UNDERSTANDING
PURE AND APPLIED
MATHS I
FOR ADVANCED LEVEL

A Dawson and R Parsons

Hutchinson

London   Melbourne   Sydney   Auckland   Johannesburg

Hutchinson Education

An imprint of Century Hutchinson Ltd
62-65 Chandos Place, London WC2N 4NW

Century Hutchinson Australia Pty Ltd
PO Box 496, 16–22 Church Street, Hawthorn,
Victoria 3122, Australia

Century Hutchinson New Zealand Limited
PO Box 40–086, Glenfield, Auckland 10,
New Zealand

Century Hutchinson South Africa (Pty) Ltd
PO Box 337, Bergvlei, 2012 South Africa

First published 1988

© A Dawson and R Parsons 1988

Phototypesetting by Thomson Press (India) Limited,
New Delhi

Printed and bound in Great Britain by
Scotprint Ltd., Musselburgh

British Library Cataloguing in Publication Data

Dawson, Tony
    Understanding pure and applied maths
    for·A level.
    Bk. 1
    1. Mathematics—1961–
    I. Title   II. Parsons, Rodney
    510   QA39.2

    ISBN   0-09-173219-0

Acknowledgements

The publishers would like to thank the following for
permission to reproduce copyright material:

Allsport UK Ltd; British Aluminium/Hudeh;
Burson–Marsteller; Camera Press; Tony Dawson;
Adrian Meredith Photography; Claire Starkey;
Topham Picture Library

Contents

Contents

4

Introduction

This is the first of two text books for students studying Advanced Level Pure and Applied Mathematics. We have written the texts bearing in mind:

(a) the needs of students and teachers;

(b) the increasing numbers of students studying advanced level mathematics;

(c) the variety of needs of advanced level mathematics students especially those who have followed GCSE syllabuses with a variety of backgrounds in the subject;

(d) its use as a possible text for AS levels.

We have tried to incorporate enough examples to assimilate and practise techniques and topics while providing enough explanation with worked examples for a student working alone. These two aims are not always compatible.

The usual style of advanced level mathematics texts contain:

      EXPOSITION

           WORKED EXAMPLES

                EXERCISES   in that order,

which may be regarded as the usual mathematics lesson. We have included 'INVESTIGATIONS' to introduce certain topics and to give the student a sense of exploration and discovery, so vital for the enjoyment of mathematics at advanced level.

The investigations might form the basis of a particular lesson or section of exposition and we assume that most students will have access to computers and computer programs. This is not essential, but graphical illustration and interpretation are so much a part of mathematics that a quick, reliable method of illustration can only help the understanding of topics and the assimilation of ideas and concepts.

Mathematics is a subject that requires students to be able to apply their knowledge to solve particular problems. This expertise cannot be acquired merely by attending lessons or even by reading a textbook, but must be gained by constantly applying the basic principles and techniques to actual questions. In other words, to be successful in mathematics the student needs a good understanding, a willingness to attempt many problems and the self-discipline to write out the solutions.

We have found that, within an A-level class, there are students who have reached different standards of achievement, have different needs and difficulties and have followed a variety of GCSE courses and so it is necessary to introduce, or maybe revise, certain basic concepts.

Chapters 1, 2 and 3 may contain material with which some students are very familiar. On the other hand, before the calculus and other important topics are attempted, a thorough knowledge of algebraic techniques is advisable. Some A-level students may need only practice, whereas others may never have come across logarithms, functions, co-ordinates, polynomials or different types of number. It may be possible for students to look at Chapters 1, 2 and 3 during the summer recess.

Some two-year courses concentrate on pure mathematics topics during the first year leaving the mechanics (or applied mathematics) until the second year. This may tie in conveniently with a set following Pure Mathematics with Statistics (where Statistics is covered during the second year) but we have found this inadvisable for two reasons: firstly, the ideas of mechanics are best assimilated over a longer period; and secondly, concentration on applied mathematics during the second year leads to a lack of variety and neglect of pure topics.

We have introduced a number of topics during the first year, many of which will be revised and consolidated during the second year as students prepare for formal examinations. Consequently, more formal examination questions and techniques will appear in Book 2, where the more difficult topics requiring a mature approach will be covered.

Vector methods appear both in pure mathematics and applied mathematics syllabuses and, consequently, we have included in Chapter 7 the various uses and notations so that students can be completely familiar with vectors from an early stage of the course.

In considering the order of chapters, our algebraic introduction enables the calculus to be covered as early as possible and so, during the first term, we would expect that two-year A-level students will have finished Chapter 7 and will complete the book in the first year.

In Book 2, in addition to covering the remaining topics of advanced level syllabuses, more attention will be given to examination techniques and the practicalities of coping with examination questions and papers.

Many of the topics and techniques learned in Book 1 will be constantly revised and practised and so Book 2 will contain revision exercises based upon the chapters of Book 1. Very often questions demand expertise from different topics and may require a recollection of techniques from different areas.

Different examination boards will use different notations, so we have tried to mention the different possibilities especially when dealing with vectors.

A list of symbols and notations is provided on page 7, but the student must be quite clear which notation is appropriate for a given examination board.

Notation

$=$	is equal to
$\neq$	is not equal to
$\equiv$	is identically equal to
$\simeq$	is approximately equal to
$>$	is greater than
$\geqslant$	is greater than or equal to
$<$	is less than
$\leqslant$	is less than or equal to
∞	infinitely large
$\Rightarrow$	implies
$\Leftarrow$	is implied by
$\Leftrightarrow$	implies and is implied by
$\rightarrow$	tends to
$:$	such that
$\parallel$	parallel to
$\perp$	perpendicular to
$+\,$ve	positive
$-\,$ve	negative
w.r.t.	with respect to
s.f.	significant figures
d.p.	decimal places

m	metres
s	seconds
kg	kilograms
N	newtons
$\mathrm{m\,s}^{-1}$	metres per second
$\mathrm{m\,s}^{-2}$	metres per second per second
M, L, T	dimensions of mass, length and time
a	acceleration
s	displacement
v	velocity
F	force
g	acceleration of gravity, $g \simeq 9.8\,\mathrm{m\,s}^{-2}$
μ	coefficient of friction
λ	angle of friction or modulus of elasticity
ω	angular speed
$i = \sqrt{-1}$	
x^{c}	x radians

$\in$	is a member of
$\notin$	is not a member of
$\subset$	A is a subset of B
$\supset$	A contains B
ϕ	empty set
ξ	universal set
A'	complement of set A
$n(A)$	number of elements in set A
$\cup$	union
$\cap$	intersection

$\mathbb{N}$	the set of natural numbers
$\mathbb{Z}$	the set of integers
$\mathbb{Q}$	the set of rational numbers
$\mathbb{R}$	the set of real numbers
$\mathbb{C}$	the set of complex numbers
$\mathbb{F}$	the set of irrational numbers

Σ	the sum of
Δ	determinant, or triangle
$\mathbf{PQ} = \overrightarrow{PQ}$	vector represented by PQ
$|\mathbf{PQ}| = PQ$	magnitude of vector $\mathbf{PQ}$
$|\mathbf{a}| = a$	magnitude of vector $\mathbf{a}$
$\mathbf{i}, \mathbf{j}, \mathbf{k}$	unit vectors along the x, y and z axes
$\mathbf{r} = x\mathbf{i} + y\mathbf{j} + z\mathbf{k}$	position vector

$n!$ n factorial

$^{n}P_r$ number of permutations of r objects chosen from the $n = \dfrac{n!}{(n-r)!}$

$\dbinom{n}{r}$ the binomial coefficient; $\dfrac{n!}{r!(n-r)!} = {}^{n}C_r$ the number of combinations of r objects chosen from n

$x \rightarrow f(x)$	x is mapped onto $f(x)$
$f : x \rightarrow x^2$	the function f mapping x onto x^2
$|x|$	the modulus of x
f^{-1}	inverse of function f
gf	function f followed by function g
$\sqrt{x}$	non-negative square root of x
$[x]$	integral part of x

$\displaystyle\lim_{x \to a} f(x)$ limit of $f(x)$ as x tends to a

δx small increment in x

$\dfrac{dy}{dx} = f'(x)$ first derivative of $y = f(x)$

$\dfrac{d^2 y}{dx^2} = f''(x)$ second derivative of $y = f(x)$

$\dot{x} = \dfrac{dx}{dt}; \quad \ddot{x} = \dfrac{d^2 x}{dt^2}$

$\displaystyle\int f(x)\,dx$ indefinite integral

$\displaystyle\int_a^b f(x)\,dx$ definite integral

e exponential $e \simeq 2.71828$ or coefficient of restitution

1 Algebra: the language of mathematics

Manipulation of formulae

Do you measure temperature in degrees Fahrenheit or Celsius (Centigrade)?

How do weather forecasters know that $5°\,C$ is $41°\,F$ or that $59°\,F$ is $15°\,C$?

Perhaps they learn a table of values:

°C	0	5	10	15	20	25	30
°F	32	41	50	59	68	77	86

Do weather forecasters learn all the intermediate values?

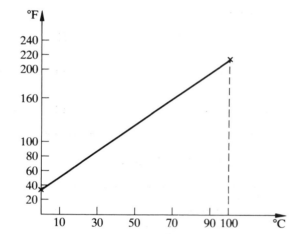

The values of °F plotted against °C on a graph, form a straight line. From the graph corresponding temperatures can be read off.

Water freezes at $0°\,C\,(=32°\,F)$ and boils at $100°\,C$ $(=212°\,F)$.

A little knowledge of straight line theory $(y = mx + c)$ gives the equation relating C and F as

$$F = \frac{9}{5}C + 32 = 1.8C + 32 \cdots\cdots (1)$$

which will give the F temperature corresponding to each C value, e.g.

$$C = 10 \quad\Rightarrow\quad F = 1.8 \times 10 + 32 = 18 + 32 = 50$$

Rearranging the formula to make C the subject, gives $\quad C = \frac{5}{9}(F - 32)$

Thus, the relation between C and F can be most conveniently summarized by an algebraic formula. Many mathematical results are expressed by an algebraic relation and mathematicians must be able to manipulate these formulae.

The author saw one weather forecaster using a simplified version of equation (1), which was $F = 2C + 30$ (double C and add 30 to get F) because he could perform the calculation quickly in his head. Although not actually true, this method was reasonably accurate over the range of temperatures he usually described.

For example,

$$C = 10 \quad\Rightarrow\quad F = 2 \times 10 + 30 = 50$$
$$C = 15 \quad\Rightarrow\quad F = 2 \times 15 + 30 = 60 \qquad (59 \text{ is the accurate value})$$
$$C = 20 \quad\Rightarrow\quad F = 2 \times 20 + 30 = 70 \qquad (68 \text{ is the accurate value})$$

What is the corresponding formula for C?

$$F = 2C + 30 \quad\Rightarrow\quad 2C = F - 30 \quad\Rightarrow\quad C = \tfrac{1}{2}(F - 30) = \tfrac{1}{2}F - 15.$$

WORKED EXAMPLE ⮞

Make y the subject of the formula $\qquad x = \sqrt{y^2 + z^2}$

Squaring both sides gives	$x^2 = y^2 + z^2$
Subtracting z^2 from both sides	$x^2 - z^2 = y^2$ or $y^2 = x^2 - z^2$
Taking the square root	$y = \sqrt{x^2 - z^2}$

Exercise 1.1

Make the letter in brackets the subject of the formula.

1 $y = 2x + 3$ $\quad (x)$

2 $C = 2\pi r$ $\quad (r)$

3 $A = \pi r^2$ $\quad (r)$

4 $S = 2\pi rh$ $\quad (h)$

5 $V = \pi r^2 h$ $\quad (h)$

6 $V = \pi r^2 h$ $\quad (r)$

7 $A = 2\pi rh + 2\pi r^2$ $\quad (h)$

8 $v = u + at$ $\quad (a)$

9 $s = ut + \frac{1}{2}at^2$ $\quad (u)$

10 $v^2 = u^2 + 2as$ $\quad (s)$

11 $\dfrac{1}{f} = \dfrac{1}{u} + \dfrac{1}{v}$ $\quad (f)$

12 $\dfrac{1}{f} = \dfrac{1}{u} + \dfrac{1}{v}$ $\quad (u)$

13 $P = \dfrac{h + 2x}{3x}$ $\quad (x)$

14 $y = \dfrac{x + 1}{x - 1}$ $\quad (x)$

15 $A = 2\pi rh + \pi r^2$ $\quad (r)$

16 $s = ut + \frac{1}{2}at^2$ $\quad (t)$

Equations and identities

Exercise 1.2

Solve the following equations.

1 $2x + 3 = 6$

2 $2x + y = 6$

3 $(x + 1)^2 = 0$

4 $2x^2 + 3 = 6$

5 $2x + y = 6$ and $x + 2y = 9$

6 $(x + 1)^2 = 1$

7 $(x + 2)^2 = 9$

8 $(x + 2)^2 = 9x$

9 $(x + 1)(x + 2) = 0$

10 $(x + 1)(x + 2) = 2$

11 $\dfrac{1}{x} + \dfrac{2}{x} = 3$

12 $\dfrac{1}{x + 1} + \dfrac{1}{x + 2} = 3$

13 $x^2 + 4x + 4 = (x + 2)^2$

Exercise 1.2 contains several different types of equation which will raise questions of recognition and methods of solution needing discussion at this stage.

Some of the questions raised will now be considered.

Identities

Question **13** in Exercise 1.2 is an identity rather than an equation. The left hand side (LHS) takes the same value as the right hand side (RHS) for all values of x, i.e. it is identically true. The RHS is the factorized form of the quadratic polynomial on the LHS.

In some cases the factorized form of an expression is most useful, while in others the expanded form is required.

WORKED EXAMPLES

1 $(x+3)^2 = (x+3)(x+3) = x^2 + 3x + 3x + 9 = x^2 + 6x + 9$

2 $(x+3)(x+4) = x^2 + 3x + 4x + 12 = x^2 + 7x + 12$

3 $(x^2 + 5x + 6) = x^2 + 2x + 3x + 6 = x(x+2) + 3(x+2) = (x+2)(x+3)$

Equations, inequalities, expressions, identities

Categorize **1** to **5** below, as an equation, inequality, expression or identity.

1 $(x+3)^2$

2 $(x+3)^2 = 2x + 6$

3 $(x+3)^2 = x^2 + 9$

4 $(x+3)^2 = x^2 + 6x + 9$

5 $(x+3)^2 > 2x + 6$

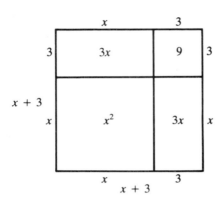

You may know that,

$(x+3)^2 = (x+3)(x+3) = x^2 + 6x + 9$
as the figure shows, so the area of the large square represents $(x+3)(x+3)$ which equals $x^2 + 6x + 9$.

So $(x+3)^2 = x^2 + 6x + 9$ whatever value x takes. We say $(x+3)^2$ is **identically equal** to $x^2 + 6x + 9$ and the equality sign should be replaced by $\equiv$ an identity sign.

4   is therefore an **identity**

1   is an **expression**, which can also be written as $x^2 + 6x + 9$

2   is **not** an **identity**. Expanding the LHS gives $x^2 + 6x + 9 = 2x + 6$, which may be true for certain values of x.

Subtracting $2x + 6$ from both sides $\Rightarrow$ $x^2 + 4x + 3 = 0$

Factorizing the LHS $\Rightarrow$ $(x+1)(x+3) = 0$ $\Rightarrow$ $x + 1 = 0$ or $x + 3 = 0$

since if two numbers multiply together to give 0, one of them must be 0.

$x + 1 = 0$ $\Rightarrow$ $x = -1$ and $x + 3 = 0$ $\Rightarrow$ $x = -3$ so there are two possible solutions: $x = -1$ or $x = -3$

2   is therefore a **quadratic equation** with two solutions.

3   is a case of 'mistaken identity'.

Some students mistakenly think that $(x+3)^2 = x^2 + 9$, forgetting $6x$ the 'middle term'.

$(x+3)^2 = x^2 + 9$ $\Rightarrow$ $x^2 + 6x + 9 = x^2 + 9$ $\Rightarrow$ $6x = 0$ which is only true for one value, $x = 0$.

5   is an **inequality**, being satisfied by many values of x, and is solved in a similar way to the equation in **2**. Expanding the LHS gives

$x^2 + 6x + 9 > 2x + 6$ $\Rightarrow$ $x^2 + 4x + 3 > 0$ $\Rightarrow$ $(x+1)(x+3) > 0$

Either $(x+1) > 0$ **and** $(x+3) > 0$ $\Rightarrow$ x must be greater than -1, $x > -1$
or $x + 1 < 0$ **and** $x + 3 < 0$ $\Rightarrow$ x must be less than -3, $x < -3$

The solution set is $\{x : x < -3\} \cup \{x : x > -1\}$

Alternatively, the graph of $y = (x + 1)(x + 3)$ is a parabola, cutting the y-axis ($y = 0$) when $x = -1$ and $x = -3$. From the graph,

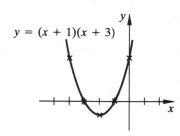
$y = (x + 1)(x + 3)$

$y = (x + 1)(x + 3) > 0$ when $x > -1$ or $x < -3$

In fact, the solution can be found without the graph by finding the crucial values $x = -1$ and $x = -3$ and testing the regions in between.

For example,
$$x < -3 \quad \Rightarrow \quad (x + 3)(x + 1) = (-) \times (-) > 0 \quad \text{inequality satisfied}$$
$$-3 < x < -1 \quad \Rightarrow \quad (x + 3)(x + 1) = (+) \times (-) < 0 \quad \text{inequality \textbf{not} satisfied}$$
$$x > -1 \quad \Rightarrow \quad (x + 3)(x + 1) = (+) \times (+) > 0 \quad \text{inequality satisfied.}$$

If you are unfamiliar with inequalities, remember $4 > 3$ and $-4 < -3$.

(a) Adding the same number to both sides of an inequality preserves the relation.

For example, adding 6 to both sides of $4 > 3$ gives $10 > 9$
and adding -6 to both sides of $4 > 3$ gives $-2 > -3$
both of which are true.

(b) Subtracting 6 is the same as adding -6, so we can subtract the same number from both sides.

(c) Multiplying both sides by $+3$ gives $12 > 9$, which is still true.

(d) Multiplying both sides by -3 gives $-12 > -9$ which is **false**. $-12 < -9$, so we must remember to alter the inequality sign.

(e) Be careful when inverting e.g. $5 < 6$ but $\dfrac{1}{5} > \dfrac{1}{6}$.

WORKED EXAMPLE

Solve $3x - 4 > 2 - x$.

$$\Rightarrow \quad 3x + x > 2 + 4 \qquad \text{Adding } x + 4 \text{ to both sides}$$
$$\Rightarrow \quad 4x > 6$$
$$\Rightarrow \quad x > 1\tfrac{1}{2} \qquad \text{Dividing both sides by 4}$$

You can check your answer by substituting back into the original equation.

$x = 1.5$ makes both sides equal
$x > 1.5$ satisfies the inequality   (try $x = 2$)
$x < 1.5$ does **not** satisfy the inequality   (try $x = 1$)

Exercise 1.3

Solve the inequalities.

1 $2x - 3 > 4$ **2** $2(x + 3) < 4$ **3** $2(x - 3) \leqslant 3(x + 4)$ **4** $2x - 3 \geqslant 4 - 5x$

5 $\dfrac{1}{x} > 3$ **6** $\dfrac{2}{x} + 3 > 4$ **7** $\dfrac{1}{x} < \dfrac{1}{4}$ **8** $\dfrac{2 + x}{x} > 3$

9 $\dfrac{2x}{x + 2} > 1$ **10** $x^2 > 4$

Algebraic mistakes

Many students make algebraic errors because they confuse the rules of algebra, or do not learn them or practise them enough to make them second nature. In many topics the practice of solving problems and performing drill exercises are essential in cementing ideas and processes which must be automatic in their execution.

1 Under the pressure of solving a difficult problem, the mistake $(a+b)^2 = a^2 + b^2$ occurs too often. This mistake can only be true if the forgotten middle term '$2ab$' is zero, which implies that either a or b is zero.

2 Another variation is $\sqrt{a^2 + b^2} = a + b$, which in general is **false**. It is only true when either $a = 0$ or $b = 0$, which makes the statement trivial.

3 Consider $\dfrac{1}{x} + \dfrac{2}{x} = 3 \;\Rightarrow\; \dfrac{1+2}{x} = 3 \;\Rightarrow\; \dfrac{3}{x} = 3 \;\Rightarrow\; \dfrac{x}{3} = \dfrac{1}{3} \;\Rightarrow\; x = 1$

Students will turn everything upside down to start with to get

$$\frac{x}{1} + \frac{x}{2} = \frac{1}{3} \;\Rightarrow\; \frac{3x}{2} = \frac{1}{3} \;\Rightarrow\; x = \frac{2}{9} \;\;(!!)$$

$$\text{or}\quad \frac{2x}{3} = \frac{1}{3} \;\Rightarrow\; x = \frac{1}{2} \;\;(?)$$

This mistake sometimes arises when using the formula $\dfrac{1}{u} + \dfrac{1}{v} = \dfrac{1}{f}$.

4 $\dfrac{1}{x-1} - \dfrac{1}{x+2} = 3$ $\quad$ Avoid the inverting mistake and express the LHS as a single fraction with $(x-1)(x-2)$ as the common denominator.

$\dfrac{(x+2)-(x-1)}{(x-1)(x+2)} = 3 \;\Rightarrow\; \dfrac{x+2-x+1}{(x-1)(x+2)}$ $\quad$ Remember the double negative.

Correctly,

$\dfrac{x+2-x+1}{(x-1)(x+2)} = 3 \;\Rightarrow\; \dfrac{3}{(x-1)(x+2)} = 3 \;\Rightarrow\; (x-1)(x+2) = 1 \cdots\cdots (1)$

Remembering that $(x-1)(x+2) = 0 \;\Rightarrow\; x-1 = 0$ or $x+2 = 0$ leads some students to do the same with **(1)** to give

$$x-1 = 1 \quad\text{or}\quad x+2 = 1 \;\Rightarrow\; x = 2 \quad\text{or}\quad x = -1$$

both of which do **not** satisfy the original equation.

You must acquire the habit of building checks into your working. It is an easy matter to see whether $x = 2$ or $x = -1$ is a solution of the original equation.

Continuing with **(1)**, $(x-1)(x+2) = 1 \;\Rightarrow\; x^2 + x - 2 = 1 \;\Rightarrow\; x^2 + x - 3 = 0$ which does not factorize; hence **(a)** the quadratic formula or **(b)** completing the square:

(a) $ax^2 + bx + c = 0 \;\Rightarrow\; x = \dfrac{-b \pm \sqrt{b^2 - 4ac}}{2a} \;\Rightarrow\; x = \dfrac{-1 \pm \sqrt{1 - 4(1)(-3)}}{2} = \dfrac{-1 \pm \sqrt{13}}{2}$

$\Rightarrow\; x = \dfrac{\sqrt{13} - 1}{2} \simeq 1.30 \quad\text{or}\quad x = \dfrac{-\sqrt{13} - 1}{2} \simeq -2.30$

(b) $x^2 + x = 3 \;\Rightarrow\; x^2 + x + \tfrac{1}{4} = 3\tfrac{1}{4} \;\Rightarrow\; (x + \tfrac{1}{2})^2 = \dfrac{13}{4} \;\Rightarrow\; x + \tfrac{1}{2} = \pm\dfrac{\sqrt{13}}{2} \;\Rightarrow\; x = \dfrac{-1 \pm \sqrt{13}}{2}$

Exercise 1.4

Find mistakes in the following solutions to the equations of Exercise 1.2.

1 $2x + 3 = 6 \Rightarrow 2x = 3 \Rightarrow x = \dfrac{2}{3}$

2 $2x + y = 6 \cdots\cdots$ **(a)** $\qquad x + 2y = 9 \cdots\cdots$ **(b)**

(A) Subtracting **(b)** from **(a)** $\Rightarrow x - y = 3 \cdots\cdots$ **(c)**

Adding **(a)** and **(c)** $\qquad\qquad \Rightarrow \qquad 3x = 9$

$\qquad\qquad\qquad\qquad\qquad\qquad \Rightarrow \qquad x = 3 \Rightarrow y = 0$

(B) Multiply **(b)** by 2 and subtract from **(a)** $\Rightarrow \quad 2x + y = 6$

$\qquad\qquad\qquad\qquad\qquad\qquad\qquad\qquad\qquad\quad 2x + 4y = 9$

$\qquad\qquad\qquad\qquad\qquad\qquad\qquad\qquad -3y = -3 \Rightarrow y = 1$

From **(a)** $y = 1 \Rightarrow \qquad 2x = 5 \Rightarrow x = 2\frac{1}{2}$
From **(b)** $y = 1 \Rightarrow \quad x + 2 = 9 \Rightarrow x = 7 \qquad$ (?)

(C) $\begin{array}{l} 2x + y = 6 \\ x + 2y = 9 \end{array} \Rightarrow \begin{pmatrix} 2 & 1 \\ 1 & 2 \end{pmatrix}\begin{pmatrix} x \\ y \end{pmatrix} = \begin{pmatrix} 6 \\ 9 \end{pmatrix} \qquad$ a matrix equation

Pre-multiply both sides by $\begin{pmatrix} 2 & -1 \\ -1 & 2 \end{pmatrix}$ to give $\begin{pmatrix} 2 & -1 \\ -1 & 2 \end{pmatrix}\begin{pmatrix} 2 & 1 \\ 1 & 2 \end{pmatrix}\begin{pmatrix} x \\ y \end{pmatrix} = \begin{pmatrix} 2 & -1 \\ -1 & 2 \end{pmatrix}\begin{pmatrix} 6 \\ 9 \end{pmatrix}$

$$\begin{pmatrix} 3 & 0 \\ 0 & 3 \end{pmatrix}\begin{pmatrix} x \\ y \end{pmatrix} = \begin{pmatrix} 3 \\ 12 \end{pmatrix}$$

$$\begin{pmatrix} 3x \\ 3y \end{pmatrix} = \begin{pmatrix} 3 \\ 12 \end{pmatrix} \Rightarrow \begin{array}{l} x = 1 \\ y = 4 \end{array}$$

(D) From **(a)** $y = 6 - 2x$, substitute for y in **(b)** to give

$x + 2(6 - 2x) = 9 \Rightarrow x + 12 - 4x = 9$

$\qquad\qquad\qquad\qquad \Rightarrow \quad -3x = 3 \Rightarrow x = -1$

$x = -1 \Rightarrow y = 6 - 2 = 4 \Rightarrow y = 4$

(E) Draw the graphs of **(a)** $2x + y = 6$
$\qquad\qquad\qquad$ and **(b)** $x + 2y = 9$

The graphs intersect at $(2, 2)$.

So $x = 2$, $y = 2$ is the solution.

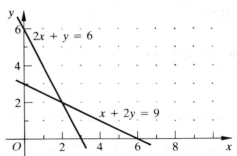

3 $(x + 2)^2 = 9 \Rightarrow x + 2 = \sqrt{9} = 3 \Rightarrow x = 1$

4 $(x + 1)(x + 2) = 2 \Rightarrow x + 1 = 2 \quad\text{or}\quad x + 2 = 2$

$\qquad\qquad\qquad\qquad\quad \Rightarrow \qquad x = 1 \quad\text{or}\qquad x = 0$

5 $(x + 2)^2 = 9 \Rightarrow (x + 2)(x + 2) = 9 \Rightarrow x + 2 = 3 \quad\text{or}\quad x + 2 = -3$

$\qquad\qquad\qquad\qquad\qquad\qquad\qquad\qquad \Rightarrow \qquad x = 1 \quad\text{or}\qquad x = -1$

6 $(x + 2)^2 = 9 \Rightarrow x^2 + 4 = 9 \Rightarrow x^2 = 5 \Rightarrow x = \pm\sqrt{5} \simeq \pm 2.24$

7 $(x + 2)^2 = 9 \Rightarrow x^2 + 4x + 4 = 9$

$\qquad\qquad\qquad\quad \Rightarrow x^2 + 4x - 5 = 0$

$\qquad\qquad\qquad\quad \Rightarrow (x + 1)(x - 5) = 0$

$\qquad\qquad\qquad\quad \Rightarrow x + 1 = 0 \quad\text{or}\quad x - 5 = 0$

$\qquad\qquad\qquad\quad \Rightarrow x = 1 \quad\text{or}\quad x = -5$

Indices, powers and logarithms

The power of powers

If you place one penny on the first square of a chessboard, two pennies on the next, four on the third and so on, doubling the value on each successive square, how much money will be placed on the last square?

What is the value of $2^3, 2^0, 2^{\frac{1}{2}}$?

$2^3 = 2 \times 2 \times 2 = 8$, although some students, knowing that 2^3 stands for three twos multiplied together, sometimes give an answer of 6   (!)

A similar misconception is that 2^0 stands for no twos multiplied together, which instinctively is given as zero!!

$2^3 = 8$ means that 2 is raised to the **power** 3 to give 8.

Another word for **power** is **index** (plural **indices**).

1	2	4	8	16	32	64	128
256	512						
							?

Rules for indices

$2^3 \times 2^4 = 8 \times 16 = 128 = 2^7$ $a^3 \times a^4 = (a \times a \times a) \times (a \times a \times a \times a) = a^7$

In general $\boxed{2^a \times 2^b = 2^{a+b}}$

When **multiplying** numbers expressed as powers, **add** the powers (indices)

$2^5 \div 2^3 = 32 \div 8 = 4 = 2^2$ $a^5 \div a^3 = \dfrac{a^5}{a^3} = \dfrac{a \times a \times a \times a \times a}{a \times a \times a} = a^2$

In general $\boxed{2^a \div 2^b = 2^{a-b}}$

When **dividing** numbers expressed as powers, **subtract** the powers (indices)

We need only try a few examples to establish these rules.

Exercise 1.5

1 $4^2 =$		**8** $(b^2)^3 =$		**15** $g^3 \div g^3 =$
2 $2^4 =$		**9** $c^7 \times c =$		**16** $h^4 \div h^5 =$
3 $4^3 =$		**10** $2 \times 2^3 =$		**17** $j^5 \times j^0 =$
4 $3^4 =$		**11** $2 \times 2^x =$		**18** $k^7 \times k^{-2} =$
5 $x^3 \times x^2 =$		**12** $d^6 \div d^2 =$		**19** $m^{\frac{1}{2}} \times m^{\frac{1}{2}} =$
6 $a^7 \times a^7 =$		**13** $e^5 \div e^3 =$		**20** $n^5 \div n^{-2} =$
7 $b^2 \times b^2 \times b^2 =$		**14** $f^9 \div f^8 =$		**21** $4^{3^2} =$

Mathematicians like to use the same rules for any numbers they are dealing with, which leads to some interesting results!

$2^4 = 16$
$2^3 = 8$
$2^2 = 4$
$2^1 = 2$
$2^0 = ?$
$2^{-1} = ?$

The sequence $16, 8, 4, 2$ continues $1, \frac{1}{2}, \frac{1}{4}, \ldots$ by a process of halving which leads to

$2^0 = 1$
$2^{-1} = \frac{1}{2}$
$2^{-2} = \frac{1}{4}$ so in general $\boxed{2^{-a} = \dfrac{1}{2^a}}$

Another way to find a meaning for $2^0, 2^{-1}, 2^{-2}$ etc. is to use $2^a \div 2^b = 2^{a-b}$

$$2^3 \div 2^3 = 2^0 \quad \text{but} \quad 2^3 \div 2^3 = 8 \div 8 = 1 \quad \Rightarrow \quad 2^0 = 1$$

$$2^3 \div 2^4 = 2^{-1} \quad \text{but} \quad 2^3 \div 2^4 = 8 \div 16 = \tfrac{1}{2} \quad \Rightarrow \quad 2^{-1} = \tfrac{1}{2}$$

$$2^3 \div 2^5 = 2^{-2} \quad \text{but} \quad 2^3 \div 2^5 = 8 \div 32 = \tfrac{1}{4} \quad \Rightarrow \quad 2^{-2} = \tfrac{1}{4}$$

Alternatively, using $2^a \times 2^b = 2^{a+b}$

$$2^5 \times 2^0 = 2^{5+0} = 2^5 \quad \Rightarrow \quad 2^0 = \frac{2^5}{2^5} = \frac{32}{32} = 1 \qquad \Rightarrow \quad 2^0 = 1$$

$$2^5 \times 2^{-2} = 2^{5+-2} = 2^3 \quad \Rightarrow \quad 2^{-2} = \frac{2^3}{2^5} = \frac{8}{32} = \frac{1}{4} = \frac{1}{2^2} \quad \Rightarrow \quad 2^{-2} = \frac{1}{2^2}$$

In general $\boxed{2^{-n} = \dfrac{1}{2^n}}$ So we have a meaning for negative powers. But still some students instinctively think that 2^{-3} is negative, perhaps -6 or -8 (!!)

Exercise 1.6

1 $p^3 \times p^{-4} =$	**10** $x^2 y \times y =$	**19** $\left(1\frac{1}{2}\right)^{-2} =$
2 $q^7 \times q^{-5} =$	**11** $xy^2 \times xy =$	**20** $x^{-1} =$
3 $q^{-5} \times q^7 =$	**12** $xy^2 \div xy =$	**21** $x^{-2} =$
4 $r^8 \div r^4 =$	**13** $3^{-2} =$	**22** $x^{-m} =$
5 $s^{-2} \div s^2 =$	**14** $4^{-1} =$	**23** $a^{\frac{1}{2}} \times a^{\frac{1}{2}} =$
6 $t^{-3} \times t^{-3}$	**15** $5^0 =$	**24** $a^{\frac{1}{3}} \times a^{\frac{1}{3}} \times a^{\frac{1}{3}} =$
7 $(t^{-3})^2 =$	**16** $\left(\frac{1}{2}\right)^2 =$	**25** $a^0 =$
8 $x^3 \div x^{-2} =$	**17** $\left(1\frac{1}{2}\right)^2 =$	**26** $0^a =$
9 $x^3 \times x^2 \times x^{-5} =$	**18** $\left(2\frac{1}{2}\right)^2 =$	**27** $0^0 =$

Exercise 1.7

1 $(2^3)^2 = 2^3 \times 2^3 =$	**4** $(a^3)^2 =$	**7** $(b^7)^3 =$
2 $(5^2)^3 = 5^2 \times 5^2 \times 5^2 =$	**5** $(x^4)^2 =$	**8** $(2^{-1})^2 =$
3 $(a^2)^3 =$	**6** $(x^2)^4 =$	**9** $(2^{-2})^3 =$

We now have another rule of indices $\boxed{(2^a)^b = (2^b)^a = 2^{a \times b} = 2^{ab}}$

Algebra

Fractional indices

We have established these rules

$$x^a \times x^b = x^{a+b} \cdots\cdots\cdots (1)$$

$$x^a \div x^b = x^{a-b} \cdots\cdots\cdots (2)$$

$$x^{-a} = \frac{1}{x^a} \cdots\cdots\cdots\cdots (3)$$

$$(x^a)^b = (x^b)^a = x^{ab} \cdots\cdots (4)$$

when a and b are **integers** (i.e. the set of whole numbers positive or negative including 0)
The integers are denoted by $\mathbb{Z}$.

The same rules apply when a and b are **fractions** (rational numbers, denoted by $\mathbb{Q}$).

$$2^{\frac{1}{2}} \times 2^{\frac{1}{2}} = 2^1 = 2 \qquad \text{so } 2^{\frac{1}{2}} \text{ stands for } \sqrt{2} \simeq \pm 1.4142\ldots$$

$$3^{\frac{1}{2}} \times 3^{\frac{1}{2}} = 3^1 = 3 \qquad \text{so } 3^{\frac{1}{2}} = \sqrt{3}$$

$$a^{\frac{1}{3}} \times a^{\frac{1}{3}} \times a^{\frac{1}{3}} = a^1 = a \qquad \text{so } a^{\frac{1}{3}} = \sqrt[3]{a}$$

and in general

$$\boxed{a^{1/n} = \sqrt[n]{a}}$$

WORKED EXAMPLES

(a) $9^{\frac{1}{2}} = \sqrt{9} = \pm 3$

(b) $\left(\dfrac{9}{16}\right)^{\frac{1}{2}} = \sqrt{\dfrac{9}{16}} = \dfrac{\sqrt{9}}{\sqrt{16}} = \pm\dfrac{3}{4}$

(c) $8^{\frac{1}{3}} = \sqrt[3]{8} = 2$ or $8^{\frac{1}{3}} = (2^3)^{\frac{1}{3}} = 2^1 = 2$ using rule (4)

Exercise 1.8

Simplify

1 $16^{\frac{1}{4}}$

2 $36^{\frac{1}{2}}$

3 $27^{\frac{1}{3}}$

4 $\left(\dfrac{8}{27}\right)^{\frac{1}{3}}$

5 $81^{\frac{1}{4}}$

6 $\sqrt[3]{x^6}$

7 $25^{\frac{1}{2}} \times 4^{\frac{1}{2}}$

8 $128^{\frac{1}{7}}$

9 $\sqrt{\dfrac{2}{98}}$

10 $4^{-\frac{1}{2}}$

11 $\left(\dfrac{1}{4}\right)^{-\frac{1}{2}}$

12 $(2\frac{1}{4})^{\frac{1}{2}}$

13 $(12.25)^{\frac{1}{2}}$

14 $\left(\dfrac{64}{125}\right)^{-\frac{1}{3}}$

15 $\left(\dfrac{32}{162}\right)^{\frac{1}{4}}$

16 $36^{\frac{1}{2}} \times 64^{\frac{1}{2}}$

17 $(36 \times 64)^{\frac{1}{2}}$

18 $16^{\frac{1}{4}} \div 81^{\frac{1}{4}}$

19 $36^{\frac{1}{2}} + 64^{\frac{1}{2}}$

20 $(36 + 64)^{\frac{1}{2}}$

21 $(9 + 16)^{\frac{1}{2}}$

22 $x^{\frac{1}{2}} \times x^{\frac{1}{3}}$

23 3^{x+2}

24 $3^x \times 3^{x+1}$

25 $3^x + 3^{x+1}$

26 $4^x + 2^{x+1}$

27 $4^x \times 2^{x+1}$

28 $4^x \div 2^{x+1}$

29 $x^{\frac{1}{3}} \times (x^2)^{\frac{1}{3}}$

30 $\left(\dfrac{1}{4}\right)^{-\frac{3}{2}}$

We have now established meanings and rules to deal with integer powers like 2^3 and 2^{-3} and fractional powers like $2^{\frac{1}{4}}$. We can now establish a definition of $x^{\frac{2}{3}}$.

$x^{\frac{2}{3}} = (x^{\frac{1}{3}})^2 = (\sqrt[3]{x})^2$ the cube root of x, squared

or $x^{\frac{2}{3}} = (x^2)^{\frac{1}{3}} = \sqrt[3]{x^2}$ the cube root of 'x squared'

For example, $8^{\frac{2}{3}} = (\sqrt[3]{8})^2 = 2^2 = 4$

or $8^{\frac{2}{3}} = \sqrt[3]{8^2} = \sqrt[3]{64} = 4$ both methods giving the same answer

Similarly, $4^{\frac{3}{2}} = (4^{\frac{1}{2}})^3 = (\pm 2)^3 = \pm 8$

or $4^{\frac{3}{2}} = (4^3)^{\frac{1}{2}} = \sqrt{64} = \pm 8$

In general $\boxed{a^{p/q} = \sqrt[q]{a^p} = (\sqrt[q]{a})^p}$

In numerical work it is usually easier to take the root first (if it can be done) to prevent the numbers becoming too large, although the results can always be checked on a calculator.

Exercise 1.9

Simplify

1   $9^{\frac{3}{2}}$

2   $16^{\frac{3}{4}}$

3   $32^{\frac{2}{5}}$

4   $32^{\frac{4}{5}}$

5   $27^{\frac{2}{3}}$

6   $81^{\frac{3}{4}}$

7   $81^{-\frac{1}{4}}$

8   $81^{-\frac{3}{4}}$

9   $64^{\frac{5}{6}}$

10   $125^{-\frac{2}{3}}$

11   $\left(\dfrac{8}{27}\right)^{\frac{2}{3}}$

12   $\left(\dfrac{8}{27}\right)^{-\frac{2}{3}}$

13   $\left(\dfrac{16}{9}\right)^{\frac{3}{2}}$

14   $(0.16)^{\frac{3}{2}}$

15   $(0.01)^{\frac{3}{2}}$

16   $(0.25)^{\frac{1}{2}}$

17   $(0.125)^{\frac{1}{3}}$

18   $\left(\dfrac{1}{8}\right)^{\frac{2}{3}}$

19   $(x^2)^{\frac{3}{2}}$

20   $(x^3)^{\frac{2}{3}}$

21   $(a^4)^{\frac{3}{2}}$

22   $(a^2 b^2)^{\frac{3}{2}}$

23   $(a^6 b^6 c^6)^{\frac{2}{3}}$

24   $(a^2 + 2ab + b^2)^{\frac{1}{2}}$

25   $x^{\frac{3}{2}} \times x^{\frac{1}{2}}$

26   $x^{\frac{3}{2}} \div x^{\frac{1}{2}}$

27   $x^{\frac{3}{2}} \times x^{-\frac{1}{2}}$

28   $x^{\frac{3}{2}} \div x^{-\frac{1}{2}}$

29   $(x+1)^{\frac{3}{2}} \times (x+1)^{-\frac{1}{2}}$

30   $(x+1)^{\frac{3}{2}} - (x+1)^{\frac{1}{2}}$

31   $\dfrac{x^{\frac{3}{2}} + x^{\frac{1}{2}}}{x^{\frac{1}{2}}}$

32   $\dfrac{x^{\frac{3}{2}} + x^{\frac{1}{2}}}{x}$

33   $\dfrac{x^{\frac{3}{2}} + x^{\frac{1}{2}}}{x^{\frac{3}{2}}}$

34   $(0.064)^{-\frac{1}{3}}$

35   $\dfrac{x^{\frac{3}{2}} - x^{-\frac{1}{2}}}{x+1}$

36   $\dfrac{x^{\frac{1}{2}} + x^{-\frac{1}{2}}}{x^{\frac{1}{2}}}$

Logarithms

$$A^B = C \quad \Rightarrow \quad B = \log_A C$$

We have established various rules for dealing with powers or indices. In certain situations we call the power (or index) a logarithm.

$2^3 = 8 \quad \Leftrightarrow \quad 3$ is the logarithm of 8 in base 2   or $3 = \log_2 8$

$3^4 = 81 \quad \Leftrightarrow \quad 4$ is the logarithm of 81 in base 3   or $4 = \log_3 81$

Rule 1 for indices gives $\quad 2^3 \times 2^4 = 2^7$
$$8 \times 16 = 128$$

which leads to $\log_2 8 + \log_2 16 = \log_2(8 \times 16) = \log_2 128$

In general $\qquad \boxed{\log_a b + \log_a c = \log_a bc}$

Similarly, $\quad 2^5 \div 2^3 = 2^{5-3} = 2^2$ leads to $\qquad \boxed{\log_a b - \log_a c = \log_a \dfrac{b}{c}}$

WORKED EXAMPLES

(a) $3^2 = 9 \iff 2 = \log_3 9$ $\qquad$ (c) $4^1 = 4 \iff 1 = \log_4 4$

(b) $5^3 = 125 \iff 3 = \log_5 125$ $\qquad$ (d) $4^0 = 1 \iff 0 = \log_4 1$

Exercise 1.10

1 Use the worked examples to express the following in their equivalent (logarithmic or power) form.

(a) $2^4 = 16$ $\qquad$ (f) $\log_7 49 = 2$ $\qquad$ (k) $\log_3 1 = 0$

(b) $4^2 = 16$ $\qquad$ (g) $\log_3 27 = 3$ $\qquad$ (l) $2^{-1} = \frac{1}{2}$

(c) $4^3 = 64$ $\qquad$ (h) $\log_{10} 100 = 2$ $\qquad$ (m) $3^{-2} = \frac{1}{9}$

(d) $2^5 = 32$ $\qquad$ (i) $\log_2 8 = x$ $\qquad$ (n) $y = e^x$

(e) $2^0 = 1$ $\qquad$ (j) $\log_a b = c$ $\qquad$ (o) $\log_e x = y$

2 Evaluate (a) $\log_3 9$ $\qquad$ (b) $\log_4 64$ $\qquad$ (c) $\log_2 16$ $\qquad$ (d) $\log_2 32$.

We can see that $\quad \log_2 81 = \log_2(3^4) = \log_2(3 \times 3 \times 3 \times 3) = \log_2 3 + \log_2 3 + \log_2 3 + \log_2 3 = 4 \times \log^3$

This suggests that $\qquad \boxed{\log_a(b^c) = c \times \log_a b} \qquad$ which we will now prove.

Put $x = \log_a(b^c) \iff b^c = a^x \iff b = a^{x/c} \iff \dfrac{x}{c} = \log_a b \iff x = c \log_a b$

$\qquad\qquad \iff c \times \log_a b = \log_a(b^c) \qquad$ You can check this on a calculator.

On calculators, logarithms in base 10 are denoted by log or lg and logarithms in base e (useful in calculus) are denoted by ln (and in books also by $\log_e$).

Change of base of logarithms

$2^x = 8 \iff x = 3 \qquad$ CAN

$\qquad\qquad\qquad\qquad$ WE

$2^x = 12 \iff x = ? \qquad\qquad$ SOLVE

$2^x = 16 \iff x = 4 \qquad\qquad\qquad 2^x = 12?$

The graph of $y = 2^x$ indicates that the solution of $2^x = 12$ lies between 3 and 4. Is it $3\frac{1}{2}$?

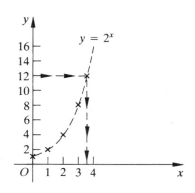

$2^{3\frac{1}{2}} = 2^3 \times 2^{\frac{1}{2}} = 8\sqrt{2} \simeq 11.31$

$2^{3.6} \simeq 12.12$ so the solution to $2^x = 12$ lies between $x = 3.5$ and $x = 3.6$.

Taking logs of both sides of $2^x = 12 \quad \Rightarrow \quad \log 2^x = \log 12$

$$\Rightarrow \quad x \log 2 = \log 12$$

$$\Rightarrow \quad x = \frac{\log_{10} 12}{\log_{10} 2} = \frac{1.079}{0.301} = 3.58$$

From the definition of logarithms $2^x = 12 \quad \Leftrightarrow \quad x = \log_2 12 = \dfrac{\log_{10} 12}{\log_{10} 2}$

This result can be proved in general to give $\boxed{\log_a b = \dfrac{\log_c b}{\log_c a}}$ used for changing the base of logarithms.

WORKED EXAMPLES

1 Simplify $\log 2 + \log 3 - \log 4$.

$\log 2 + \log 3 = \log (2 \times 3) = \log 6$; $\quad \log 2 + \log 3 - \log 4 = \log 6 - \log 4 = \log \frac{6}{4} = \log \frac{3}{2} = \log 1.5$

2 Solve $9^x + 3^{x+1} - 4 = 0$.

Use $9^x = (3^2)^x = 3^{2x} = (3^x)^2$ and also $3^{x+1} = 3^x \times 3^1 = 3 \times 3^x$

Then $9^x + 3^{x+1} - 4 = 0 \quad \Rightarrow \quad (3^x)^2 + (3 \times 3^x) - 4 = 0 \quad \Rightarrow \quad y^2 + 3y - 4 = 0 \quad$ if $y = 3^x$

$y^2 + 3y - 4 = 0 \quad \Rightarrow \quad (y-1)(y+4) = 0 \quad \Rightarrow \quad (3^x - 1)(3^x + 4) = 0$

$$\Rightarrow \quad 3^x = 1 \quad \text{or} \quad 3^x = -4$$

$$3^x = 1 \quad \Rightarrow \quad x = 0 \quad \text{or} \quad 3^x = -4 \quad \Rightarrow \quad \text{no solution for } x$$

Exercise 1.11

1 Use logarithm rules to express the following as logarithms of a single number.

(a) $\log 2 + \log 3 + \log 4$ (d) $\log 8 - \log 4$ (g) $\frac{1}{2}\log 18 - \frac{1}{2}\log 2$

(b) $\log 2 - \log 3 + \log 4$ (e) $\log 16 - \log 8$ (h) $2\log 4 + 3\log 3 - 4\log 2$

(c) $2\log 3 - 3\log 2$ (f) $\log 25 - \log 5$ (i) $\frac{1}{3}\log 64 - \frac{1}{5}\log 32$

2 Express as a single logarithm

(a) $\log a^3 - \log a^2$ (d) $\log xy - \log x - \log y$ (g) $\log a^2 - 2\log a$

(b) $\log 2a + \log 3a$ (e) $\log xy^2 - \log xy$ (h) $\frac{1}{2}\log a^4 - \frac{1}{3}\log b^3$

(c) $\log(1+x) + \log(1-x)$ (f) $\log(a+b) - \log(a^2 - b^2)$ (i) $5\log b + \frac{1}{3}\log b^6$

3 Solve   (a) $2^x = 3$ (b) $3^x = 15$ (c) $4^x = 20$ (d) $5^x = 25$ (e) $6^x = 5$ (f) $2^{x+1} = 10$

4 Solve   (a) $2^{2x} = 2^x$ (c) $2^{2x} = 2^{x+1} - 1$ (e) $4^x = 3 \times 2^x - 2$

          (b) $2^{2x} = 2^{x+1}$ (d) $4^x = 2^{2x}$ (f) $4^x - 2^x - 1 = 0$

Algebra

Surds

$$x^2 = 9 \quad \Rightarrow \quad x = \pm 3 \text{ and } \sqrt{9} \text{ is an integer (whole number)}$$

$$\text{But } x^2 = 3 \quad \Rightarrow \quad x = \pm \sqrt{3} \text{ which is irrational (see Chapter 2).}$$

$$\sqrt{3} \simeq 1.732\,050\,8\ldots \text{ but } \sqrt{3} \text{ has no exact decimal equivalent form.}$$

In much of advanced level mathematics we leave $\sqrt{3}$ (the positive square root of 3) in its exact form, $\sqrt{3}$, so we often have to deal with $\sqrt{2}, \sqrt{3}, \sqrt{5}$ etc. as **surds**.

WORKED EXAMPLE

Remember that $\sqrt{12}$ can be positive or negative. It is usually clear in the context of the question whether to write $\sqrt{12} = \pm 2\sqrt{3}$ but in practice we omit both signs and remember that $\sqrt{12}$ is positive or negative.

1 $\sqrt{12} = \sqrt{4 \times 3} = \sqrt{4} \times \sqrt{3} = 2\sqrt{3}$

4 $(2 + \sqrt{3})(2 - \sqrt{3}) = 4 - (\sqrt{3})^2 = 4 - 3 = 1$

2 $\sqrt{18} = \sqrt{9 \times 2} = \sqrt{9} \times \sqrt{2} = 3\sqrt{2}$

5 $\dfrac{2}{\sqrt{3}} = \dfrac{2}{\sqrt{3}} \times \dfrac{\sqrt{3}}{\sqrt{3}} = \dfrac{2\sqrt{3}}{3}$

3 $(2 + \sqrt{3})^2 = (2 + \sqrt{3})(2 + \sqrt{3})$
$$= 4 + 2 \times 2\sqrt{3} + (\sqrt{3})^2$$
$$= 4 + 4\sqrt{3} + 3$$
$$= 7 + 4\sqrt{3}$$

6 $\dfrac{2 - \sqrt{3}}{2 + \sqrt{3}} = \dfrac{(2 - \sqrt{3})(2 - \sqrt{3})}{(2 + \sqrt{3})(2 - \sqrt{3})} = \dfrac{4 - 4\sqrt{3} + 3}{4 - 3} = 7 - 4\sqrt{3}$

Exercise 1.12

1 Simplify

   (a) $\sqrt{8}$ **(d)** $\sqrt{20}$ **(g)** $\sqrt{98}$ **(j)** $\sqrt{2a^2}$

   (b) $\sqrt{27}$ **(e)** $\sqrt{48}$ **(h)** $\sqrt{125}$ **(k)** $\sqrt{a^3}$

   (c) $\sqrt{50}$ **(f)** $\sqrt{72}$ **(i)** $\sqrt{128}$ **(l)** $\sqrt{\pi r^2}$

2 Expand and simplify

   (a) $\sqrt{2}(\sqrt{2} + 1)$ **(d)** $(2 + \sqrt{3})(3 + \sqrt{2})$ **(g)** $(\sqrt{6} - \sqrt{5})(\sqrt{6} + \sqrt{5})$ **(j)** $\sqrt{2}(1 + \sqrt{2})^2$

   (b) $(\sqrt{2} + 1)^2$ **(e)** $(\sqrt{3} + \sqrt{2})(\sqrt{3} - \sqrt{2})$ **(h)** $(3 - \sqrt{5})^2$ **(k)** $(\sqrt{3} + 1)(\sqrt{3} - 1)$

   (c) $(\sqrt{2} + 1)(\sqrt{2} - 1)$ **(f)** $(12 - \sqrt{3})(12 + \sqrt{3})$ **(i)** $(1 + \sqrt{2})(2 + \sqrt{2})$ **(l)** $(3 + \sqrt{3})(3 - \sqrt{3})$

3 Rationalize (as in Worked Examples **5** and **6**) i.e. with surds only in the numerator.

   (a) $\dfrac{1}{\sqrt{2}}$ **(e)** $\dfrac{1}{\sqrt{3} + 2}$ **(i)** $\dfrac{6}{3 + \sqrt{3}}$ **(m)** $\left(\dfrac{1}{2}\right)^2 + \left(\dfrac{\sqrt{3}}{2}\right)^2$

   (b) $\dfrac{1}{\sqrt{3}}$ **(f)** $\dfrac{1}{3 + \sqrt{2}}$ **(j)** $\dfrac{3 + \sqrt{3}}{\sqrt{3} + 1}$ **(n)** $\left(\dfrac{1}{\sqrt{2}}\right)^2 + \left(\dfrac{1}{\sqrt{2}}\right)^2$

   (c) $\dfrac{1}{\sqrt{2} - 1}$ **(g)** $\dfrac{1}{\sqrt{5} - 2}$ **(k)** $\dfrac{2 + \sqrt{2}}{2 - \sqrt{2}}$ **(o)** $\left(\dfrac{2}{3}\right)^2 + \left(\dfrac{\sqrt{5}}{3}\right)^2$

   (d) $\dfrac{1}{\sqrt{2} + 1}$ **(h)** $\dfrac{2}{\sqrt{3} - 1}$ **(l)** $\dfrac{2 + \sqrt{2}}{\sqrt{2} + 1}$ **(p)** $\left(\dfrac{3}{4}\right)^2 + \left(\dfrac{\sqrt{7}}{4}\right)^2$

Partial fractions

$$\frac{3}{4} = \frac{1}{2} + \frac{1}{4} \cdots (1) \qquad \frac{7}{8} = \frac{1}{2} + \frac{1}{4} + \frac{1}{8} \cdots (2) \qquad \frac{5}{6} = \frac{1}{2} + \frac{1}{3} \cdots (3)$$

$\dfrac{1}{x+1} + \dfrac{1}{x+2} = \dfrac{2x+3}{(x+1)(x+2)}$ We can easily build up fractions because there are rules with methods (i.e. algorithms) to achieve this.

Can we break down fractions into components? i.e. into their partial fractions? $\dfrac{2}{(x+1)(x+2)} = \dfrac{?}{(x+1)} + \dfrac{?}{(x+2)}$ $\cdots (A)$

Why do **(1)** and **(3)** break down into 2 fractional components, yet **(2)** breaks down into 3?

Discuss possible solutions for expressing **(A)**, **(B)**, **(C)** and **(D)** in partial fractions.

(B) $\dfrac{2x}{(x+1)(x+2)}$ **(C)** $\dfrac{2x+3}{(x+1)(x+2)}$ **(D)** $\dfrac{x^2}{(x+1)(x+2)}$

Perhaps it is easier to start with a possible solution for the partial fractions.

$$\frac{a}{x+1} + \frac{b}{x+2} = \frac{a(x+2) + b(x+1)}{(x+1)(x+2)} = \frac{ax + 2a + bx + b}{(x+1)(x+2)} = \frac{(a+b)x + 2a + b}{(x+1)(x+2)}$$

To produce an answer to **(A)** $\dfrac{2}{(x+1)(x+2)}$ we need $a + b = 0$ and $2a + b = 2$

$$\Rightarrow \quad a = 2 \text{ and } b = -2$$

Solution to **(A)** is $\dfrac{2}{(x+1)(x+2)} = \dfrac{2}{x+1} + \dfrac{-2}{x+2} = \dfrac{2}{x+1} - \dfrac{2}{x+2}$

(B) $\dfrac{2x}{(x+1)(x+2)}$ requires $a + b = 2$ and $2a + b = 0 \Rightarrow a = -2, b = 4$

Solution to **(B)** is $\dfrac{2x}{(x+1)(x+2)} = \dfrac{-2}{x+1} + \dfrac{4}{x+2}$

(C) $\dfrac{2x+3}{(x+1)(x+2)}$ requires $a + b = 2$ and $2a + b = 3 \Rightarrow a = 1, b = 1$

Solution to **(C)** is $\dfrac{2x+3}{(x+1)(x+2)} = \dfrac{1}{x+1} + \dfrac{1}{x+2}$

(D) $\dfrac{x^2}{(x+1)(x+2)}$ requires $x^2 = (a+b)x + 2a + b$ so if a and b are numbers, we will not find values of a, b which work for all x values.

Consider $\dfrac{11}{6} = \dfrac{?}{2} + \dfrac{?}{3}$. We would firstly write $\dfrac{11}{6} = 1 + \dfrac{5}{6} = 1 + \dfrac{1}{2} + \dfrac{1}{3}$

$\frac{11}{6}$ is an **improper** fraction (top-heavy or vulgar) so we first change its form into a whole number plus a **proper** fraction. This is the method to use with **(D)**.

$$\frac{x^2}{(x+1)(x+2)} = \frac{x^2}{x^2+3x+2} = \frac{x^2+3x+2-3x-2}{x^2+3x+2} = \frac{x^2+3x+2}{x^2+3x+2} - \frac{3x+2}{x^2+3x+2} = 1 - \frac{3x+2}{(x+1)(x+2)}$$

$$\frac{3x+2}{(x+1)(x+2)} = \frac{a}{x+1} + \frac{b}{x+2} = \frac{a(x+2)+b(x+1)}{(x+1)(x+2)} = \frac{ax+bx+2a+b}{(x+1)(x+2)}$$

So we require $3x + 2 = (a+b)x + 2a + b \quad \Rightarrow \quad a + b = 3,\ 2a + b = 2 \quad \Rightarrow \quad a = -1$ and $b = 4$

$$\frac{x^2}{(x+1)(x+2)} = 1 - \left(\frac{-1}{x+1} + \frac{4}{x+2}\right) = 1 + \frac{1}{x+1} - \frac{4}{x+2}$$

WORKED EXAMPLE

Separate $\dfrac{2x+3}{(x+2)(x+3)}$ into partial fractions.

Method 1   Assume $\dfrac{2x+3}{(x+2)(x+3)} = \dfrac{a}{x+2} + \dfrac{b}{x+3} = \dfrac{a(x+3)+b(x+2)}{(x+2)(x+3)} = \dfrac{(a+b)x+3a+2b}{(x+2)(x+3)}$

$\Rightarrow \quad 2x + 3 = (a+b)x + 3a + 2b \quad \Rightarrow \quad a + b = 2 \cdots\cdots \textbf{(1)}$

$\qquad\qquad\qquad\qquad\qquad\qquad\qquad\qquad 3a + 2b = 3 \cdots\cdots \textbf{(2)}$

$\textbf{(2)} - 2 \times \textbf{(1)}$ gives $\quad a = -1 \quad \Rightarrow \quad b = 3 \quad$ by substituting $a = -1$ in equation $\textbf{(1)}$

So, $\quad \dfrac{2x+3}{(x+2)(x+3)} = \dfrac{-1}{x+2} + \dfrac{3}{x+3}$

Method 2 $\dfrac{2x+3}{(x+2)(x+3)} = \dfrac{a}{x+2} + \dfrac{b}{x+3} = \dfrac{a(x+3)+b(x+2)}{(x+2)(x+3)}$

requiring $\quad 2x + 3 = a(x+3) + b(x+2) \quad$ for all values of x

Put $x = -3 \quad \Rightarrow \quad$ LHS $= 2(-3) + 3 = -3$ and RHS $= a(0) + b(-1) \quad \Rightarrow \quad -3 = -b \quad \Rightarrow \quad b = 3$

Put $x = -2 \quad \Rightarrow \quad 2(-2) + 3 = a(-2+3) + b(-2+2) \quad \Rightarrow \quad -1 = a \quad \Rightarrow \quad a = -1$

So, $\quad \dfrac{2x+3}{(x+2)(x+3)} = \dfrac{-1}{x+2} + \dfrac{3}{x+3}$

Exercise 1.13

Separate into partial fractions using both methods.

1 $\dfrac{1}{(x+1)(x+4)}$

2 $\dfrac{1}{(x+1)(x-2)}$

3 $\dfrac{x+5}{(x-2)(x-3)}$

4 $\dfrac{2x+1}{(x-2)(x+2)}$

5 $\dfrac{x^2+1}{(x+2)(x-1)}$

6 $\dfrac{3x}{(2x+1)(x+2)}$

Covering-up rule

This is based on Method **2**.

Consider $\dfrac{x+5}{(x-2)(x-3)} = \dfrac{a}{x-2} + \dfrac{b}{x-3} = \dfrac{a(x-3)+b(x-2)}{(x-2)(x-3)} \quad$ requiring $\quad x + 5 = a(x-3) + b(x-2)$

$x = 3$ gives $3 + 5 = b(3 - 2)$ so b is the value of $\dfrac{x + 5}{x - 2}$ with $x = 3$ i.e. $b = \dfrac{3 + 5}{3 - 2} = 8$

$x = 2$ gives $2 + 5 = a(2 - 3)$ so a is the value of $\dfrac{x + 5}{x - 3}$ with $x = 2$ i.e. $a = \dfrac{2 + 5}{2 - 3} = \dfrac{7}{-1} = -7$

Hence $\dfrac{x + 5}{(x - 2)(x - 3)} = \dfrac{-7}{x - 2} + \dfrac{8}{x - 3}$

'a' is found by 'covering-up' $(x - 2)$ in LHS and putting $x = 2$ in the rest of expression $\dfrac{x + 5}{x - 3}$

'b' is found by 'covering-up' $(x - 3)$ in LHS and putting $x = 3$ in the rest $\dfrac{x + 5}{x - 2}$

With practice we can use this method with very little working, but care will have to be taken with the more complicated examples we shall meet later.

However the next example will show that time can be saved by using the 'covering-up' method.

WORKED EXAMPLE

Separate $\dfrac{2x + 4}{(x - 1)(x - 2)(x + 3)}$ into $\dfrac{a}{x - 1} + \dfrac{b}{x - 2} + \dfrac{c}{x + 3}$

$$\frac{a}{x - 1} + \frac{b}{x - 2} + \frac{c}{x + 3} = \frac{a(x - 2)(x + 3) + b(x - 1)(x + 3) + c(x - 1)(x - 2)}{(x - 1)(x - 2)(x + 3)}$$

requiring $2x + 4 = a(x - 2)(x + 3) + b(x - 1)(x + 3) + c(x - 1)(x - 2)$

$x = 1$ gives $2 + 4 = a(-1)(4) + b(0)(4)\ \ \ + c(0)(-1)$ $\Rightarrow$ $6 = -4a$ $\Rightarrow$ $a = \frac{-3}{2}$

$x = 2$ gives $4 + 4 = a(0)(5)\ \ \ + b(1)(5)\ \ \ + c(1)(0)$ $\Rightarrow$ $8 = 5b$ $\Rightarrow$ $b = \frac{8}{5}$

$x = -3$ gives $-6 + 4 = a(-5)(0) + b(-4)(0) + c(-4)(-5)$ $\Rightarrow$ $-2 = 20c$ $\Rightarrow$ $c = \frac{-1}{10}$

Thus, $\dfrac{2x + 4}{(x - 1)(x - 2)(x + 3)} = \dfrac{\frac{-3}{2}}{x - 1} + \dfrac{\frac{8}{5}}{x - 2} + \dfrac{\frac{-1}{10}}{x + 3} = \dfrac{-3}{2(x - 1)} + \dfrac{8}{5(x - 2)} - \dfrac{1}{10(x + 3)}$

The covering-up rule can be used again.

Put $x = 1$ to find a; $a = \dfrac{2x + 4}{(x - 2)(x + 3)} = \dfrac{2 + 4}{(1 - 2)(1 + 3)} = \dfrac{6}{(-1)(4)} = \dfrac{6}{-4} = \dfrac{-3}{2}$

Put $x = 2$ to find b; $b = \dfrac{2x + 4}{(x - 1)(x + 3)} = \dfrac{4 + 4}{(2 - 1)(2 + 3)} = \dfrac{8}{(1)(5)} = \dfrac{8}{5}$

Put $x = -3$ to find c; $c = \dfrac{2x + 4}{(x - 1)(x - 2)} = \dfrac{-6 + 4}{(-3 - 1)(-3 - 2)} = \dfrac{-2}{(-4)(-5)} = \dfrac{-2}{20} = \dfrac{-1}{10}$

Exercise 1.14

Separate into partial fractions

1 $\dfrac{3x + 1}{(x - 1)(x + 1)}$ **3** $\dfrac{11x - 9}{(3x - 1)(x - 3)}$ **5** $\dfrac{3x - 2}{(x + 1)(x - 2)(x + 3)}$ **7** $\dfrac{(x + 1)^2}{(x - 1)(x - 2)}$

2 $\dfrac{4x - 4}{x^2 - 4}$ **4** $\dfrac{(2x - 1)(x - 1)}{(x - 2)(x + 1)}$ **6** $\dfrac{9x}{(2x - 1)(x - 2)(x + 1)}$ **8** $\dfrac{(x + 1)^3}{(x - 1)(x - 2)}$

2 Functions

Number sets $\mathbb{N}, \mathbb{Z}, \mathbb{Q}, \mathbb{R}, \mathbb{C}$

Solve the equations
$$x + 3 = 5 \cdots\cdots (1)$$
$$x + 5 = 3 \cdots\cdots (2)$$
$$2x = 7 \cdots\cdots (3)$$
$$x^2 = 2 \cdots\cdots (4)$$
$$x^2 = -4 \cdots (5)$$

and state to which set of numbers each answer belongs.

Equation **(1)** is trivial, the answer being 2, a positive whole number, which belongs to the set of **natural numbers** $\mathbb{N} = \{1, 2, 3, 4, \ldots,\}$ also known as the counting numbers.

Equation **(2)** has solution $x = -2$, a negative whole number, so to solve equations such as this we need to expand our number set to include whole numbers positive and negative including zero. This set is called the set of **integers** $\mathbb{Z} = \{0, \pm 1, \pm 2, \pm 3, \ldots\}$.

Equation **(3)** has no whole number solution. $x = 3\frac{1}{2}$, which belongs to the **rational numbers**, or fractions. These are numbers which can be expressed as a ratio of two integers. So $3\frac{1}{2}$ can be expressed as $\frac{7}{2}$, where 2 and 7 belong to $\mathbb{Z}$.

In general, the set $\mathbb{Q}$ of rational numbers consists of numbers of the form p/q where $p, q \in \mathbb{Z}$ and $q \neq 0$. The word rational comes from the ratio p/q.

$\mathbb{Q}$ contains $\mathbb{Z}$ and $\mathbb{N}$ since -6 can be expressed as $\frac{-6}{1}$, so $\mathbb{N}$ has been extended to $\mathbb{Z}$ and $\mathbb{Z}$ to $\mathbb{Q}$. Thus, $\mathbb{Q} \supset \mathbb{Z} \supset \mathbb{N}$.

In trying to solve equations **(2)** and **(3)** we have extended our basic set of natural numbers by inventing a new type of number. We still require our numbers to obey the same laws of combination $(+, -, \times, \div)$ so that the **rationals** include the previous numbers considered.

Each new equation may require a different type of number in order that the equation can be solved.

Equations like type **(4)** may have a solution in $\mathbb{N}$, $\mathbb{Z}$ or $\mathbb{Q}$.

For example, $x^2 = 4 \Rightarrow x = 2$ $(2 \in \mathbb{N})$ or $x = -2$ $(-2 \in \mathbb{Z})$

and $x^2 = 2.25 \Rightarrow x = 1\frac{1}{2}$ or $-1\frac{1}{2}$ (both belonging to $\mathbb{Q}$)

However, $x^2 = 2$ poses a problem; $x^2 = 2 \Rightarrow x = \sqrt{2}$ the square root of 2.

A calculator gives $\sqrt{2} = 1.4142136$ but $1.4142136^2 = 2.0000001$, so 1.4142136 is not the **exact** square root of 2.

Does $\sqrt{2}$ have an exact square root? What do we mean by an exact square root? $1.41^2 = 1.9881$; $1.414^2 = 1.999396$; $1.4142^2 = 1.9999616$ and the more decimal places we consider for $\sqrt{2}$ the closer, on squaring, we approach 2.

Can we ever find an exact decimal which when squared produces 2 exactly?

If you can, find it!!

If you cannot, prove it!

1.41, 1.414 and 1.4142 are all rational numbers since $1.41 = \dfrac{141}{100}$, $1.414 = \dfrac{1414}{1000}$ and $1.4142 = \dfrac{14\,142}{10\,000}$, so any terminating decimal is a rational number. Some rational numbers can be expressed in a decimal form which terminates.

For example, $\quad \dfrac{3}{8} = 0.375 \qquad \dfrac{1}{4} = 0.25 \qquad \dfrac{3}{5} = 0.6$

Other rational numbers have a decimal form which recurs but is infinite.

For example, $\quad \dfrac{1}{3} = 0.333\ldots$ which is written in shorthand form as $0.\dot{3}$

$\dfrac{1}{6} = 0.166\ldots$ which is written in shorthand form as $0.1\dot{6}$

$\dfrac{1}{7} = 0.142\,857\,142\,857\ldots$ in shorthand form $0.\dot{1}42\,85\dot{7}$ the bank of six repeating digits within the dots being the recurring set.

Do all recurring decimals have a fractional (rational number) form?

Here is a method (algorithm) for changing a recurring decimal into a fraction.

$x = 0.121\,212\ldots = 0.\dot{1}\dot{2}$
$100x = 12.1212\ldots$
Subtracting $\quad 99x = 12$

$y = 0.123\,123\,123\ldots$
$1000y = 123.123\,123\ldots$
Subtracting $\quad 999y = 123$

(the decimal tails disappearing)

$\Rightarrow \quad x = \dfrac{12}{99} = \dfrac{4}{33} = 0.\dot{1}\dot{2}$

$\Rightarrow \quad y = \dfrac{123}{999} = \dfrac{41}{333}$

Similarly, $0.\dot{1}23\dot{4} = \dfrac{1234}{9999} \quad$ and $\quad 0.\dot{4} = \dfrac{4}{9}$.

Some students make the mistake of writing $\dfrac{2}{3} = 0.66$ instead of $0.\dot{6}$.

$$0.66 = \dfrac{66}{100} \quad \text{but} \quad \dfrac{2}{3} = \dfrac{66}{99}$$

So pure mathematicians will write $\dfrac{8}{3} = 2\frac{2}{3}$, not 2.66. But $\dfrac{8}{3} = 2.\dot{6}$.

Exercise 2.1

With the aid of a calculator:

1 Find the recurring decimal forms for

(a) $\dfrac{1}{9}, \dfrac{2}{9}, \dfrac{3}{9}, \dfrac{4}{9}, \dfrac{5}{9}, \dfrac{6}{9}, \dfrac{7}{9}, \dfrac{8}{9}, \dfrac{9}{9}$

(b) $\dfrac{1}{7}, \dfrac{2}{7}, \dfrac{3}{7}, \dfrac{4}{7}, \dfrac{5}{7}, \dfrac{6}{7}$

(c) $\dfrac{1}{11}, \dfrac{2}{11}, \dfrac{3}{11}, \dfrac{4}{11}, \dfrac{5}{11}, \dfrac{6}{11}, \dfrac{7}{11}, \dfrac{8}{11}, \dfrac{9}{11}, \dfrac{10}{11}$

2 (a) On your calculator, $\pi =$ (b) $\dfrac{22}{7} =$ (c) $\dfrac{355}{113} =$

3 Find a proof that $\sqrt{2}$ cannot be written as a rational number i.e. in the form p/q.

Irrational numbers

$\sqrt{2} \simeq 1.4142$ is an **irrational** number, $\mathbb{F}$; its decimal form does not terminate nor does it recur. It is an example of an algebraic number, being the root of an algebraic equation like

$$a_n x^n + a_{n-1} x^{n-1} + \cdots + a_1 x + a_0 = 0$$

There are also irrational numbers which are not algebraic, the best known being $\pi \simeq 3.14159\ldots$. The decimal form for π does not recur or terminate.

A rational number approximation for π is $\dfrac{22}{7} = 3.\dot{1}42\,85\dot{7}$

A better approximation is $\dfrac{355}{113} \simeq 3.141\,592\,9$

In decimal form **rational numbers** terminate or recur.

In decimal form **irrational numbers** are infinite and non-recurring.

We can invent irrational numbers, such as $0.4040040004\ldots$, which have a pattern but do not recur. Later you will meet e, the exponential base number.

Real numbers

All the numbers mentioned so far, naturals $\mathbb{N}$, integers $\mathbb{Z}$, rationals $\mathbb{Q}$ and irrationals make up the set of **real numbers**, $\mathbb{R}$, which consist of all the numbers on the real number line.

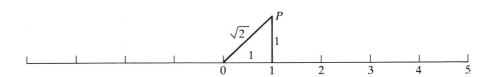

$\sqrt{2}$ has been constructed using a right-angled triangle (sides $1, 1, \sqrt{2}$) from which a length $\sqrt{2}$ has been constructed (OP) and by rotating OP through $45°$ clockwise, a length $\sqrt{2}$ can be pinpointed on the real number line.

This method of construction can be extended to produce $\sqrt{3}, \sqrt{4} = 2, \sqrt{5}$ etc. as in this figure.

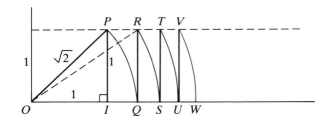

$$OP = \sqrt{2} = OQ$$
$$OR^2 = OQ^2 + QR^2 = (\sqrt{2})^2 + 1^2 = 2 + 1 = 3 \quad \Rightarrow \quad OR = \sqrt{3} = OS$$
$$OT^2 = OS^2 + ST^2 = (\sqrt{3})^2 + 1^2 = 3 + 1 = 4 \quad \Rightarrow \quad OT = \sqrt{4} = 2 = OU$$
$$OV^2 = OU^2 + UV^2 = (\sqrt{4})^2 + 1^2 = 4 + 1 = 5 \quad \Rightarrow \quad OV = \sqrt{5} = OW$$

Complex numbers

From our original set of five equations at the beginning of the chapter, we can now solve the first four types.

Type **(4)** $x^2 = 2$ has 2 solutions $x = +\sqrt{2}$ or $-\sqrt{2}$

Similarly, $x^2 = 4 \Rightarrow x = 2$ or -2

Type **(5)** $x^2 = -4$ causes difficulties. Our first guess is $x = -2$, but $(-2)^2 = +4$.

We use a symbol i to stand for $\sqrt{-1}$, so that $i^2 = -1$, and express all square roots of negative whole numbers in terms of i (some books use j).

$$x^2 = -1 \Rightarrow x = i \quad \text{or} \quad x = -i \quad \text{since } (-i) \times (-i) = +i^2 = -1$$

$$x^2 = -4 \Rightarrow x = 2i \quad \text{or} \quad x = -2i$$

$$x^2 = -2 \Rightarrow x = \sqrt{2}i \quad \text{or} \quad x = -\sqrt{2}i$$

The numbers involving i are called **complex numbers**, $\mathbb{C}$, and are dealt with in Chapter 6 of Book 2. They are also mentioned later in this chapter, in the section on quadratic equations.

Exercise 2.2

To which of the following sets of numbers do the solutions of the following equations belong?

naturals,   integers,   rationals,   irrationals,   reals,   complex
$\quad \mathbb{N} \qquad\quad \mathbb{Z} \qquad\quad \mathbb{Q} \qquad\quad \mathbb{F} \qquad\quad \mathbb{R} \qquad\quad \mathbb{C}$

1 $3x + 2 = 1$

2 $x^2 = 7$

3 $x^2 = 9$

4 $(x + 1)^2 = 9$

5 $x^2 = -9$

6 $4x^2 = 9$

7 $x^2 = 6.25$

8 $x^2 = 3.24$

9 $x^2 + 4^2 = 5^2$

10 $x^2 + 4^2 = 3^2$

11 $x^3 = 8$

12 $x^3 = -8$

13 $x^2 = x$

14 $x^2 = 1$

15 $x^3 = x$

16 Which of the following are finite decimals?

$$\frac{1}{3}, \frac{1}{6}, \frac{1}{8}, \frac{3}{25}, \frac{4}{50}, \frac{5}{16}, \frac{7}{75}, \frac{4}{125}$$

17 Express these numbers in rational form

$0.\dot{6}, 0.3\dot{6}, 0.0\dot{3}\dot{7}, 0.\dot{8}5714\dot{2}$

18 Prove that $\sqrt[2]{5}$ is an irrational number.

19 Draw a Venn diagram to represent the sets $\mathbb{N}, \mathbb{Z}, \mathbb{Q}$ and $\mathbb{R}$.

20 Simplify the following sets

(a) $\mathbb{N} \cap \mathbb{Z}$ (b) $\mathbb{N} \cup \mathbb{Q}$ (c) $\mathbb{Z} \cap \mathbb{R}$

Mappings

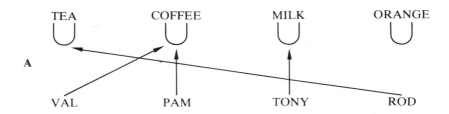

It's time to have a break, so Rod drinks tea, Tony milk for energy, while Pam and Val drink coffee. The arrows represent the relation between the four people and their drink.

Each person could drink more than one drink and then the mapping diagram would have more arrows.

If only one cup of each drink were available, the diagram could look like **B**.

One person has one drink, which gives a

> ONE to ONE mapping

In diagram **A**, more than one person (many) have one drink so **A** represents a

> MANY to ONE mapping

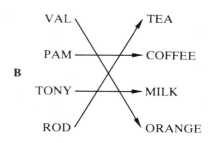

Diagram **C** shows the result when the men are thirsty.
MANY persons have ONE drink (coffee)
ONE person has MANY drinks (the men have two)
This provides an example of a

> MANY to MANY mapping

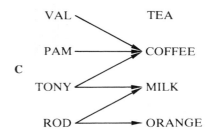

If the drinking set is restricted to Tony and Rod, diagram **D** shows the result.
ONE person has MANY (more than one) drinks; a

> ONE to MANY mapping

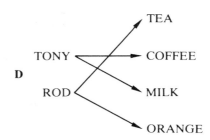

The type of mapping may depend on the number in the starting set (**domain**) and the number in the finishing set (**codomain**).

The domain is the set of people, the codomain the set of drinks.

In diagram **A**, the domain is {Val, Pam, Tony, Rod} the codomain is {Tea, Coffee, Milk, Orange} and the set of drinks actually taken {Tea, Coffee, Milk} is called the **range**.

An **inverse** mapping reverses the process, mapping drinks onto the people who drank them. In diagram **D**, orange and tea are drunk by Rod, and the inverse mapping is MANY to ONE.

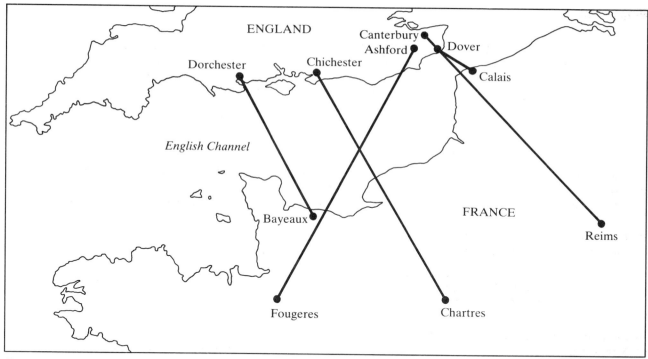

Five pairs of twinned towns – an example of one to one mapping

WORKED EXAMPLE

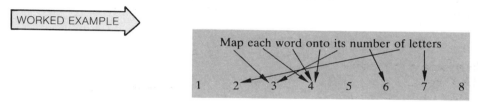

The domain has 8 members: the codomain could have as many as 8 members. The range has 5 members.

The mapping is MANY to ONE.

Exercise 2.3

For each mapping state the domain, codomain, range and specify the type of mapping.

1 Map each person in the class onto (a) their age
  (b) their birth-date
  (c) the day of their birthday this year
  (d) their last homework mark
  (e) their house number

2 Map the first five natural numbers onto (a) the first five odd numbers (b) the first five prime numbers.

3 Map the first nine natural numbers onto their prime factors.

4 With domain $\mathbb{N}$, classify the mapping (a) $x \to 2x + 1$
  (b) $x \to x^2$
  (c) $x \to \sqrt{x}$

5 With domain $\mathbb{Z}$, classify the mapping (a) $x \to x^2$
  (b) $x \to x^3$

Functions

Mappings which are ONE to ONE or MANY to ONE are called **functions**.
   A function specifies a rule by which a member of the domain (object) is mapped onto its corresponding member in the codomain (image).

For the function $f: x \to 2x$ the image of 1 is 2 and the image of 2 is 4.

Different notations are used to describe a function.

   $f: x \to 2x$ is read as 'the function f such that x goes to $2x$'.

   $f(x) = 2x$ reads 'f of x equals $2x$', which means $f(1) = 2$ and $f(2) = 4$.

Representing functions

Different diagrams can be used to represent a function, depending on the context and perhaps the domain and codomain.

(A)   function $f: x \to 2x$

   domain $\{1, 2, 3, 4, 5\}$

$$
\begin{array}{c}
\overset{f}{} \\
1 \longrightarrow 2 \\
2 \longrightarrow 4 \\
3 \longrightarrow 6 \\
4 \longrightarrow 8 \\
5 \longrightarrow 10 \\
\cdots \\
x \longrightarrow 2x
\end{array}
$$

(B)   The domain and codomain can be represented by vertical number lines.

   If the domain is represented by x values and the codomain by y values,

   x is mapped onto y

   and the function $f(x) = 2x$ can be denoted by $y = 2x$

   In this diagram it is easy to extend the domain to include all real numbers i.e. $x \in \mathbb{R}$.

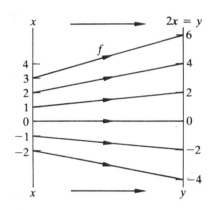

(C) $y = 2x$ leads to another diagram, where x and corresponding y values are plotted as co-ordinates on a graph:

$$(1, 2)\ (2, 4)\ (3, 6)\ (4, 8)$$

   The domain forms the x-axis.
   The codomain forms the y-axis.

   Points satisfying the relation (function) lie on a straight line $y = 2x$.

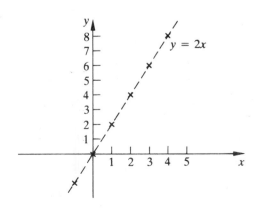

Inverse functions

The drinking function F in the diagram tells us that $F(\text{VAL}) = \text{ORANGE}$.

$$F$$
$$\text{VAL} \longrightarrow \text{ORANGE}$$
$$\text{PAM} \longrightarrow \text{COFFEE}$$
$$\text{TONY} \longrightarrow \text{MILK}$$
$$\text{ROD} \longrightarrow \text{TEA}$$

Its **inverse function** F^{-1} specifies that ORANGE was drunk by VAL.

$$F^{-1}(\text{ORANGE}) = \text{VAL}$$
$$F^{-1}(\text{COFFEE}) = \text{PAM}$$

An inverse reverses the direction of the function.

For $f(x) = 2x$, $f(1) = 2$, $f(2) = 4$ where each number is **doubled**.

Consequently $f^{-1}(2) = 1$, $f^{-1}(4) = 2$ and the inverse function will **halve** each number.

So, $f^{-1}(x) = \dfrac{1}{2}x = \dfrac{x}{2}$

Difficulties will arise if the function is MANY to ONE.

Consider $g(x) = x^2$ $\qquad g(2) = 4 \qquad$ and $g(-2) = 4$

So $g^{-1}(4)$ could be 2 or -2, and the inverse mapping is not a function.
If we restrict the domain to positive numbers no problems arise, and $g^{-1}(x) = \sqrt{x}$ the positive square root of x.

In practice with $g(x) = x^2$ it is easy to consider both the positive and negative square root so $g^{-1}(x) = \pm\sqrt{x}$, although strictly the inverse mapping is not a function.

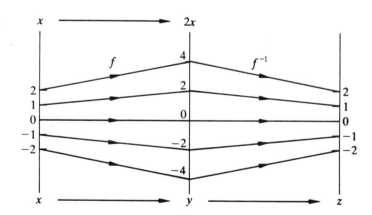

The figure shows $f(x) = 2x$ represented on the x and y lines. To find the inverse function f^{-1}, either we reverse the arrows to take y numbers back to the x values, or we can find another function f^{-1} which will map y values onto z values knowing that the z values must be the same as the x values.

f^{-1} takes y values to x values, $\quad y = 2x \;\Rightarrow\; x = \frac{1}{2}y$ so f^{-1} takes y to $\frac{1}{2}y$ i.e. $\;f^{-1}(y) = \frac{1}{2}y$ or $f^{-1}(x) = \frac{1}{2}x$

It is easy (in this case) to see that the z values are half the y values, so $z = \frac{1}{2}y$

$$\text{and } f^{-1}(y) = \tfrac{1}{2}y \;\Rightarrow\; f^{-1}(x) = \tfrac{1}{2}x$$

Functions

The figure illustrates two important facts

(A) $\left.\begin{array}{l} f(x)=2x \\ \text{or} \quad y=2x \\ \text{or} \quad y=f(x) \end{array}\right\} \Rightarrow x=f^{-1}(y) \quad \text{or} \quad \begin{array}{l} f^{-1}(y)=\frac{1}{2}y \\ f^{-1}(x)=\frac{1}{2}x \end{array}$

The inverse function $f^{-1}(x)$ reverses the arrows on the mapping diagram.

(B) If f^{-1} is performed after f, the result is x i.e. the application of a function f followed by its inverse f^{-1} takes each value back to its starting value.

$$f^{-1}f(x)=x$$

Exercise 2.4

For the functions in **1** to **9**

    (a) sketch (or describe) the mapping diagram (as in diagram **(B)**, Representing Functions)
    (b) describe the function in words
    (c) sketch (or describe) its inverse function (f^{-1})
    (d) describe the inverse function in words
    (e) express the inverse function in the form $f^{-1}(x)=$

1 $f(x)=3x$ **4** $F(x)=\frac{1}{4}x$ **7** $V(x)=\dfrac{1}{x}$

2 $t(x)=x+2$ **5** $I(x)=-x$ **8** $E(x)=\dfrac{6}{x}$

3 $h(x)=x-3$ **6** $S(x)=6-x$ **9** $N(x)=-x+5$

10 to 18 For each function in questions **1** to **9**, sketch the graph, as in diagram **(C)**, Representing Functions. On the same graph sketch in a different colour the graph of the corresponding inverse function. What do you notice?

Composite functions

Consider the function $f(x)=2x$ followed by $g(x)=x+1$.

$$f(1)=2 \text{ and } g(2)=3 \Rightarrow g[f(1)]=3 \quad gf(1)=3$$
$$f(2)=4 \text{ and } g(4)=5 \Rightarrow g[f(2)]=5 \text{ or } gf(2)=5$$
$$f(3)=6 \text{ and } g(6)=7 \Rightarrow g[f(3)]=7 \quad gf(3)=7$$

$gf(1)$ means do f first then do g to the result.
gf stands for f first then g, and this can be confusing!

However the composite function gf has to act on a number like $gf(1)$ and f has to act on the 1 first, then g on the result $f(1)$.

Thinking of gf as **doubling** then **adding one** leads to $gf(x)=2x+1$

Alternatively, $gf(x)=g(2x)=2x+1$

$$fg(1)=f(2)=4; \quad fg(2)=f(3)=6; \quad fg(3)=f(4)=8$$

$$fg(x)=f(x+1)=2(x+1)=2x+2; \quad \text{so } fg\neq gf$$

32

WORKED EXAMPLES

Using $f(x) = 2x$ and $g(x) = x + 1$

(a) express $h(x) = x + 3$ in terms of f and/or g
(b) express $k(x) = fg^2$ in terms of x

(a) $x + 3$ can be seen as adding 1 three times
   once $g(x) = x + 1$
   twice $gg(x) = g(x + 1) = x + 1 + 1 = x + 2$
   thrice $ggg(x) = g(x + 2) = x + 2 + 1 = x + 3$ $h(x) = x + 3 = ggg(x) = g^3(x)$

(b) $fg^2(x) = fgg(x) = fg(x + 1) = f(x + 2) = 2(x + 2) = 2x + 4$

Exercise 2.5

Express the following in terms of $f(x) = 2x$ and $g(x) = x + 1$.

1 $W(x) = 2x + 2$ 3 $H(x) = 2(x + 1)$ 5 $I(x) = 2x - 2$ 7 $V(x) = \frac{1}{2}(x + 1)$ 9 $N(x) = \frac{1}{2}(2x + 2)$

2 $T(x) = 2(x + 2)$ 4 $F(x) = x - 1$ 6 $S(x) = \frac{1}{2}x$ 8 $E(x) = 2x - 1$

Express the following composite functions in the form $C(x) =$

10 $f^2g = ffg$ 13 gfg 16 $g^{-1}g$ 19 $(fg)^{-1}$

11 $gf^2 = gff$ 14 fgf 17 ff^{-1} 20 $f^{-1}g^{-1}$

12 $g^2f = ggf$ 15 $f^3 = fff$ 18 $g^{-1}f^{-1}$ 21 $(gf)^{-1}$

Using $f(x) = 2x$, $g(x) = x + 1$, $h(x) = 3x + 2$ and $k(x) = x^2$, find the following functions in terms of x.

22 fh 26 kg 30 hk 34 $(gh)^{-1}$

23 hf 27 gk 31 kh 35 $(fh)^{-1}$

24 gh 28 fk 32 h^{-1} 36 $f^{-1}h^{-1}$

25 hg 29 kf 33 gh^{-1} 37 $h^{-1}f^{-1}$

38 For $S(x) = 5 - x$, find   (a) $SS(1)$ (b) $SS(2)$ (c) $SS(3)$ (d) $SS(x)$ (e) What do you deduce about $S(x)$?

39 For $R(x) = \dfrac{5}{x}$, find   (a) $RR(1)$ (b) $RR(2)$ (c) $RR(3)$ (d) $RR(x)$

Inverse of a composite function

Exercise 2.5, questions **18** and **19**, lead us to the result $(fg)^{-1} = g^{-1}f^{-1}$.

The logic is simple. fg stands for g first then f.

START with $x \rightarrow$ | DO g | to get $g(x)$, then | DO f | to get $fg(x)$ —|

REVERSE

$g^{-1}f^{-1}(x)$ | UNDO g / DO g^{-1} | $\leftarrow f^{-1}(x)$ | UNDO f / DO f^{-1} | $\leftarrow x$ $\leftarrow$

Reverse the instructions (or flow diagram) to get back to where you started.

Using $f(x) = 2x \Rightarrow f^{-1}(x) = \frac{1}{2}x$ and $g(x) = x + 1 \Rightarrow g^{-1}(x) = x - 1$

$$fg(x) = f(x + 1) = 2(x + 1) \qquad g^{-1}f^{-1}(x) = g^{-1}(\tfrac{1}{2}x) = \tfrac{1}{2}x - 1$$

There is a simpler (algebraic) process which saves times and is easier to apply with more complicated composite functions.

If $y = fg(x)$, then $x = (fg)^{-1}(y)$, so $(fg)^{-1}$ takes y back to x.

$$y = 2(x + 1) \Rightarrow \tfrac{1}{2}y = x + 1 \Rightarrow x = \tfrac{1}{2}y - 1$$

$(fg)^{-1}$ takes y back to $\frac{1}{2}y - 1$, so $(fg)^{-1}(y) = \frac{1}{2}y - 1 \Rightarrow (fg)^{-1}(x) = \frac{1}{2}x - 1$

i.e. **1** Put fg in the form $y = 2(x + 1)$

    2 Find x in terms of y i.e. $x = \frac{1}{2}y - 1$

    3 Change y for x i.e. $(fg)^{-1} = \frac{1}{2}x - 1$

Exercise 2.6

Find the inverses of the following functions.

1 $W(x) = x + 7$ **5** $I(x) = 3(x + 2)$ **9** $N(x) = 9 - x$

2 $T(x) = 2x - 1$ **6** $S(x) = 2(x - 3)$ **10** $X(x) = 12 \div x = \dfrac{12}{x}$

3 $H(x) = 2x + 3$ **7** $V(x) = -x$ **11** $L(x) = \dfrac{12}{9 - x}$

4 $F(x) = 3x + 2$ **8** $E(x) = \dfrac{1}{x}$ **12** $D(x) = 9 - \dfrac{12}{x}$

13 For the function $f(x) = 3x + 2$ **(a)** find $f^{-1}(x)$ **(b)** find in terms of x, $f^{-1}f(x)$ **(c)** $ff^{-1}(x)$
    (d) can you prove this in general? i.e. that $ff^{-1}(x) = x$.

Rational functions

The composite functions so far considered have been built up using successive applications of simple one-step functions. We need the algebraic confidence to cope with all types of function.

Can we find the inverse of $f(x) = \dfrac{x + 2}{x + 1}$?

Method 1 $\dfrac{x + 2}{x + 1} = \dfrac{x + 1 + 1}{x + 1} = \dfrac{x + 1}{x + 1} + \dfrac{1}{x + 1} = 1 + \dfrac{1}{x + 1}$

Flow diagram:

START $x \rightarrow$ | ADD 1 | $\xrightarrow{x+1}$ | DIVIDE INTO 1 | $\xrightarrow{\frac{1}{x+1}}$ | ADD 1 | $\rightarrow \dfrac{1}{x+1} + 1$ ⎞ REVERSE

$\dfrac{1 - (x - 1)}{x - 1} = \dfrac{1}{x - 1} - 1$ ← | SUBTRACT 1 | $\xleftarrow{\frac{1}{x-1}}$ | DIVIDE INTO 1 | $\xleftarrow{x-1}$ | SUBTRACT 1 | $\leftarrow x$

So, $f^{-1}(x) = \dfrac{1-x+1}{x-1} = \dfrac{2-x}{x-1}$

Method 2 Put $y = \dfrac{x+2}{x+1}$ Cross multiply $y(x+1) = x+2$

$$\Rightarrow \quad xy + y = x + 2$$

Isolate x $\quad xy - x = 2 - y$

$$\Rightarrow \quad x(y-1) = 2 - y$$

$$\Rightarrow \quad x = \dfrac{2-y}{y-1}$$

So, $f^{-1}(y) = \dfrac{2-y}{y-1} \quad \Rightarrow \quad f^{-1}(x) = \dfrac{2-x}{x-1}$

Both methods involve work with rational functions (the ratio of two functions) and every student must develop the ability and confidence to cope with the necessary algebraic manipulation. The work on partial fractions (Chapter 1) gave practice in adding and subtracting rational functions and the work on polynomials (later in this Chapter) will provide examples for multiplication and division.

WORKED EXAMPLE

Use both methods above to find the inverse function for $y = f(x) = \dfrac{2x+3}{x+4}$.

Method 1   Divide out the algebraic function $\dfrac{2x+3}{x+4} = \dfrac{2x+8-8+3}{x+4} = \dfrac{2x+8}{x+4} - \dfrac{5}{x+4} = 2 - \dfrac{5}{x+4}$

Flow diagram:

| START x | | ADD 4 | $x+4$ | DIVIDE INTO 5 | $\dfrac{5}{x+4}$ | SUBTRACT FROM 2 | $\rightarrow 2 - \dfrac{5}{x+4}$ |

REVERSE

| $\dfrac{5}{2-x} - 4$ | SUBTRACT 4 | $\dfrac{5}{2-x}$ | DIVIDE INTO 5 | $2-x$ | SUBTRACT FROM 2 | $\leftarrow x \leftarrow$ |

So, $f^{-1}(x) = \dfrac{5}{2-x} - 4 = \dfrac{5 - 4(2-x)}{2-x} = \dfrac{5 - 8 + 4x}{2-x} = \dfrac{4x-3}{2-x}$

Method 2   Put $y = \dfrac{2x+3}{x+4}$.   Cross multiply $y(x+4) = 2x+3 \quad \Rightarrow \quad xy + 4y = 2x + 3$

Isolate x $\quad xy - 2x = 3 - 4y \quad \Rightarrow \quad x(y-2) = 3 - 4y$

$$\Rightarrow \quad x = \dfrac{3-4y}{y-2} = \dfrac{4y-3}{2-y}$$

Change y for $x \quad \Rightarrow \quad f^{-1}(x) = \dfrac{4x-3}{2-x}$.

Exercise 2.7

Use both methods to find the inverse functions to those given.

1 $\dfrac{3x+2}{x+4}$

2 $\dfrac{2x-3}{x-4}$

3 $\dfrac{2x}{x-1}$

4 $\dfrac{3}{x-1}$

5 $\dfrac{2x+3}{x}$

6 $\dfrac{2x+3}{4}$

7 $5+\dfrac{2}{x+1}$

8 $3-\dfrac{1}{x-1}$

9 $\dfrac{1}{2-x}$

10 $7-\dfrac{3}{x-1}$

11 $\dfrac{7x-10}{x-1}$

12 $\dfrac{3x-4}{x-1}$

13 **(a)** Find the inverse of the function $y=f(x)=\dfrac{3x+4}{5x+7}$

(b) Can you spot the pattern of coefficients change from $f(x)$ to $f^{-1}(x)$?

(c) Make a table of results so far obtained:

$f(x)$	$f^{-1}(x)$
$\dfrac{x+2}{x+1}$	$\dfrac{-x+2}{x-1}=\dfrac{+x-2}{-x+1}$
$\dfrac{2x+3}{x+4}$	$\dfrac{4x-3}{-x+2}$
$\dfrac{3x+4}{5x+7}$	

(d) Guess the inverse of $y=f(x)=\dfrac{ax+b}{cx+d}$ and check it using Method **2**.

(e) Express $\dfrac{3}{x-1}$ in the form $\dfrac{ax+b}{cx+d}$ and check that the 'switching' method agrees with your answer to question **4**.

(f) Repeat part **(e)** for $\dfrac{1}{2-x}$; $3x+2$; $6-x$; $\dfrac{12}{x}$ etc.

(g) How is this method related to inverse matrices?

Polynomials

1 2 3 4

What does 1234 mean to you? Are we counting? Is there a space (or comma) between 1 and 2?

ONE THOUSAND, TWO HUNDRED AND THIRTY FOUR

1×10^3 $+$ 2×10^2 $+$ 3×10 $+$ 4

1 five cubed 2 fives squared 3 fives + four

1×5^3 $+2\times5^2$ $+3\times5$ $+4$ if we are using base 5

or $1\times x^3$ $+2\times x^2$ $+3x$ $+4$ if we are using base x

or $(x-1)^3+2(x-1)^2+3(x-1)+4$ if we are using base $x-1$

Can you change 1234_{10} (base 10) into $abcd_5$ (base 5)? Or is it $abcde_5$?

Can you change $(x-1)^3 + 2(x-1)^2 + 3(x-1) + 4$, a **polynomial** in powers of $x-1$, into a polynomial in powers of x?

A **polynomial** is an expression of terms of positive powers of a number, so $x^3 + 2x^2 + 3x + 4$ is a polynomial in terms of x whose **degree** is 3 (its highest power being x^3).
   In fact, degree 3 polynomials are also called **cubics** or **trinomials**.

Addition and subtraction of polynomials

$$f(x) = x^2 + x + 2 \text{ and } g(x) = x + 3 \quad \Rightarrow \quad f(x) + g(x) = x^2 + x + 2 + x + 3 = x^2 + 2x + 5$$

$$f(x) - g(x) = x^2 + x + 2 - x - 3 = x^2 - 1$$

Polynomials are added and subtracted by combining corresponding terms (powers of x).

Multiplication

$$g(x) \times f(x) = (x+3)(x^2 + x + 2) = x(x^2 + x + 2) + 3(x^2 + x + 2)$$
$$= x^3 + x^2 + 2x + 3x^2 + 3x + 6$$
$$= x^3 + 4x^2 + 5x + 6$$

$f(x) \times g(x)$ gives the same result.

Do not confuse $f(x) \times g(x)$ with $fg(x) = f[g(x)]$ the composite function.

   In this case $fg(x) = (x+3)^2 + (x+3) + 2 = x^2 + 6x + 9 + x + 3 + 2 = x^2 + 7x + 14$

Division

$$f(x) \div g(x) = \frac{x^2 + x + 2}{x+3} = \frac{x^2 + 6x + 9 - 5x - 7}{x+3} = \frac{(x+3)^2 - 5(x+3) + 8}{x+3} = \frac{(x+3)^2}{x+3} - \frac{5(x+3)}{x+3} + \frac{8}{x+3}$$

$$= x + 3 - 5 + \frac{8}{x+3} = x - 2 + \frac{8}{x+3}$$

Thus, when $x^2 + x + 2$ is divided by $x + 3$, the answer (quotient) is $x - 2$ with remainder 8.

i.e. $x^2 + x + 2 = (x+3)(x-2) + 8$

Compare this with $35 \div 8 = 4$ remainder 3.

$$35 = 8 \times 4 + 3 \qquad \text{i.e.} \qquad \frac{35}{8} = 4 + \frac{3}{8}$$

An **easier process** achieves the same result by **long division**.

Look at x into x^2, answer x

$x \times (x+3) = x^2 + 3x$

Subtract and bring down 2.

x into $-2x$ goes -2; $-2(x+3) = -2x - 6$

Subtract to give remainder 8.

$$\begin{array}{r} x - 2 \\ x+3\overline{)x^2 + x + 2} \\ \underline{x^2 + 3x} \\ 0 - 2x + 2 \\ \underline{-2x - 6} \\ 0 + 8 \end{array}$$

$$\text{c.f.} \quad \begin{array}{r} 21 \text{ rem. } 3 \\ 12\overline{)255} \\ \underline{24} \\ 15 \\ \underline{12} \\ 3 \end{array}$$

37

So $\quad x^2 + x + 2 = (x+3)(x-2) + 8$

$$\Rightarrow \quad \frac{x^2+x+2}{x+3} = \frac{(x+3)(x-2)}{x+3} + \frac{8}{x+3} = x - 2 + \frac{8}{x+3}$$

WORKED EXAMPLES

1 Write $M(x) = (x-1)^3 + 2(x-1)^2 + 3(x-1) + 4$ as a polynomial in x.

$$M(x) = (x-1)(x-1)^2 + 2(x^2 - 2x + 1) + 3(x-1) + 4$$
$$= (x-1)(x^2 - 2x + 1) + 2x^2 - 4x + 2 + 3x - 3 + 4$$
$$= x(x^2 - 2x + 1) - 1(x^2 - 2x + 1) + 2x^2 - 4x + 2 + 3x - 3 + 4$$
$$= x^3 - 2x^2 + x - x^2 + 2x - 1 + 2x^2 - 4x + 2 + 3x - 3 + 4$$
$$= x^3 - 3x^2 + 3x - 1 + 2x^2 - 4x + 2 + 3x + 1$$
$$= x^3 - x^2 + 2x + 2 \qquad\qquad \text{Note:} \quad (x-1)^3 = x^3 - 3x^2 + 3x - 1$$

2 Divide $M(x) = x^3 - x^2 + 2x + 2$ by $x - 1$.

From the original form of $M(x)$ in terms of $x - 1$, the remainder should be 4.

Long division:

x into x^3 goes x^2

$x^2(x-1) = x^3 - x^2$

Subtract and bring down $2x + 2$

x into $2x$ goes 2; $\quad 2(x-1) = 2x - 2$

$$\begin{array}{r} x^2 \qquad\quad + 2 \\ x-1 \overline{\smash{)}\, x^3 - x^2 + 2x + 2} \\ \underline{x^3 - x^2 \qquad\qquad} \\ 0 \qquad 0 \;+ 2x + 2 \\ \underline{2x - 2} \\ + 4 \text{ remainder} \end{array}$$

So, $\quad \dfrac{x^3 - x^2 + 2x + 2}{x+3} = x - 2 + \dfrac{4}{x-1}$

Exercise 2.8

For $f(x) = x^2 + 2x + 3$ and $g(x) = x + 1$ evaluate:

1 $fg(x)$	**6** $f(x) \div g(x)$	**11** $[g(x)]^4$	**16** $[g(x)]^3 \div f(x)$
2 $gf(x)$	**7** $[g(x)]^2$	**12** $g^{-1}(x)$	**17** $f(x) \div [g(x)]^2$
3 $f(x) + g(x)$	**8** $[f(x)]^2$	**13** $f^{-1}(x)$	**18** $f(x) \div g(x^2)$
4 $f(x) - g(x)$	**9** $f^2(x)$	**14** $g^{-1}f(x)$	**19** $f(x) \div (x+2)$
5 $f(x) \times g(x)$	**10** $[g(x)]^3$	**15** $fg^{-1}(x)$	**20** $f(x) \div (x-2)$

21 Expand $(ax+b)^2$, $(ax+b)^3$ and use these to guess $(ax+b)^4$

22 Compare $(a+b)^2$ with $(a+b+c)^2$ in their expanded forms.

23 Compare $(a+b)^3$ with $(a+b+c)^3$ in their expanded forms.

Remainder theorem

When 1234 is divided by 12, the answer is 102 remainder 10.

$$1234 = 102 \times 12 + 10$$

What is the remainder when $S(x) = x^2 + x + 2$ is divided by $x + 3$?

$$S(x) = x^2 + x + 2 = (x + 3)(x - 2) + 8 \text{ so the remainder is 8.}$$

What is the remainder when $M(x) = x^3 - x^2 + 2x + 2$ is divided by $x - 1$?

$$M(x) = (x - 1)^3 + 2(x - 1)^2 + 3(x - 1) + 4 = (x - 1)\{(x - 1)^2 + 2(x - 1) + 3\} + 4 \quad \text{so the remainder is 4.}$$

An easier way to find the remainder uses the factor form of the polynomial:

$$S(x) = (x + 3)(x - 2) + 8 \quad \Rightarrow \quad S(2) = (2 + 3)(2 - 2) + 8 = 5 \times 0 + 8 = 8 \quad \text{which is the remainder when } S(x) \text{ is divided} \\ \text{by } x - 2.$$

If $S(x)$ were divided by $(x + 3)$ the remainder would be $S(-3)$, the value of S when $x = -3$.

$$S(-3) = 0 \times (-5) + 8 = 8 \text{ the same remainder.}$$

To find the remainder when $M(x)$ is divided by $x - 1$, work out $M(1)$.

$$M(1) = 1^3 - 1^2 + 2 \times 1 + 2 = 1 - 1 + 2 + 2 = 4$$

When $M(x)$ is divided by $(x - 2)$, the remainder will be $M(2)$.

$$M(2) = 2^3 - 2^2 + 2 \times 2 + 2 = 8 - 4 + 4 + 2 = 10 \qquad \text{check this by long division}$$

In general, if a polynomial $P(x)$ is divided by $x - a$ to give a quotient $Q(x)$ with a remainder R, then

$$P(x) = (x - a) \times Q(x) + R$$

$x = a \quad \Rightarrow \quad P(a) = 0 \times Q(a) + R = R$ so the remainder $R = P(a)$.

> **Remainder Theorem:** The remainder on dividing a polynomial $P(x)$ by a linear factor $(ax + b)$ is given $P\left(\dfrac{-b}{a}\right)$.

For $S(x) = x^2 + x - 2$, $\quad S(1) = 1 + 1 - 2 = 0$ so the remainder is 0 when dividing by $x - 1$
i.e. $x - 1$ is a factor of S.

$$S(-2) = 4 - 2 - 2 = 0 \quad \Rightarrow \quad x + 2 \text{ is a factor of } S$$

$$S(x) = x^2 + x - 2 = (x - 1)(x + 2)$$

This special form of the Remainder Theorem (Factor Theorem) helps us to factorize polynomials.

> **Factor Theorem:** If on dividing a polynomial $P(x)$ by a linear factor $(ax + b)$ the remainder is **zero**, then $(ax + b)$ is a **factor** of $P(x)$.

WORKED EXAMPLE

Factorize $F(x) = x^3 + 2x^2 - x - 2$.

$$F(1) \quad = 1 + 2 - 1 - 2 \quad = 0 \quad \Rightarrow \quad x - 1 \text{ is a factor}$$
$$F(-1) = -1 + 2 + 1 - 2 = 0 \quad \Rightarrow \quad x + 1 \text{ is a factor}$$
$$F(2) \quad = 8 + 8 - 2 - 2 \quad = 12 \quad \Rightarrow \quad x - 2 \text{ is \textbf{not} a factor}$$
$$F(-2) = -8 + 8 + 2 - 2 = 0 \quad \Rightarrow \quad x + 2 \text{ is a factor}$$

A cubic has at most three linear factors $\Rightarrow$
$$F(x) = (x - 1)(x + 1)(x + 2)$$

Note 1 It is sometimes quicker to divide $F(x)$ by the first factor found i.e. $x - 1$ in this case.

So, $F(x) = x^3 + 2x^2 - x - 2 = (x - 1)(x^2 + 3x + 2) = (x - 1)(x + 1)(x + 2)$

Note 2 The possible factors are $(x \pm 1)$ or $(x \pm 2)$, so only try $F(\pm 1)$ or $F(\pm 2)$.

Note 3 Some gifted students can spot $F(x) = x^2(x + 2) - 1(x + 2)$
$$= (x + 2)(x^2 - 1)$$
$$= (x + 2)(x - 1)(x + 1)$$

Note 4 The factors are **unique**; whichever way round you factorize, you get the same 3 factors.

Exercise 2.9

1 Find the remainder when $M(x) = x^3 - x^2 + 2x + 2$ is divided by

    (a) $x + 1$ **(b)** $x + 2$ **(c)** $x - 2$ **(d)** $x - 3$ **(e)** x

    Has $M(x)$ any factors?

2 Find the remainder when $S(x) = x^2 + x + 2$ is divided by

    (a) $x + 1$ **(b)** $x + 2$ **(c)** $x - 1$ **(d)** $x - 2$ **(e)** x

    Has $S(x)$ any factors?

3 Find the remainder when $R(x) = x^3 - 6x^2 + 11x - 6$ is divided by

    (a) $x + 1$ **(b)** $x - 1$ **(c)** $x + 2$ **(d)** $x - 2$ **(e)** $x + 3$ **(f)** $x - 3$

    Factorize $R(x)$.

4 Factorize   **(a)** $x^3 - x$ **(c)** $x^3 - 3x^2 + 3x - 1$ **(e)** $x^2 - 5x + 6$ **(g)** $2a^2 + ab - b^2$

              (b) $x^3 - 8$ **(d)** $x^3 + 2x^2 + 4x + 8$ **(f)** $6x^2 - 5x + 1$ **(h)** $a^2 - b^2$

5 Is the second polynomial a factor of the first? If so find the other factor(s).

    (a) $x^2 + 7x + 6, \quad x + 1$ **(e)** $x^4 - 3x^2 + 2, \quad x + 1$ **(i)** $x^2 + 2x + 4, \quad x + 2$

    (b) $x^3 + 3x + 2, \quad x + 2$ **(f)** $x^3 - x, \quad x - 1$ **(j)** $x^3 + 2x^2 + 3x + 4, \quad x + 2$

    (c) $x^4 + 3x + 2, \quad x + 1$ **(g)** $x^2 + 2x + 1, \quad x - 1$ **(k)** $x^2 + x + 1, \quad x + 1$

    (d) $x^4 - 3x^2 + 2, \quad x - 1$ **(h)** $x^3 + 3x^2 + 3x + 1, \quad x - 1$ **(l)** $x^2 + 2x + 3, \quad x + 3$

6 Divide the first polynomial by the second, giving the quotient and the remainder

    (a) $x^2 + 2x + 1, \quad x + 1$ **(d)** $x^3 + 2x^2 + 3x + 4, \quad x + 2$ **(g)** $x^3 - 2x + 1, \quad x - 1$

    (b) $x^3 + 3x^2 + 3x + 1, \quad x - 1$ **(e)** $x^3 + x + 1, \quad x + 1$ **(h)** $x^3 - 8, \quad x - 2$

    (c) $x^2 + 2x + 4, \quad x + 2$ **(f)** $x^2 + 2x + 3, \quad x + 3$ **(i)** $x^3 + 27, \quad x + 3$

Quadratics

Quadratic polynomials

Can the product of two consecutive integers equal 12?
But of course; you can think of the answer in your head: 3 and 4.
You may also come up with -3 and -4.

The problem can be expressed as $x \times (x+1) = 12$ where x and $x+1$ are consecutive integers.

$$x(x+1) = 12 \quad \Rightarrow \quad x^2 + x = 12 \quad \Rightarrow \quad x^2 + x - 12 = 0 \quad \Rightarrow \quad (x-3)(x+4) = 0$$

$$\Rightarrow \quad x - 3 = 0 \quad \textbf{or} \quad x + 4 = 0$$

$$\Rightarrow \quad x = 3 \qquad \textbf{or} \quad x = -4$$

Can the product of two consecutive integers be 6? Yes: 2 and 3 or -3 and -2.
Can the product of two consecutive integers be 3? Perhaps, though an answer does not readily spring to mind.

$$x(x+1) = 3 \quad \Rightarrow \quad x^2 + x = 3 \quad \Rightarrow \quad x^2 + x - 3 = 0 \text{ which does not factorize,}$$

Is there a solution?

The quadratic formula $x = \dfrac{-b \pm \sqrt{b^2 - 4ac}}{2a}$ for solving $ax^2 + bx + c = 0$ leads to,

for $x^2 + x - 3 = 0$, $\quad x = \dfrac{-1 \pm \sqrt{1 - 4(1)(-3)}}{2} = \dfrac{-1 \pm \sqrt{13}}{2} = \dfrac{\sqrt{13} - 1}{2}$ or $\dfrac{-\sqrt{13} - 1}{2} = 1.303$ or -2.303.

There are no integer solutions.

Plotting the graph of $q(x) = x^2 + x - 12$ gives the **parabola** in the figure.

The graph cuts the x-axis at $x = 3$ and $x = -4$ which are the solutions of $x^2 + x - 12 = 0$.

The graph looks symmetrical, having a line of symmetry at $x = -\frac{1}{2}$ the values either side of this line being the same

i.e. $\quad q(0) = q(-1) = -12$
$\quad\quad q(1) = q(-2) = -10$
$\quad\quad q(2) = q(-3) = -6$ etc.

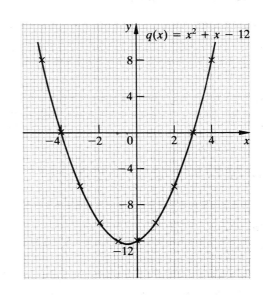

Plotting the graph of $p(x) = x^2 + x - 3$ gives the parabola in the lower figure.

The graph has an axis of symmetry at $x = -\frac{1}{2}$.
The graph cuts the x-axis at

$$x \simeq 1.3 \text{ and } x \simeq -2.3$$

from our calculations above, on this page.

The lowest point (minimum) is at $x = -\frac{1}{2}$.

$$y(\min) = p\left(-\tfrac{1}{2}\right) = \left(-\tfrac{1}{2}\right)^2 + \left(-\tfrac{1}{2}\right) - 3$$
$$= 0.25 - 0.5 - 3$$
$$= -3.25$$

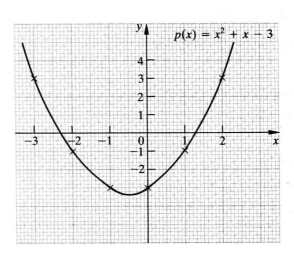

Exercise 2.10

Complete the following table.

	$q(x)$	Solutions (roots) of $q(x) = 0$	Sum of roots	Product of roots
1	$x^2 - 5x + 6$	$x = 2$ or $x = 3$	5	6
2	$x^2 - 3x + 2$			
3	$x^2 + x - 12$			
4		$x = 3$ or $x = 4$		
5		$x = \frac{1}{3}$ or $x = \frac{1}{4}$		
6	$x^2 + x - 3$			
7			12	20
8			8	15
9			-5	6
10			-3	4
11		$x = \frac{1}{2}$ or $x = 3$		
12		$x = 2$ or $x = \frac{1}{3}$		
13	$x^2 - 4x + 5$			

Roots of quadratics

If the roots (solutions) of a quadratic equation are α and β then the quadratic equation can be expressed as $(x - \alpha)(x - \beta) = 0$ in its factorized form.

$$(x - \alpha)(x - \beta) = 0 \quad \Rightarrow \quad x^2 - \alpha x - \beta x + \alpha\beta = 0$$
$$\Rightarrow \quad x^2 - (\alpha + \beta)x + \alpha\beta = 0$$

Thus, the sum of the roots is the negative of the coefficient of x and the product of the roots is the constant term, provided the coefficient of x^2 is unity.

If $ax^2 + bx + c = 0$ has roots α and β, then $\alpha + \beta = \dfrac{-b}{a}$ and $\alpha\beta = \dfrac{c}{a}$

For example, $2x^2 + 5x - 3 = 0 \quad \Rightarrow \quad (2x - 1)(x + 3) = 0$ which has roots $\alpha = \frac{1}{2}$ or $\beta = -3$

Sum $\alpha + \beta = \frac{1}{2} - 3 = -2\frac{1}{2} = \dfrac{-5}{2}$ **Product** $\alpha\beta = -\frac{1}{2} \times 3 = -1\frac{1}{2} = \dfrac{-3}{2}$

The quadratic formula for solving $2x^2 + 5x - 3 = 0$ leads to

$$x = \frac{-5 \pm \sqrt{25 - 4 \times (2)(-3)}}{4} = \frac{-5 \pm \sqrt{25 + 24}}{4} = \frac{-5 \pm \sqrt{49}}{4} = \frac{-5 + 7}{4} \text{ or } \frac{-5 - 7}{4} = \frac{1}{2} \text{ or } -3$$

Where does the quadratic formula come from?
  It certainly helps when quadratic equations do not factorize with whole numbers.

Completing the square

$$\text{If } f(x) = x^2 + 2x - 3 \text{ what is } f^{-1}(x)?$$

We need to express $f(x)$ as a composite function, so that f can be built up in easy stages starting from x. The key process is to work on x before squaring.

$$(x + 1)^2 = x^2 + 2x + 1$$

So $f(x) = x^2 + 2x - 3 = x^2 + 2x + 1 - 4 = (x + 1)^2 - 4$

$$y = (x + 1)^2 - 4 \quad \Rightarrow \quad (x + 1)^2 = y + 4 \quad \Rightarrow \quad x + 1 = \sqrt{y + 4} \quad \Rightarrow \quad x = -1 \pm \sqrt{y + 4}$$

$f^{-1}(x) = -1 \pm \sqrt{x + 4}$ which is not strictly a function, but we can use it if we remember to consider both positive and negative square roots.

The process of expressing $f(x) = x^2 + 2x - 3 = (x + 1)^2 - 4$ is called **completing the square** and enables us to solve $x^2 + 2x - 3 = 0$ in a different way.

$$x^2 + 2x - 3 = 0 \quad \Rightarrow \quad (x + 1)^2 - 4 = 0 \quad \Rightarrow \quad (x + 1)^2 = 4 \quad \Rightarrow \quad x + 1 = \sqrt{4} = +2 \text{ or } -2$$

$$x + 1 = 2 \quad \Rightarrow \quad x = 1 \quad \text{and} \quad x + 1 = -2 \quad \Rightarrow \quad x = -3$$

In this case $x^2 + 2x - 3 = 0$ factorizes into $(x - 1)(x + 3) = 0 \quad \Rightarrow \quad x = 1 \text{ or } x = -3$

However, $x^2 + 2x - 4 = 0 \quad \Rightarrow \quad x^2 + 2x + 1 - 5 = 0 \quad \Rightarrow \quad (x + 1)^2 = 5 \quad \Rightarrow \quad x = -1 \pm \sqrt{5}$

WORKED EXAMPLES

1 Complete the square for **(a)** $q(x) = x^2 + x - 12$ and **(b)** $p(x) = x^2 + x - 3$, interpreting your results.

(a) $q(x) = x^2 + x - 12$

$\quad = x^2 + x + \frac{1}{4} - 12\frac{1}{4}$

$\quad = (x + \frac{1}{2})^2 - 12\frac{1}{4}$

$(x + \frac{1}{2})^2 \geqslant 0$ for all values of x so the least value (minimum) of q occurs when $x = -\frac{1}{2}$ (line of symmetry) and $q(\text{min}) = -12\frac{1}{4}$

(b) $p(x) = x^2 + x - 3 = x^2 + x + \frac{1}{4} - 3\frac{1}{4}$

$\quad = (x + \frac{1}{2})^2 - 3\frac{1}{4}$

$p(\text{min}) = -3\frac{1}{4}$ occurs when $x = -\frac{1}{2}$ (line of symmetry)

2 Complete the square for $r(x) = 7 - 6x - x^2$ and $s(x) = 2x^2 + 12x + 6$.

$r(x) = 7 - 6x - x^2$

$\quad = 7 - (6x + x^2)$

$\quad = 7 + 9 - (x^2 + 6x + 9)$

$\quad = 16 - (x + 3)^2$

$s(x) = 2x^2 + 12x + 6 = 2(x^2 + 6x + 3)$

$\quad = 2(x^2 + 6x + 9 - 6)$

$\quad = 2[(x^2 + 6x + 9) - 6]$

$\quad = 2(x + 3)^2 - 12$

Exercise 2.11

Solve the quadratics by **(a)** factors and **(b)** completing the square.

1 $x^2 + 2x - 8 = 0$ **4** $2x^2 - x - 3 = 0$ **7** $x^2 - 4 = 0$ **10** $6x^2 - 5x - 6 = 0$

2 $x^2 + 4x - 5 = 0$ **5** $2(x + 1)^2 = 18$ **8** $5x^2 + 4x - 1 = 0$ **11** $6x^2 + x - 15 = 0$

3 $x^2 + 6x - 7 = 0$ **6** $3x^2 + 8x - 3 = 0$ **9** $6x^2 - 5x + 1 = 0$ **12** $ax^2 + bx + c = 0$

13 For each of the equations in **1** to **11**, check that the sum of the roots is $-b/a$ and the product of the roots is c/a when you regard the equation as $ax^2 + bx + c = 0$.

Functions

Quadratic formula for the solution of $ax^2 + bx + c = 0$

Divide the equation through by a
$$x^2 + \frac{b}{a}x + \frac{c}{a} = 0$$

Add $\left(\frac{b}{2a}\right)^2$ to both sides
$$x^2 + \frac{b}{a}x + \left(\frac{b}{2a}\right)^2 + \frac{c}{a} = \left(\frac{b}{2a}\right)^2$$

$$\left(x + \frac{b}{2a}\right)^2 + \frac{c}{a} = \frac{b^2}{4a^2}$$

$$\left(x + \frac{b}{2a}\right)^2 = \frac{b^2}{4a^2} - \frac{c}{a} = \frac{b^2 - 4ac}{4a^2}$$

Take the square root of both sides
$$x + \frac{b}{2a} = \sqrt{\frac{b^2 - 4ac}{4a^2}} = \frac{\sqrt{b^2 - 4ac}}{2a}$$

Subtract $\frac{b}{2a}$ from both sides
$$x = \frac{-b}{2a} \pm \frac{\sqrt{b^2 - 4ac}}{2a} = \frac{-b \pm \sqrt{b^2 - 4ac}}{2a}$$

Solving quadratics

Method 1   Factorization

$$x^2 + 2x - 8 = 0$$
$$(x - 2)(x + 4) = 0$$
$$x - 2 = 0 \text{ or } x + 4 = 0$$
$$x = 2 \text{ or } x = -4$$

Method 2   Completing the square

$$x^2 + 2x - 8 = 0$$
$$x^2 + 2x + 1 - 9 = 0$$
$$(x + 1)^2 = 9; \quad x + 1 = \pm 3$$
$$x + 1 = 3 \text{ or } x + 1 = -3$$
$$x = 2 \text{ or } x = -4$$

Method 3   Quadratic formula

$$x^2 + 2x - 8 = 0 \qquad \begin{array}{l} a = 1 \\ b = 2 \\ c = -8 \end{array} \qquad x = \frac{-b \pm \sqrt{b^2 - 4ac}}{2a} = \frac{-2 \pm \sqrt{4 - 4(1)(-8)}}{2}$$

$$= \frac{-2 \pm \sqrt{4 + 32}}{2} = \frac{-2 \pm \sqrt{36}}{2}$$

$$= \frac{-2 + 6}{2} \text{ or } \frac{-2 - 6}{2} = 2 \text{ or } -4$$

WORKED EXAMPLES

1 Solve $x^2 - 6x + 7 = 0$.

Some equations do not factorize. The formula gives $x = \dfrac{6 \pm \sqrt{36 - 4 \times 1 \times 7}}{2} = \dfrac{6 \pm \sqrt{8}}{2}$

$$= \frac{6 \pm 2\sqrt{2}}{2} = 3 + \sqrt{2} \text{ or } 3 - \sqrt{2}$$

$$x \simeq 4.14 \text{ or } 1.59$$

2 Solve $x^2 + 4x + 5 = 0$.

The formula gives $x = \dfrac{-4 \pm \sqrt{16 - 4(1)(5)}}{2} = \dfrac{-4 \pm \sqrt{16 - 20}}{2} = \dfrac{-4 \pm \sqrt{-4}}{2}$

$\sqrt{-4}$ is a **complex** number; $\quad \sqrt{-4} = 2i$ where $i = \sqrt{-1} \quad \Rightarrow \quad i^2 = -1$

$$x = \frac{-4 \pm 2i}{2} = \frac{-4 + 2i}{2} \text{ or } \frac{-4 - 2i}{2} = -2 + i \text{ or } -2 - i$$

Exercise 2.12

If practice is needed in using the quadratic formula, use it to solve the quadratic equations in Exercise 2.11.

Roots of quadratic equations

The nature of the roots of $ax^2 + bx + c = 0$ can be deciphered by using the formula $x = \dfrac{-b \pm \sqrt{b^2 - 4ac}}{2a}$.

The **discriminant** '$b^2 - 4ac$' can be   **(A)**   a perfect square, as in $x^2 + 2x - 8 = 0$ leading to rational roots
                                                        (B)   positive,   leading to real roots
                                                        (C)   zero,   leading to equal roots
                                                        (D)   negative,   leading to complex roots.

> WORKED EXAMPLES

1   (B) $x^2 - 6x + 7 = 0$

The upper figure is the graph of $y = x^2 - 6x + 7$.

'$b^2 - 4ac$' $= 36 - 28 = 8 > 0$ hence real roots.

The real roots of $x^2 - 6x + 7 = 0$ occur where the graph cuts the line $y = 0$ (x-axis).

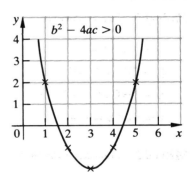

2   (C) $x^2 - 6x + 9 = 0$

The middle figure is the graph of $y = x^2 - 6x + 9$.

'$b^2 - 4ac$' $= 36 - 4 \times 9 = 0 \quad \Rightarrow \quad$ equal roots.

The solutions are $x = 3$ (repeated) and the graph 'touches' the x-axis ($y = 0$) at $x = 3$.

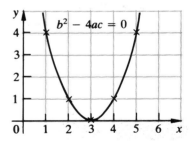

3   (D) $x^2 - 6x + 10 = 0$

The lower figure is the graph of $y = x^2 - 6x + 10$.

'$b^2 - 4ac$' $= 36 - 4 \times 10 = -4 < 0 \quad \Rightarrow \quad$ complex roots.

No real solutions $\Rightarrow$ the graph does not cut the x-axis.

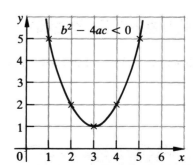

The graph of the general quadratic function $y = ax^2 + bx + c$

Different values of a, b and c will produce different graphs.

   It would be helpful to use a graph-plotting program on a computer to draw different quadratic graphs by varying the values of a, b and c.

Investigation 1

Keeping $b = c = 0$ what is the effect of varying 'a'?

(a) Consider $y_1 = x^2$, $\quad y_2 = 2x^2$, $\quad y_3 = 3x^2$ etc. What is the effect on the graph?

(b) Make 'a' smaller; i.e. compare $y_{\frac{1}{2}} = \frac{1}{2}x^2$, $\quad y_{\frac{1}{3}} = \frac{1}{3}x^2$ etc. with $y = x^2$,

(c) Make 'a' negative; draw $y_{-1} = -x^2$, $\quad y_{-2} = -2x^2$ etc.

Investigation 2

What is the effect of varying 'c'? You can probably answer this without too much trouble.

Investigation 3

Vary 'b'; consider $y = x^2 + x$, $\quad y = x^2 + 2x$, $\quad y = x^2 - 2x$ etc.

We shall see later that the ideas contained in these Investigations can be applied to graphs of other functions.

As an extension, compare with the basic $y = x^2$ the graphs of $y = (x + 1)^2$, $\quad y = (x + 2)^2$, $\quad y = (x - 1)^2$ and $y = (x - 2)^2$. Then try plotting $x = y^2$ and $x = (y + 1)^2$ etc.

You may have noticed that all quadratics have a line of symmetry.

WORKED EXAMPLE ▷

Sketch the graph $y = x^2 - 6x + 7$ and find its line of symmetry. What else does this tell you?

The previous Worked Examples show the graph of $y = x^2 - 6x + 7$.

Completing the square $\Rightarrow$ $y = x^2 - 6x + 9 - 2 = (x - 3)^2 - 2$

The line of symmetry is given by $x - 3 = 0$ $\Rightarrow$ $x = 3$ and this value of x gives the minimum value of y.

$y(\min) = (3 - 3)^2 - 2 = -2$ and $(3, -2)$ is the **minimum point** on the graph.

Comparing the graph of $y = (x - 3)^2 - 2$ with $y = x^2$, we can see that $y = (x - 3)^2 - 2$ is the translation of $y = x^2$, 3 units to the right and 2 units down.

If the coefficient of x^2 is negative, the graph is 'upside down' and the line of symmetry leads to a **maximum point**.

Completing the square for $y = ax^2 + bx + c$ gives $y = a\left(x^2 + \dfrac{bx}{a}\right) + c$

$$\Rightarrow \quad y = a\left(x^2 + \frac{bx}{a} + \frac{b^2}{4a^2}\right) + c - \frac{b^2}{4a}$$

$$= a\left(x + \frac{b}{2a}\right)^2 + \frac{4ac - b^2}{4a}$$

The minimum point is given by $x + \dfrac{b}{2a} = 0$ $\Rightarrow$ $x + \dfrac{-b}{2a}$ $\Rightarrow$ $y(\min) = \dfrac{4ac - b^2}{4a}$

Can you deduce the nature of the roots of $ax^2 + bx + c = 0$ from the value of $y(\min)$?

Exercise 2.13

1 For the following quadratic functions find **(i)** the line of symmetry **(ii)** the maximum or minimum point and **(iii)** sketch the curve.

  (a) $y = x^2 + 4x - 5$ **(b)** $y = 12 - 3x^2$ **(c)** $y = (x - 2)(x - 4)$ **(d)** $y = 2(x - 1)^2 - 3$

2 Which of the following equations have real roots, equal roots or complex roots?

  (a) $x^2 + x + 1 = 0$ **(b)** $x^2 + 4x + 4 = 0$ **(c)** $x^2 + 8x + 8 = 0$ **(d)** $x^2 + 8x + 16 = 0$

3 For what values of k do the following equations have equal roots?

  (a) $x^2 + kx + k = 0$ **(b)** $x^2 + x + k = 0$ **(c)** $(x + 2)(x + 4) = k$ **(d)** $(x + 2)(x + 8) = kx$

4 Show that the following quadratic expressions are positive for all real values of x.

  (a) $x^2 + 2x + 2$ **(b)** $x^2 - 2x + 2$ **(c)** $x^2 - 4x + 5$ **(d)** $x^2 - 6x + 10$

5 For what values of k do the following equations have real roots?

  (a) $x^2 + x + k = 0$ **(b)** $x^2 + kx + 4 = 0$ **(c)** $(x + 3)(x + 5) = k$ **(d)** $(x + 1)(x + 9) = kx$

Odd and even functions

Investigation 4

Plot the graphs of $y = x$, $y = x^2$, $y = x^3$, $y = x^4$ for $-2 \leqslant x \leqslant +2$.

(A) A computer graph plotting program would help. Can you superimpose $y = x^2$ and $y = x^3$ on the same picture for comparison?

(B) Without the computer use the table here with the aid of a calculator and the x^y button. Use more values, such as 1.2 and 1.4, if required.

x	-2	-1	0	0.5	1	1.5	2
x^2	4	1	0	0.25	1	2.25	4
x^3							
x^4							

1 As the power of x increases, what happens for **(a)** $0 \leqslant x \leqslant +1$ **(b)** $1 \leqslant x \leqslant 2$ **(c)** $-2 \leqslant x \leqslant -1$?

2 Are the graphs symmetric in any way? In which way?

3 What can you say about $y = x^2$ and $y = x^4$, the **even powers** of x?

4 What can you say about $y = x$ and $y = x^3$, the **odd powers** of x?

5 How would you describe the graphs of $y = x^5$ and $y = x^6$?

6 What can you say about $f(-x)$ for **(a)** **even** powers **(b)** **odd** powers?

7 Describe the symmetry of the graphs for **(a)** **even** powers **(b)** **odd** powers.

Exercise 2.14

1 Sketch the graphs of **(a)** $y = -x$ **(b)** $y = -x^2$ **(c)** $y = -x^3$ **(d)** $y = -x^4$. Which are odd or even?

2 Sketch the graphs of **(a)** $x + x^2$ **(b)** $x + x^3$ **(c)** $x^3 - x$ **(d)** $x^4 - x^2$. Which are odd or even?

3 Compare $f(+2)$ and $f(-2)$ for each of the functions in question **2**.

4 Can you tell whether a function is odd or even before sketching the graph?

5 Complete the following graphs if the functions are **(a)** odd **(b)** even.

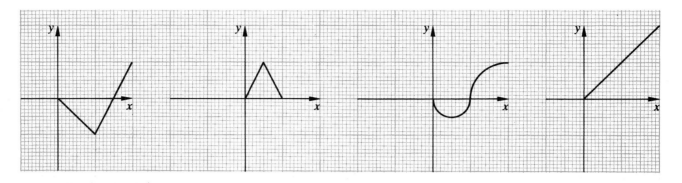

6 Which function is both odd and even?

7 We have considered the functions x^4, x^3, x^2 and x. What about x^0, x^{-1}, x^{-2}?

8 If $f(x)$ and $g(x)$ are both **even**, what can you say about **(a)** $f(x) + g(x)$ **(b)** $f(x) \times g(x)$?

9 If $f(x)$ and $g(x)$ are both **odd**, what can you say about **(a)** $f(x) + g(x)$ **(b)** $f(x) \times g(x)$?

3 Co-ordinates

Introduction

How does the pilot know where the plane is?

How is the plane's position specified?

What is the landing procedure?

How does the control tower 'talk' the plane down?

Certain facts are vital for the pilot to know at any time i.e. the plane's position, speed, altitude (height), direction of flight and distance from destination. The instrument panel will convey all this information, enabling the pilot to adjust when necessary. (In fact the 'automatic pilot' will probably do this.)

A pilot on a flight from Norwich to London, for example, could specify the plane's position in different ways:

(A) by giving his distance East and North of London;

(B) by giving his straight line distance and bearing (direction) from London;

(C) by specifying latitude and longitude.

In all cases, of course, he would specify the height (altitude) of the plane.

(A) Cartesian Co-ordinates

Relative to London:

Norwich is 63 miles East, 96 miles North
  $(63, 96)$
Doncaster is 53 miles West, 162 miles North
  $(-53, 162)$
Dover is 71 miles East, 33 miles South
  $(71, -33)$
Plymouth is 203 miles W, 74 miles South
  $(-203, -74)$

(B)   Polar Co-ordinates

Relative to London:

Norwich is 115 miles in direction North 33° East
   i.e. 33° to the East of North;

Doncaster is 170 miles in direction N 18° W;

Dover is 78 miles in direction S 65° E;

Plymouth is 216 miles in direction S 70° W.

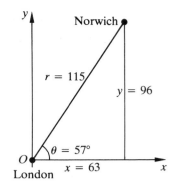

(C)   Spherical Polar Co-ordinates

Norwich is Latitude 52.6° N, Longitude 1.3° E.

(A)   Cartesian Co-ordinates (x, y) specify distances to the east (x) and to the north (y) of a fixed origin $O\,(0, 0)$. For Norwich, $x = 63$ and $y = 96$, when the origin is London.

To avoid confusion, towns to the west of London will have a negative x co-ordinate and towns to the south a negative y co-ordinate.
   So Doncaster is $(-53, 162)$ and Dover $(71, -33)$.

The Cartesian system can be extended to three dimensions by including the z co-ordinate to represent height above ground level (origin O).

(B)   Polar Co-ordinates (r, θ) specify the direct distance from the origin (r) and the direction (θ) made with a fixed direction Ox.
   For Norwich, $r = 115$ and $\theta = 57°$.

For Polar Co-ordinates the angle (direction) is measured anti-clockwise from the x-axis in contrast to a bearing which is measured **(a)**   clockwise from North or **(b)**   with reference to the compass points. Norwich is on a bearing 033° (or N 33° E or E 57° N). Doncaster has a bearing 342° (or N 18° W), but its polar co-ordinates are $(170, 108°)$.

(C)   Spherical Polar Co-ordinates   are used in higher mathematics. For navigational purposes these are the most accurate (since the earth closely resembles a sphere).

Cartesian co-ordinates

These are named after the French mathematician, Descartes.

The co-ordinates of a point fix its position relative to a fixed origin. In the figure:

The point O is the origin $(0, 0)$.

The co-ordinates of A are $(2, 3)$, 2 across, 3 up from O.

The co-ordinates of B are $(-2, 1)$, 2 back, 1 up from O.

The co-ordinates of C are $(-3, -4)$, 3 back, 4 down from O.

The co-ordinates of D are $(4, -2)$, 4 across, 2 down from O.

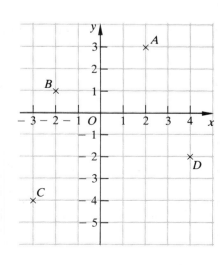

In general, for the point $P(x, y)$, x gives the displacement in the x direction and y gives the displacement in the y direction.

The vertical line through A consists of points whose x co-ordinate is 2, and is called the line $x = 2$.

The horizontal line through A consists of points whose y co-ordinate is 3, and is called the line $y = 3$.

The point A is the intersection of the line $x = 2$ with the line $y = 3$ and has co-ordinates $(2, 3)$.

Ox is called the x-axis whose equation is $y = 0$, each point on this line having y co-ordinate O.
Oy is called the y-axis with equation $x = 0$, each point on this line having x co-ordinate O.

Gradient

The gradient of a line measures its slope, the steeper the line, the greater its gradient.

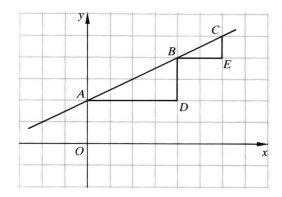

In the figure, the mathematical gradient of the line joining A to B is given by $\dfrac{BD}{AD} = \dfrac{2}{4} = \dfrac{1}{2}$.

Gradient $BC = \dfrac{CE}{BE} = \dfrac{1}{2}$

AB and BC form the same line ABC and the gradients of AB and BC both equal $\frac{1}{2}$.

Investigation 1

In the figure, right, the gradient of $PQ = \dfrac{2}{4} = \dfrac{1}{2}$ but the gradient of OS is $\dfrac{-1}{4}$ (OS slopes downwards as x increases)

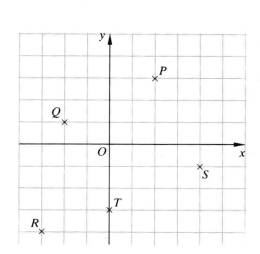

PQ, RQ and RS have positive gradients.

PS, OS and QO have negative gradients.

1. Find the gradients of OP, OQ, OR and OS.
    Can you relate each gradient to the co-ordinates of P, Q, R and S?

2. Find the gradients of $PQ, PS, TS, QT, RT, RS, RQ, RP, PT$ and QS.

3. What do the gradients of QP and TS tell you? Also QT and PS?

4. What shape is $PQTS$?

5. Compare the gradients of **(a)** QT and TS **(b)** QP and PS **(c)** QS and PT. Comment!

6. Can you find the gradient of HJ where H is (x_1, y_1) and J is (x_2, y_2)?

7. What is the gradient of Ox and TO?

Distance

Investigation 2   (figure, right)

1   Find the distance *PC*.

2   Find the distance *PB*, and comment.

3   Find the distance *OP, OQ, OR, OS, OT.*

Can you relate these distances to the co-ordinates of *P, Q, R, S* and *T*?

4   Find *PQ, QT, TS* and *PS*.

5   What shape is *PQTS*?

6   Find *PC, TS, TP* and *SC*. What shape is *PTSC*?

7   Insert the point *F*(1,0) and find *FP, FQ, FS, FT*.

What can you deduce?

8   Can you find the distance *HJ* where *H* is (x_1, y_1) and *J* is (x_2, y_2)?

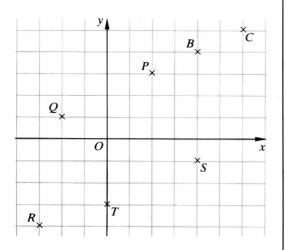

Mid-points

Investigation 3

In the figure for Investigation 2,

The mid-point of *P* (2, 3) and *C* (6, 5) is *B* (4, 4)

The mid-point of *Q* (− 2, 1) and *C* (6, 5) is *P* (2, 3)

The mid-point of *P* (2, 3) and *B* (4, 4) is ?

The mid-point of *P* (2, 3) and *S* is ?

1   Complete a table of values using as many mid-points from the diagram as you need to discover the general rule for mid-points.

2   Find the mid-point of *PT* and *QS*; comment!

3   Find the mid-point of *PS* and *TC*; comment!

4   Can you find the co-ordinates of the mid-point of *HJ* where *H* is (x_1, y_1) and *J* is (x_2, y_2)?

5   Join the mid-points of *PQ, PS, ST* and *TQ* in that order.

What is the resulting shape?

Straight line equations

Investigation 4

The points $(1, 1)$ $(2, 2)$ $(3, 3)$ $(-2, -2)$ all lie on the line $y = x$.
The points $(1, 2)$ $(2, 4)$ $(-1, -2)$ all lie on the line $y = 2x$ because for each point the y co-ordinate is double the x co-ordinate (see figure, right).

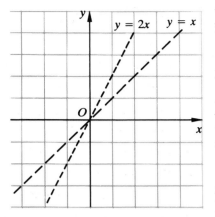

1 Find enough points on each line to draw the lines:
   $y = 3x$, $y = 4x$, $y = \frac{1}{2}x$, $y = \frac{1}{3}x$. What do you notice?

2 Find points on the lines $y = -x$, $y = -2x$, $y = -3x$, $y = -\frac{1}{2}x$.
   What do you notice?

3 What is the relation between $y = 2x$ and $y = -2x$?

4 What is the relation between $y = 2x$ and $y = -\frac{1}{2}x$?

5 What is the equation of the line through the origin O and (a) $(3, 4)$ (b) (a, b)?

6 What is the gradient of the line $y = 6x$?

7 What is the equation of the line through O perpendicular to $y = 6x$?

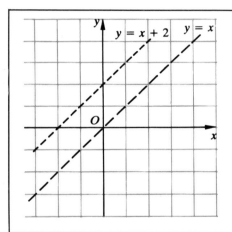

Investigation 5

In the figure, left, points $(0, 2)$ $(1, 3)$ $(2, 4)$ $(-1, 1)$ all lie on the line $y = x + 2$, because each y co-ordinate is 2 more than the x co-ordinate.

1 Find enough points to draw the lines $y = x + 1$, $y = x + 4$, $y = x - 1$, $y = x - 3$. What do you notice?

2 Where would you draw the lines $y = x + 10$ and $y = x - 8$?

3 Find enough points to draw the lines $y = -x$, $y = -x + 2$, $y = -x + 3$, $y = 4 - x$, $x + y = 3$. What do you notice?

Investigation 6

1 Find points which satisfy $y = 2x + 1$, $y = -3x + 2$, $y = \frac{1}{2}x + 2$ and draw their graphs.
   What do you notice?

2 What can you say about the straight line $y = mx + c$?

Through two given points

If (see figure) $P(x, y)$ is any point on the line AB, we can

express the gradient of PB as $\dfrac{y - 4}{x - 5}$.

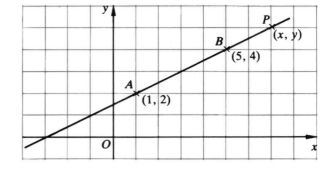

The gradient of AB is $\dfrac{4 - 2}{5 - 1} = \dfrac{2}{4} = \dfrac{1}{2}$.

Gradient of AB = gradient of PB $\Rightarrow$

$$\dfrac{y - 4}{x - 5} = \tfrac{1}{2} \quad \Rightarrow \quad y - 4 = \tfrac{1}{2}(x - 5)$$

$$\Rightarrow \quad y = \tfrac{1}{2}x - 2\tfrac{1}{2} + 4$$

$$\Rightarrow \quad y = \tfrac{1}{2}x + 1\tfrac{1}{2} \quad \text{which is the equation of } AB.$$

$x = 0 \quad \Rightarrow \quad y = 1\tfrac{1}{2}$, so AB cuts the y-axis at $(0, 1\tfrac{1}{2})$; the intercept on the y-axis is $1\tfrac{1}{2}$.

The gradients AP and AB being equal $\quad \Rightarrow \quad \dfrac{y - 2}{x - 1} = \tfrac{1}{2} \quad \Rightarrow \quad y - 2 = \tfrac{1}{2}(x - 1) \quad \Rightarrow \quad y = \tfrac{1}{2}x + 1\tfrac{1}{2}$.

Through a given point with a certain gradient

To find the equation of the straight line through the point $A(1, 2)$ with gradient 3, if $P(x, y)$ is a general point on

the line, the gradient of AP is $\dfrac{y - 2}{x - 1}$, which equals 3.

$$\dfrac{y - 2}{x - 1} = 3 \quad \Rightarrow \quad y - 2 = 3(x - 1) \quad \Rightarrow \quad y = 3x - 3 + 2 \quad \Rightarrow \quad y = 3x - 1$$

The equation of the straight line through $A(1, 2)$ with gradient m is $y - 2 = m(x - 1)$.

The equation of the straight line through $H(x_1, y_1)$ with gradient m is $\boxed{y - y_1 = m(x - x_1)}$

The equation $\boxed{y = mx + c}$ represents a straight line with **gradient** m and **intercept** c.

Summary of results

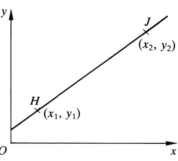

The **gradient** of the line HJ is $m = \dfrac{y_2 - y_1}{x_2 - x_1} = \dfrac{y_1 - y_2}{x_1 - x_2}$.

The **distance** $HJ = \sqrt{(x_2 - x_1)^2 + (y_2 - y_1)^2} = \sqrt{(x_1 - x_2)^2 + (y_1 - y_2)^2}$.

The **mid-point** of HJ has co-ordinates $\left(\dfrac{x_1 + x_2}{2}, \dfrac{y_1 + y_2}{2} \right)$.

The **straight line** HJ has equation $\qquad y - y_1 = m(x - x_1)$, where m = gradient of HJ,

which can be written as $\dfrac{y - y_1}{x - x_1} = \dfrac{y_2 - y_1}{x_2 - x_1}$ or $y - y_2 = m(x - x_2)$.

The gradient of the line **perpendicular** to a line of gradient m, has gradient $\dfrac{-1}{m}$.

Two lines with gradients m_1 and m_2 are perpendicular if $m_1 m_2 = -1$.

Exercise 3.1

1 For the following straight line equations, find **(a)** the gradient **(b)** the intercept on the y-axis **(c)** the point of intersection of the line with the x-axis.

(i) $y = 2x + 4$ **(ii)** $2y = 3x - 6$ **(iii)** $y = 5 - x$ **(iv)** $x + y = 3$ **(v)** $2x + 3y = 6$ **(vi)** $3y = x - 2$.

2 Find the equation of the straight lines through the points:
(a) $(0, 3)$ and $(4, 5)$ **(b)** $(1, 4)$ and $(5, 2)$ **(c)** $(4, -1)$ and $(-2, 1)$ **(d)** $(-2, 3)$ and $(-4, -2)$.

3 Find the equations of the lines through $(2, 1)$ with gradient **(a)** 1 **(b)** 2 **(c)** -3 **(d)** 0.

4 Find the equations of the lines through $(3, 4)$ parallel and perpendicular to $y = 2x + 1$.

5 **(a)** For the following points, $A(1, 2)$, $B(4, 3)$, $C(5, 7)$, $D(2, 6)$, find the lengths and gradients of AB, BC, CD, DA. What do you conclude?
  (b) Find the mid-point M of AC and the mid-point N of BD. What does this mean?
  (c) What shape is $ABCD$? Draw a diagram to confirm this.
  (d) Find the area of $ABCD$.

6 For $A(1, 2)$, $E(5, 3)$, $F(6, 7)$, $D(2, 6)$, repeat question **5**.
  Find the gradients of AF and DE. What do you notice?

7 Repeat question **5** for $A(1, 2)$, $J(4, 5)$, $G(6, 3)$ and $H(3, 0)$.
  Draw the points to confirm the shape. What else must you do to prove $AJGH$ is the true shape? Find the equation of JG. Can this be used to write down the equation of AH?

8 Find the equations of the sides and diagonals of a square of centre $(3, 2)$ when one side is $y = \frac{1}{2}x + 3$.

9 **(a)** For $O(0, 0)$, $A(6, 2)$, $B(1, 7)$, find OA, OB and AB. What kind of triangle is $\triangle AOB$?
  (b) Find $\angle AOX$, $\angle BOX$ and hence $\angle AOB$.
  (c) Find the other angles of the triangle? Is their sum $180°$?
  (d) Can you design an equilateral triangle with whole number co-ordinates? If so, write to the authors; if not, prove that it cannot be done!
  (e) Find the mid-point M of OA and the gradient BM. How should this compare with the gradient of OA? Does it?

10 Using $\triangle OAB$ from question **9** and M, the midpoint of OA,
  (a) Find N, the mid-point of AB, and the equation of ON.
  (b) Find the equation of BM and G the point of intersection of ON and BM.
  (c) Find the equation of AG and the point of intersection, K, of AG with OB.
  (d) Find the mid-point of OB. What do you notice?
  (e) Find the lengths of ON and OG and hence GN. Comment!
  (f) Prove that $BG = 2 \times GM$ (no calculation is needed; look for similar triangles).
  (g) How is G related to the triangle OAB? What is it called?

11 For $O(0, 0)$ and $D(6, 2)$ find the gradient of OD. Rotate D through $90°$ clockwise to D_1, state the co-ordinates of D_1 and the gradient OD_1. Rotate D through $90°$ anticlockwise to D_2 and state the co-ordinates of D_2 and the gradient OD_2. Compare the gradients of OD and OD_1.

12 Repeat question **11**, starting with the point (a, b). What can you say about the gradients of lines which are at right angles to each other?

13 What can you say about the gradients of lines which are parallel to each other?

Dividing in a given ratio

From the figure, with $C(1, 3)$ and $D(7, 6)$, the mid-point of CD is

$$(\tfrac{1}{2}(1 + 7), \tfrac{1}{2}(3 + 6)) = (4, 4\tfrac{1}{2})$$

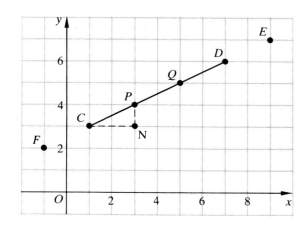

P and Q are the points of trisection (P nearer C) of CD so that $CP = PQ = QD$ and

$$CP:PD = 1:2.$$

Given the co-ordinates of C and D, can we calculate the co-ordinates of P and Q?

Remember: A **vector** is a quantity having magnitude and direction, e.g. force, or velocity.

The displacement (vector) $\mathbf{CD} = \begin{pmatrix} 6 \\ 3 \end{pmatrix}$ and $\mathbf{CP} = \tfrac{1}{3}\mathbf{CD} = \tfrac{1}{3}\begin{pmatrix} 6 \\ 3 \end{pmatrix} = \begin{pmatrix} 2 \\ 1 \end{pmatrix}$

Denoting the co-ordinates of P by X_P and Y_P,

$$\left.\begin{array}{l} X_P = X_C + CN = 1 + 2 = 3 \\ Y_P = Y_C + NP = 3 + 1 = 4 \end{array}\right\} \quad \text{So } P \text{ is } (X_P, Y_P) = (3, 4)$$

In terms of displacements (vectors) $\quad \begin{pmatrix} X_P \\ Y_P \end{pmatrix} = \begin{pmatrix} X_C \\ Y_C \end{pmatrix} + \begin{pmatrix} 2 \\ 1 \end{pmatrix} \quad$ or $\quad P = C + \begin{pmatrix} 2 \\ 1 \end{pmatrix}$

Similarly $\quad \begin{pmatrix} X_Q \\ Y_Q \end{pmatrix} = \begin{pmatrix} X_C \\ Y_C \end{pmatrix} + \tfrac{2}{3}\mathbf{CD} = \begin{pmatrix} X_C \\ Y_C \end{pmatrix} + \tfrac{2}{3}\begin{pmatrix} 6 \\ 3 \end{pmatrix} = \begin{pmatrix} 1 \\ 3 \end{pmatrix} + \begin{pmatrix} 4 \\ 2 \end{pmatrix} = \begin{pmatrix} 5 \\ 5 \end{pmatrix}$ so Q is $(5, 5)$.

In terms of co-ordinates, $\quad (X_P, Y_P) = (X_C, Y_C) + (2, 1) = (1, 3) + (2, 1) = (3, 4)$

$$(X_Q, Y_Q) = (1, 3) + (4, 2) = (5, 5)$$

The co-ordinates of the point P which divides CD in the ratio 1:2 can be found by adding $\tfrac{1}{3}$ of the difference between the C and D co-ordinates to the co-ordinates of C.

For example, $\qquad X_P = X_C + \tfrac{1}{3}(X_D - X_C) \quad$ and $\quad Y_P = Y_C + \tfrac{1}{3}(Y_D - Y_C)$

$$\Rightarrow \quad X_P = X_C + \tfrac{1}{3}X_D - \tfrac{1}{3}X_C \quad \text{and} \quad Y_P = Y_C + \tfrac{1}{3}Y_D - \tfrac{1}{3}Y_C$$

$$\Rightarrow \quad X_P = \tfrac{2}{3}X_C + \tfrac{1}{3}X_D \qquad\qquad \text{and} \quad Y_P = \tfrac{2}{3}Y_C + \tfrac{1}{3}Y_D$$

You may prefer to remember $\qquad \boxed{P = \tfrac{2}{3}C + \tfrac{1}{3}D} \quad \ldots\ldots\ldots (1)$

Similarly, $Q = \tfrac{1}{3}C + \tfrac{2}{3}D$. In co-ordinates $(X_Q, Y_Q) = \tfrac{1}{3}(1, 3) + \tfrac{2}{3}(7, 6) = (5, 5)$.

N.B. $\qquad P$ divides CD in the ratio 1:2 $\quad \Rightarrow \quad P = \tfrac{2}{3}C + \tfrac{1}{3}D$
$\qquad\qquad Q$ divides CD in the ratio 2:1 $\quad \Rightarrow \quad Q = \tfrac{1}{3}C + \tfrac{2}{3}D$

For mid-point M,
$\qquad\qquad M$ divides CD in the ratio 1:1 $\quad \Rightarrow \quad M = \tfrac{1}{2}C + \tfrac{1}{2}D \quad \Rightarrow \quad (X_M, Y_M) = (4, 4\tfrac{1}{2})$

If $\qquad T$ divides CD in the ratio 3:1 $\quad \Rightarrow \quad T = \tfrac{1}{4}C + \tfrac{3}{4}D$

T is three-quarters of the way from C to D $\Rightarrow$ $T = \frac{1}{4}(1,3) + \frac{3}{4}(7,6) = \left(\dfrac{1+21}{4}, \dfrac{3+18}{4} \right) = (5\frac{1}{2}, 5\frac{1}{4})$

The ratio for T is 3:1 $\Rightarrow$ ratios are $\frac{3}{4}$ and $\frac{1}{4}$ and you take $\frac{3}{4}D$ (not C) since T is nearer D than C.

Exercise 3.2

1 Find the co-ordinates of S, the point dividing CD (in the figure opposite) in the ratio 1:5.

2 Find the co-ordinates of J, the point dividing CD in the ratio 1:4

3 Find the co-ordinates of G, the point dividing CD in the ratio 4:1

4 Find the co-ordinates of H, the point dividing CD in the ratio 2:3

5 Triangle OAB has $O(0,0)$, $A(9,0)$ and $B(3,6)$.

   (a) Find M, the mid-point of OB and then G, which divides AM in the ratio 2:1.

   (b) Find L, the mid-point of AB and then H, which divides OL in the ratio 2:1.

   (c) Find N, the mid-point of OA and then I, which divides BN in the ratio 2:1.

   (d) What do you notice? What does this prove? Look at your diagram.

6 (a) For $O(0,0)$, $A(9,0)$, $B(6,3)$ find the co-ordinates of P which divides OA in the ratio 2:1 and Q which divides AB in the ratio 1:2.

   (b) Find the gradients of PQ and OB, and comment.

   (c) Find the lengths of PQ and OB, and comment.

   (d) What would be the result if the ratios were 3:1 and 1:3?

7 For the quadrilateral $O(0,0)$, $A(9,0)$, $C(9,6)$, $B(3,6)$, answer the following.

   (a) Find the centroid of $\triangle OAB$. (This is G in question 5. G is on AM such that $AG = 2GM$.)

   (b) Find X on GC where $GX = \frac{1}{4}GC$ (X divides GC in the ratio 1:3).

   (c) Find D the centroid of $\triangle ABC$ (D divides LC in the ratio 1:2; L is $(6,3)$).

   (d) Find Y on OD where $OY:YD = 3:1$ (i.e. Y divides OD in the ratio 3:1).

   (e) What do you notice about X and Y?

   (f) N is the mid-point of OA, M of OB, L of AB, E of AC, F of BC, H of OC. Find the mid-points of ME, FN and HL, and comment. Have you found the centre of the quadrilateral? If so which centre?

   (g) What are the various centres of a triangle?

   (h) Has a quadrilateral a centre, or perhaps more than one?

   (i) Find out about the centre of mass of a tetrahedron.

This work is connected with vector geometry which is dealt with in Chapter 7, on Vectors.

8 (a) Draw the triangle $A(0,4)$, $B(8,0)$, $C(3,10)$.

   (b) Find G, where $CG:GB = 2:3$; H, where $BH:HA = 3:1$; I, where $AI:IC = 1:2$.

   (c) Ceva's theorem states that if the ratios $\dfrac{CG}{GB} \times \dfrac{BH}{HA} \times \dfrac{AI}{IC} = 1$ as they do, then AG, BI and CH are concurrent (meet in a single point). Check that they do and find the co-ordinates of the meeting point.

   (d) Repeat for $CD:DB = 3:2$, $BE:EA = 1:3$, $AF:FC = 2:1$.

External division

In the figure at the beginning of this section, consider the point E lying on CD produced.

$CE:ED = 4:1$ numerically, but does not divide CD in the ratio 4:1 because there is a point near Q (G in question **3**) which does that. We have to say that E divides CD externally in the ratio 4:1 **or** in the ratio 4: -1.

Using the notation of equation **(1)** $E = D + \frac{1}{3}CD = D + \frac{1}{3}(D - C) = \frac{4}{3}D - \frac{1}{3}C$

You will notice that if the ratio is 4: -1, the fractions are $\dfrac{4}{4-1}$ and $\dfrac{-1}{4-1}$ i.e. $\dfrac{4}{3}$ and $\dfrac{-1}{3}$.

You may prefer to regard D as dividing CE in the ratio 3:1.

Then, $CD:DE = 3:1 \;\Rightarrow\; D = \frac{1}{4}C + \frac{3}{4}E \;\Rightarrow\; 4D = C + 3E \;\Rightarrow\; 3E = 4D - C \;\Rightarrow\; E = \frac{4}{3}D - \frac{1}{3}C$

We can check this result using co-ordinates: C is $(1,3)$, D is $(7,6)$

$$E = \frac{4}{3}D - \frac{1}{3}C = \frac{4}{3}(7,6) - \frac{1}{3}(1,3) = \left(\frac{27}{3}, \frac{21}{3}\right) = (9,7) \quad \text{which is correct.}$$

Consider F which divides CD externally (outside C) in the ratio $-1:4$.

The fractions are $\dfrac{-1}{-1+4}$ and $\dfrac{4}{-1+4}$ or $\dfrac{-1}{3}$ and $\dfrac{4}{3}$.

So, $$F = -\frac{1}{3}D + \frac{4}{3}C$$

You must remember which fraction goes with C and which with D. Remember that F is nearer to C than D, so the greater fraction goes with C.

More easily, regard C as dividing FD in the ratio 1:3.

Then, $C = \dfrac{3}{4}F + \dfrac{1}{4}D \quad \left(F \text{ is nearer to } C \text{ than } D, \text{ so } \dfrac{3}{4}F \text{ and only } \dfrac{1}{4}D\right)$

$$\Rightarrow \;\; 4C = 3F + D \;\;\Rightarrow\;\; 3F = 4C - D \;\;\Rightarrow\;\; F = \frac{4}{3}C - \frac{1}{3}D$$

Exercise 3.3

1 Find the co-ordinates of K, the point dividing CD externally in the ratio 2: -1

2 Find the co-ordinates of L, the point dividing CD externally in the ratio $-1:2$

3 Find the co-ordinates of R, the point dividing CD externally in the ratio $-2:5$

4 Find the co-ordinates of T, the point dividing CD externally in the ratio 3: -2

5 Draw the triangle $O\,(0,0)$, $P\,(6,0)$, $Q\,(2,2)$.
  Find R, where $QR:RP = 1:3$; S, where $PS:SO = 2:1$; and T, where $OT:TQ = 3:-2$.

  Menelaus' theorem states that if $\dfrac{QR}{RP} \times \dfrac{PS}{SO} \times \dfrac{OT}{TQ} = -1$, then S, R and T are in the same straight line (collinear).

  In this case, $\dfrac{QR}{RP} \times \dfrac{PS}{SO} \times \dfrac{OT}{TQ} = \dfrac{1}{3} \times \dfrac{2}{1} \times \dfrac{3}{-2} = -1$, so check that S, R, T are collinear.

6 Use the same information as in question **5** but make all the ratios negative. There are several possibilities, so take care with the order of the letters.

Inequalities

The line $y = x$, in the figure, divides the xy plane into two regions.
  For the shaded region containing $A(-1, 4)$, $B(1, 3)$, $C(3, 4)$, for each point the y co-ordinate is greater than the x co-ordinate.

For the shaded region, $y > x$.

To include the line we write $y \geqslant x$ (y is greater than or equal to x).

Similarly for points 'below' the line, $y < x$.

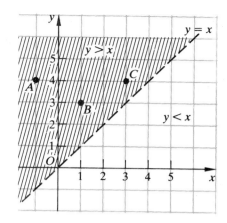

> WORKED EXAMPLE

Find the area of the region in the positive quadrant which satisfies both $x + y < 5$ and $y > x - 1$.

For points in the positive quadrant, $x > 0$ and $y > 0$.

To represent $x + y < 5$, first draw the line $x + y = 5$, which cuts the axes at $E(5, 0)$ and $D(0, 5)$. The point $(1, 1)$ clearly satisfies $x + y < 5$ so $(1, 1)$ lies in the required region which is 'below' (south-west of) the line $x + y = 5$.

To represent $y > x - 1$, first draw the line $y = x - 1$, which has gradient 1, passing through $(0, -1)$, $A(1, 0)$ and $B(3, 2)$.

To decide which side of the line $y = x - 1$ is required, substitute the point $(1, 1)$ in the inequality $y > x - 1$. Gives $1 > 0$ which is true, so $(1, 1)$ lies in the region required.

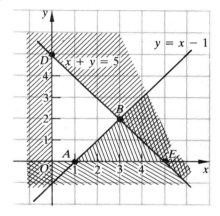

The lower figure shows each inequality with the **unwanted** region shaded so the final region required is that remaining **unshaded**. The area of this region $OABD$ is $8\frac{1}{2}$ units^2, being found in this easy example by counting squares.

Exercise 3.4

1 Sketch the regions satisfying the following inequalities.

  (a) $-3 \leqslant x \leqslant +3$ i.e. $|x| \leqslant 3$ **(b)** $-2 \leqslant y \leqslant +2$ i.e. $|y| \leqslant 2$ **(c)** $|x| \leqslant 3$ and $|y| \leqslant 2$
  (d) $y < 2x + 3$ **(e)** $y > 6 - x$ **(f)** $3y \leqslant 2x$ **(g)** $2x > y + 4$ **(h)** $2x + 3y \leqslant 6$

2 Within the restrictions of the upper figure (region $OABD$), what is the maximum value of **(a)** $x + 3y$
  (b) $3x + y$.

3 Sketch the following regions,

  (a) $x + y \leqslant 5$ **(b)** $x + y > -5$ **(c)** $|x + y| \leqslant 5$ **(d)** $|x| + |y| \leqslant 5$.

4 Find points satisfying $y = x^2$ and draw the graph of $y = x^2$.
  Label the regions **(a)** $y > x^2$ and **(b)** $y < x^2$.

5 Find points satisfying $xy = 6$ and draw its graph. Label the region $xy < 6$.

Reduction to linear form

Many relations between two variables are tested by experiments to obtain corresponding values. We may need to find the algebraic relation between the variables.

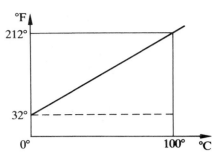

To find the relation between Celsius and Fahrenheit, we know that freezing point is $0° C \equiv 32° F$ and boiling point is $100° C \equiv 212° F$.
    What equation connects F and C?

We know that the graph of F against C is a straight line because an increase of $1° C$ gives the same increase in °F, whatever the temperature.

From the figure above, using $y = mx + c$, $c = 32$ and the gradient is $\dfrac{212 - 32}{100 - 0} = 1.8$.

$$F = 1.8C + 32 = \tfrac{9}{5}C + 32$$

On a warm day, $C = 20°$ $\Rightarrow$ $F = 1.8 \times 20 + 32 = 36 + 32 = 68° F.$

Body temperature $F = 98.4°$ $\Rightarrow$ $C = \tfrac{5}{9}(F - 32) = \tfrac{5}{9}(98.4 - 32) = \tfrac{5}{9}(66.4) \simeq 36.9 \simeq 37° C.$

If an experiment produces a straight line graph, the equation connecting the variables can easily be found using this method.

Experimental results may produce sets of figures relating two quantities, which produce a curve when plotted against each other on a graph. Can we find the equation of the curve?

WORKED EXAMPLES

1   Curves like that in the figure, right, might be quadratic, (like $y = x^2$) cubic ($y = x^3$) or any power of x.

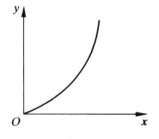

We may suspect that $y = kx^m$ e.g. $y = 3x^2$.

Taking logarithms, $\log y = \log kx^m = \log k + \log x^m$

$\Rightarrow$ $\log y = m \log x + \log k$

Compare this with $Y = mX$ $+ c$ so $Y = \log y$, $X = \log x$.

We now plot values of $\log x$ against values of $\log y$. This should give a straight line whose gradient and intercept can now be calculated from the graph. The gradient of the graph $\log x$ against $\log y$ gives m and the intercept $\log k$.
    For example, if you find $m = 3$ and $\log k = 2$, then if your logs are in base 10, $k = 10^2 = 100$ and $y = kx^m = 100x^3$.

2   If the graph looks like that in the figure, right, the relation could be $y = \dfrac{1}{x}$ or $\dfrac{6}{x}$ or $\dfrac{k}{x}$.

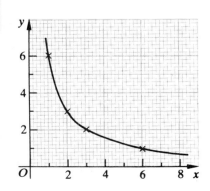

$y = \dfrac{k}{x}$ $\Rightarrow$ $xy = k$, so multiplying each pair of values of x and y together should give the constant value k.

In the figure,

x	1	2	3	6
y	6	3	2	1

For each pair of values, $x \times y = 6$, so $y = \dfrac{6}{x}$.

3 A graph similar to Worked Example **2**, may have a relation of the form

$$\frac{a}{x}+\frac{b}{y}=k$$

where a, b and k are constants (see figure, right).

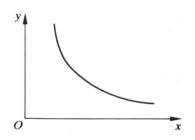

Plotting values of $\frac{1}{x}=X$ and $\frac{1}{y}=Y$ will produce a straight line since

$$\frac{a}{x}+\frac{b}{x}=k \quad \Rightarrow \quad aX+bY=k$$

$$\Rightarrow \quad Y=-\frac{a}{b}X+\frac{k}{b}.$$

From the X, Y graph the gradient $m=-\frac{a}{b}$ and the intercept $\frac{k}{b}$ can be calculated.

If $\frac{a}{b}=\frac{1}{2}$ and $\frac{k}{b}=1$ then $Y=-\frac{1}{2}X+1 \quad \Rightarrow \quad 2Y=-X+2 \quad \Rightarrow \quad \frac{2}{y}=\frac{-1}{x}+2.$

So the relation is $\frac{1}{x}+\frac{2}{y}=2.$

4 If the graph is similar to that for Worked Example **1** (but not through the origin) and the method of this Example does not work, the relation may be of the form $y=ab^x$. Remember, only b is raised to the power x.

Taking logarithms, $\log y = \log ab^x = \log a + \log b^x = x\log b + \log a$.

Plotting values of $\log y$ against x now produces a straight line whose gradient is $\log b$ and whose intercept is $\log a$.

If a is found to be 2 and $b=3$, then originally $y=2 \times 3^x$.

In all of these Examples the aim is to produce a straight line graph from which the gradient and the intercept give the constants in the x, y relation.

The original graph of x against y should provide a clue to your next move.

You could try   **(a)** $\log x$ against $\log y$

        (b) $\frac{1}{x}$ against $\frac{1}{y}$

        (c) x against $\log y$,   or many other combinations.

A simple graph-plotting computer program may cut down some numerical work, enabling you to see whether you have a combination producing a straight line.

Parametric form

Sometimes it is more convenient to specify straight lines and curves in terms of a third variable, called a **parameter**.

For example, if $x=1+t$ and $y=4-2t$,

$$t=0 \quad \Rightarrow \quad x=1 \quad \text{and} \quad y=4$$
$$t=1 \quad \Rightarrow \quad x=2 \quad \text{and} \quad y=2$$

Co-ordinates

t	0	1	2	3	4
x	1	2	3	4	5
y	4	2	0	-2	-4

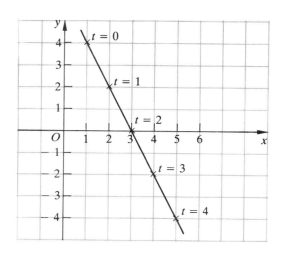

Each value of t gives a point on the line.

In Mechanics, x and y may represent displacements (or distances) at time t.

Eliminating t gives the x, y equation.

$$2x + y = 2(1 + t) + 4 - 2t = 2 + 2t + 4 - 2t = 6$$

$2x + y = 6 \implies y = 6 - 2x$, which is a straight line of gradient -2 and intercept 6.

WORKED EXAMPLE

For values of t from $t = \frac{1}{2}$ to $t = 3$, plot the graph represented by the parametric equations,

$$x = \frac{3}{t}, \quad y = 2t$$

Which point is closest to the origin? Find the x, y equation.

t	$\frac{1}{2}$	1	$1\frac{1}{2}$	2	3
x	6	3	2	$1\frac{1}{2}$	1
y	1	2	3	4	6

The curve is symmetrical in x and y. The nearest points from the table of values are $(3, 2)$ and $(2, 3)$.

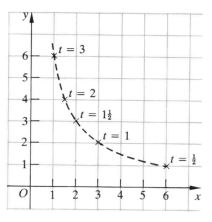

The nearest point to the origin from the curve is that for which $x = y$, i.e.
$$x = \frac{3}{t}, \quad y = 2t.$$

Eliminating t $\implies$ $xy = 6$ and if $x = y$ then $x^2 = 6$ $\implies$ $x = \sqrt{6} \simeq 2.45$.

Closest point is $(\sqrt{6}, \sqrt{6}) \simeq (2.45, 2.45)$.

Exercise 3.5

1 Plot the graphs for the following parametric equations and eliminate the parameter to obtain the x, y equation.

    (a) $x = 2t, y = 4 + t$ **(b)** $x = 3 - \theta, y = 1 + 3\theta$ **(c)** $x = 4s, y = 3s$
    (d) $x = 2(1 + t), y = 3(1 - t)$ **(e)** $x = 3\theta - 4, y = 2\theta + 1$ **(f)** $x = t^2 + 1, y = 3 - t^2$

2 Find the gradient and intercept for each of the lines in question **1**. Can you find these from the parametric equations without plotting the graph?

3 Plot the graph for the equations $x = t + 1$, $y = \frac{1}{2}t^2$. Find the x, y equation.

---

4 Differentiation 1

Introduction

In Chapter 3 you discovered that the gradient of a straight line graph is constant and defines the slope or inclination of the line. The value is calculated by finding the increase or decrease in the y co-ordinate per unit increase in the x co-ordinate (see figure).

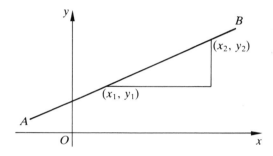

Gradient of $AB = \dfrac{y_2 - y_1}{x_2 - x_1}$

However, if the graph is a curve, the process is not quite as simple for, although you can still choose two points, the gradient of the line joining them will clearly depend upon where the points are chosen.

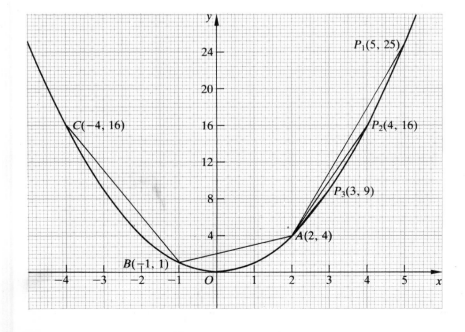

For the curve $y = x^2$ as shown left:

Gradient of $BA = \dfrac{4-1}{2-(-1)} = 1$

Gradient of $CB = \dfrac{16-1}{-4-(-1)} = -5$

Thus the gradient of a curve is **not** constant and can only be specified at a given point of the curve.

By fixing this chosen point it is possible to draw chords to a number of other points on the curve calculating the gradient of each. For example, for the chord AP_1,

$$\text{Gradient of } AP_1 = \frac{25 - 4}{5 - 2} = 7$$

Taking further points $P_2, P_3 \ldots$ from the figure, we can tabulate the results as follows.

x co-ordinate of P	y co-ordinate of P $(= x^2)$	Difference in x co-ordinate of A and P	Difference in y co-ordinate of A and P	Gradient of AP
5	25	3	21	7
4	16	2	12	6
3	9	1	5	5
2.5	6.25	0.5	2.25	4.5
2.1	4.41	0.1	0.41	4.1
2.01	4.040 1	0.01	0.040 1	4.01
2.001	4.004 001	0.001	0.004 001	4.001

The values of the gradients clearly form a sequence which will tend to a limiting value – in this case 4. This is the gradient of the tangent at the chosen point.

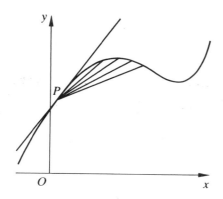

> The gradient of a curve at a given point is the gradient of the tangent at that point.

Limits

The limit of a function can be found in other ways as the following examples illustrate.

WORKED EXAMPLES

1 Find the limits of   **(a)** $f(x) = x^2 + 2x - 3$ as $x \rightarrow 0$

   (b) $f(x) = \dfrac{x + 2}{x}$ as $x \rightarrow 1$.

(a)   As x tends to zero the terms x^2 and $2x$ tend to zero and hence,

$$(x^2 + 2x - 3) \rightarrow -3$$

$$\therefore \quad \lim_{x \to 0} (x^2 + 2x - 3) = -3$$

(b)   As x tends to 1, $x + 2$ tends to 3 and hence,

$$\frac{x+2}{x} \rightarrow \frac{3}{1} = 3 \quad \text{or} \quad \lim_{x \to 1}\left(\frac{x+2}{x}\right) = 3$$

2   Find $\lim_{x \to -2}\left(\dfrac{x^2 - 4}{x + 2}\right)$.

In this Example, $x = -2$ would produce $\dfrac{0}{0}$ when substituted into $f(x) = \dfrac{x^2 - 4}{x + 2}$, and $\dfrac{0}{0}$ has no unique value (i.e. it is indeterminate). In such cases we need an alternative approach.

Consider values of $f(x)$ as $x \to -2$. You will find that $f(x)$ takes a sequence of finite values which approach a limit.

If $x = -2.1$, $\quad f(x) = \dfrac{(-2.1)^2 - 4}{-2.1 + 2} = \dfrac{4.41 - 4}{-0.1} = -\dfrac{0.41}{0.1} = -4.1$

If $x = -2.01$, $\quad f(x) = \dfrac{(-2.01)^2 - 4}{-2.01 + 2} = \dfrac{4.0401 - 4}{-0.01} = -\dfrac{0.0401}{0.01} = -4.01$

If $x = -2.001$, $\quad f(x) = \dfrac{(-2.001)^2 - 4}{-2.001 + 2} = \dfrac{4.004\,001 - 4}{-0.001} = -\dfrac{0.004\,001}{0.001} = -4.001$

Clearly this pattern will continue and as $x \to -2$, the value of $f(x) \to -4$.

This process is usually performed algebraically. Consider a value of x near -2, say $x = -2 + h$ where h is small.

$$f(-2 + h) = \frac{(-2+h)^2 - 4}{(-2+h) + 2} = \frac{4 - 4h + h^2 - 4}{h}$$

$$= \frac{-4h + h^2}{h} = -4 + h$$

Now as $h \to 0$, the value of x approaches -2 and the value of $f(x) \to -4$.

$$\text{Thus,} \quad \lim_{x \to -2}\left(\frac{x^2 - 4}{x + 2}\right) = -4$$

3   Find $\lim_{x \to \infty}\left(\dfrac{4x}{x + 1}\right)$.

Again, it is not possible just to substitute $x = \infty$ into the function $f(x) = \dfrac{4x}{x + 1}$ since this gives $\dfrac{\infty}{\infty}$ which is also indeterminate.

However, $f(x)$ can be rearranged by dividing the numerator and denominator by x.

$$\lim_{x \to \infty}\left(\frac{4x}{x + 1}\right) = \lim_{x \to \infty}\left(\frac{4}{1 + (1/x)}\right)$$

Now, as $x \to \infty$, $\dfrac{1}{x} \to 0$ and so $\lim_{x \to \infty}\left(\dfrac{4x}{x + 1}\right) = 4$

Alternatively, this can be achieved by dividing the numerator by the denominator,

$$f(x) = \frac{4x}{x+1}$$

$$\begin{array}{r} 4 \\ x+1\overline{)4x} \\ \underline{4x+4} \\ -4 \end{array}$$

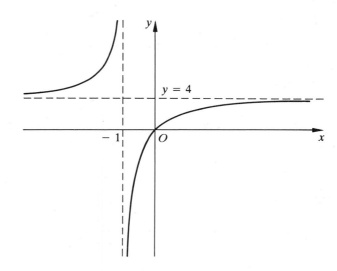

Hence $f(x) = \dfrac{4x}{x+1} = 4 - \dfrac{4}{x+1}$

As $x \to \infty$, $\dfrac{4}{x+1} \to 0$ and

$f(x) \to 4$ (from below).

As $x \to -\infty$, $y \to 4$ (from above). Also, as $x \to -1$, $y \to \infty$, and $x = 0$ when $y = 0$.

The curve sketch is shown in the figure with the curve tending to the line $y = 4$ as $x \to \infty$. The line $y = 4$ is called an **asymptote** to the curve.

Exercise 4.1

1 Find the limits of the following functions as $x \to 1$.

    (a) $3x + 4$ **(b)** $2 - x$ **(c)** $\dfrac{1}{x+3}$ **(d)** $1 + \dfrac{4}{x}$

2 Find the limits of the following functions as $x \to 3$.

    (a) $1 - x - x^2$ **(b)** $\dfrac{x-3}{x+3}$ **(c)** $\dfrac{x^2-9}{x-3}$

3 Find the limits of the following functions as $x \to 0$.

    (a) $2 - x + x^2$ **(b)** $\dfrac{1+x^2}{1-x^2}$ **(c)** $\dfrac{3x^2-x^3}{2x^2}$

4 Find the behaviour of the functions as $x \to \infty$.

    (a) $\dfrac{1}{x}$ **(b)** $\dfrac{3}{x+1}$ **(c)** $\dfrac{2x}{3x-1}$ **(d)** $\dfrac{x^2}{x-1}$

5 If $f(x) = \dfrac{2x}{x-1}$, find

    (a) the behaviour of $f(x)$ as $x \to \pm\infty$

    (b) the behaviour of $f(x)$ as $x \to 1$

    (c) where the curve cuts the axes.

    Hence sketch the curve.

6 Repeat question **5**, for **(i)** $f(x) = \dfrac{3x}{2x+1}$ **(ii)** $f(x) = \dfrac{x-1}{x+1}$.

7 Find the behaviour of the curve $y = \dfrac{3x-2}{x}$ **(a)** as $x \to \pm\infty$ and **(b)** as $x \to 0$. Hence sketch the curve.

8 Find

(a) $\displaystyle\lim_{x\to1}\left(\frac{x^2-1}{x-1}\right)$ (b) $\displaystyle\lim_{x\to-1}\left(\frac{3x^2+11x+10}{x^2+x-2}\right)$ (c) $\displaystyle\lim_{x\to-2}\left(\frac{x^3+8}{x+2}\right)$

(d) $\displaystyle\lim_{x\to\infty}\left(\frac{2x^2+1}{4x^2-3}\right)$

9 By using the substitution $x=1+h$ where $h\to0$, find

(a) $\displaystyle\lim_{x\to1}\left(\frac{4-3x}{x-1}\right)$ (b) $\displaystyle\lim_{x\to1}\left(\frac{x^3-1}{x-1}\right)$

10 Find

(a) $\displaystyle\lim_{x\to\infty}\left(\frac{4x^2-1}{x^2-9}\right)$ (b) $\displaystyle\lim_{x\to\infty}\left(\frac{x-3}{2x-1}\right)$ (c) $\displaystyle\lim_{x\to\infty}\left(\frac{2x+1}{x+3}\right)$

Use your results to show that, in this case,

$$\lim_{x\to a}[f(x)\times g(x)]=\left[\lim_{x\to a}f(x)\right]\times\left[\lim_{x\to a}g(x)\right]$$

11 If $f(x)=\dfrac{3x}{x+1}$ and $g(x)=\dfrac{2x}{x-1}$, find the limits of $f(x)$ and $g(x)$ as $x\to\infty$.

Find also $f(x)+g(x)$ and its limit as $x\to\infty$.

Use your results to show that, in this case,

$$\lim_{x\to a}[f(x)+g(x)]=\left[\lim_{x\to a}f(x)\right]+\left[\lim_{x\to a}g(x)\right]$$

Differentiation from first principles

The gradient of a curve at a given point was defined, earlier in this chapter, as the gradient of the tangent to the curve at that point. It can be evaluated by finding the limit of the sequence, formed from the gradients of chords drawn from the given point. The arithmetical process used to evaluate the gradient of $y=x^2$ at $A(2,4)$ is tedious and the result can be achieved more readily by an **algebraic method**.

The following method shows the algebraic process and will find the gradient of the curve $y=3x^2$ at $P(x,3x^2)$.

Let $P(x,3x^2)$ and $Q(x+h,3(x+h)^2)$ be neighbouring points on the curve (see figure, right).

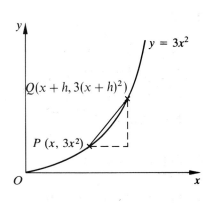

The gradient of $PQ=\dfrac{3(x+h)^2-3x^2}{(x+h)-x}$

$\qquad=\dfrac{3(x^2+2xh+h^2)-3x^2}{h}=\dfrac{3x^2+6xh+3h^2-3x^2}{h}$

$\qquad=\dfrac{6xh+3h^2}{h}=6x+3h$

67

Now as $h \to 0$, the point Q approaches P and the gradient of the chord PQ approaches the gradient of the tangent at P.

The gradient of the curve $= \lim\limits_{h \to 0} (6x + 3h) = 6x$

Notice that the term $6x$ remains fixed and is not affected by allowing h to tend to zero.

Thus the gradient of the curve $y = 3x^2$ at a point whose x co-ordinate is x, is $6x$. The function giving the gradient is called the **derived function** or the **differential**.

The process for finding the derived function is called **differentiation**.

Note that for $y = 3x^2$, the gradient at $(1, 3) = 6(1) = 6$

and the gradient at $(-3, 27) = 6(-3) = -18$.

Generally, if we consider the function $y = f(x)$, then P will be the point $(x, f(x))$ and Q will be the point $(x + h, f(x + h))$.

The gradient of $PQ = \dfrac{f(x + h) - f(x)}{(x + h) - x} = \dfrac{f(x + h) - f(x)}{h}$

The gradient of the curve is the limit of the gradient of PQ as $h \to 0$ and is written $f'(x)$.

Hence, $f'(x) = \lim\limits_{h \to 0} \left[\dfrac{f(x + h) - f(x)}{h} \right]$ provided the limit exists.

An alternative form can be written as

$$f'(x) = \lim\limits_{h \to 0} \left[\dfrac{f(x + h) - f(x - h)}{2h} \right]$$

which is particularly useful for differentiating the trigonometrical functions (see page 178).

The δx notation

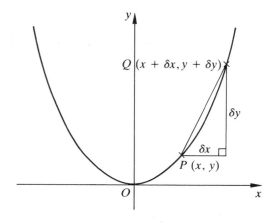

A special notation can be used instead of the h of the previous section (see figure, left).

The Greek letter δ means 'a small increase or decrease' in a stated variable.

So δx means a small increase or decrease (called an increment) in the variable x, and δy means an increment in the variable y.

In this context, the δ and the x (or y) cannot be separated and must be regarded as a single entity.

Consider the curve $y = 3x^2$.

Let $P(x, y)$ and $Q(x + \delta x, y + \delta y)$ be neighbouring points of the curve.

Since Q lies on the curve, $y + \delta y = 3(x + \delta x)^2$ ...... **(1)**

Since P lies on the curve, $y = 3x^2$ .... **(2)**

Subtracting **(2)** from **(1)** $\delta y = 3(x + \delta x)^2 - 3x^2$

Dividing by δx gives $\dfrac{\delta y}{\delta x} = \dfrac{3(x + \delta x)^2 - 3x^2}{\delta x}$

Now δy is the difference in the y co-ordinates of P and Q and δx is the difference in the x co-ordinates. Thus, $\dfrac{\delta y}{\delta x}$ represents the gradient of the chord PQ.

$$\therefore \text{Gradient of } PQ = \frac{\delta y}{\delta x} = \frac{3[x^2 + 2x(\delta x) + (\delta x)^2] - 3x^2}{\delta x}$$

$$= \frac{3x^2 + 6x(\delta x) + 3(\delta x)^2 - 3x^2}{\delta x}$$

$$= \frac{6x(\delta x) + 3(\delta x)^2}{\delta x} = 6x + 3(\delta x)$$

The gradient of the chord $PQ = \dfrac{\delta y}{\delta x} = 6x + 3(\delta x)$

The gradient of the curve is the limit of the gradient of the chord PQ as $\delta x \to 0$ and is denoted by $\dfrac{dy}{dx}$.

The gradient of the curve is $\dfrac{dy}{dx} = \lim\limits_{\delta x \to 0} \left(\dfrac{\delta y}{\delta x} \right) = \lim\limits_{\delta x \to 0} [6x + 3(\delta x)]$

$$\therefore \frac{dy}{dx} = 6x \qquad \text{(as before)}$$

Notes

1   $\dfrac{dy}{dx}$ is a special notation for the gradient of the curve and in this context cannot be separated into dy and dx.

2   Both $\dfrac{dy}{dx}$ and $f'(x)$ notations are used widely, but are not usually mixed.

   i.e.   If $y = 3x^2$, then write $\dfrac{dy}{dx} = 6x$,

   but if $f(x) = 3x^2$, then write $f'(x) = 6x$.

3   It is also possible to write $\dfrac{d}{dx}(3x^2) = 6x$, the $\dfrac{d}{dx}$ being read as 'the differential of'.

4   The above methods are known as **differentiation from first principles**.

1 Differentiate from first principles $y = x^2 - 4x$ at $P(x, y)$.

Let $P(x, y)$ be a given point on the curve and let $Q(x + \delta x, y + \delta y)$ be a neighbouring point of the curve.

Since Q lies on the curve, $\quad y + \delta y = (x + \delta x)^2 - 4(x + \delta x)$

Since P lies on the curve, $\quad y = x^2 - 4x$

Subtracting,
$$\delta y = [(x + \delta x)^2 - 4(x + \delta x)] - [x^2 - 4x]$$
$$= x^2 + 2x(\delta x) + (\delta x)^2 - 4x - 4(\delta x) - x^2 + 4x$$
$$= 2x(\delta x) - 4(\delta x) + (\delta x)^2$$

Dividing by δx,
$$\frac{\delta y}{\delta x} = \frac{2x(\delta x) - 4(\delta x) + (\delta x)^2}{\delta x} = 2x - 4 + \delta x$$

The gradient of the curve is $\dfrac{dy}{dx} = \lim\limits_{\delta x \to 0}\left(\dfrac{\delta y}{\delta x}\right)$

$$= \lim_{\delta x \to 0} [2x - 4 + \delta x] = 2x - 4$$

Hence the gradient of the curve $y = x^2 - 4x$ is $2x - 4$. i.e. If $y = x^2 - 4x$, then $\dfrac{dy}{dx} = 2x - 4$.

2 Differentiate $f(x) = \dfrac{1}{x}$ from first principles.

Let $P\left(x, \dfrac{1}{x}\right)$ be a given point of the curve and let $Q\left(x + h, \dfrac{1}{x + h}\right)$ be a neighbouring point of the curve.

The gradient of the curve, $f'(x) = \lim\limits_{h \to 0}\left[\dfrac{f(x + h) - f(x)}{h}\right]$

$$f'(x) = \lim_{h \to 0}\left[\frac{\dfrac{1}{(x + h)} - \dfrac{1}{x}}{h}\right] = \lim_{h \to 0}\left[\frac{x - (x + h)}{x(x + h)} \times \frac{1}{h}\right]$$

$$= \lim_{h \to 0}\left[\frac{-h}{x(x + h)} \times \frac{1}{h}\right] = \lim_{h \to 0}\left[\frac{-1}{x(x + h)}\right]$$

$$\therefore f'(x) = -\frac{1}{x^2} \quad \text{or} \quad -x^{-2}$$

Investigation 1

By letting $P(x, y)$ and $Q(x + \delta x, y + \delta y)$ be neighbouring points on the curve $y = \dfrac{1}{x}$, show that the result

$\dfrac{dy}{dx} = -\dfrac{1}{x^2}$ is obtained using the $\delta x, \delta y$ notation of Worked Example **1**.

Investigation 2

Differentiate the following functions from first principles.

(a) $3x$ **(b)** $8x + 7$ **(c)** x^2 **(d)** $5x^2$ **(e)** $2x^2 + 3x$ **(f)** $x^2 + 7x - 5$

(g) x^3 **(h)** x^4 **(i)** $\dfrac{3}{x}$ **(j)** $\dfrac{1}{x^2}$ **(k)** $x - 3$ **(l)** 6 **(m)** $-2x^3$

Make a table of your results. The gradient functions found in the worked examples and text are already inserted below.

Function	$3x^2$	$x^2 - 4x$	$\dfrac{1}{x}$			
Gradient	$6x$	$2x - 4$	$-x^{-2}$			

Do you notice any relation between the original function and the results obtained for the gradients?

General rule for differentiation

Did you manage to find the relationship asked for in Investigation 2?

It seems that, if $y = ax^n$ where a and n are constants, then $\dfrac{dy}{dx} = nax^{n-1}$.

i.e. if $y = 3x^4$, then $\dfrac{dy}{dx} = 12x^3$

and if $y = -\dfrac{2}{x^2} = -2x^{-2}$, then $\dfrac{dy}{dx} = 4x^{-3} = \dfrac{4}{x^3}$

The result $\dfrac{d}{dx}(ax^n) = nax^{n-1}$ can be proved, but requires an application of the binomial expansion when n is a positive integer, and other techniques of differentiation if n is negative or fractional.
    You should attempt the proofs later, but we shall use the result.

$$\text{If } y = ax^n \text{ then } \frac{dy}{dx} = nax^{n-1}$$

WORKED EXAMPLES

Differentiate **(a)** $3x^7$ **(b)** $\dfrac{2}{x^3}$ and **(c)** $\sqrt[3]{x}$

(a) If $y = 3x^7$ then $\dfrac{dy}{dx} = 21x^6$

(b) If $y = \dfrac{2}{x^3} = 2x^{-3}$ then $\dfrac{dy}{dx} = -6x^{-4} = -\dfrac{6}{x^4}$

(c) If $y = \sqrt[3]{x} = x^{\frac{1}{3}}$ then $\dfrac{dy}{dx} = \dfrac{1}{3}x^{-\frac{2}{3}} = \dfrac{1}{3\sqrt[3]{x^2}}$

Differentiation 1

Powers, polynomials, sums

The differential of the function $x^2 + 7x - 5$ in Investigation 2 gave the gradient as $2x + 7$.
   In fact, the derivative of the sum of several terms is equal to the sum of the derivatives of the separate terms.

If y is a function which is the sum of two terms u and v, both of which are functions of x, then an increment δx in x will cause increments of δu, δv and δy in u, v and y respectively.

$\therefore$ If $y = u + v$,   then $y + \delta y = (u + \delta u) + (v + \delta v)$

Subtracting gives $\delta y = \delta u + \delta v$

Dividing by δx $\dfrac{\delta y}{\delta x} = \dfrac{\delta u}{\delta x} + \dfrac{\delta v}{\delta x}$

Now if $\delta x \to 0$, $\delta u, \delta v$ and δy will all tend to zero and

$$\frac{dy}{dx} = \lim_{\delta x \to 0}\left(\frac{\delta y}{\delta x}\right) = \lim_{\delta x \to 0}\left(\frac{\delta u}{\delta x} + \frac{\delta v}{\delta x}\right) = \lim_{\delta x \to 0}\left(\frac{\delta u}{\delta x}\right) + \lim_{\delta x \to 0}\left(\frac{\delta v}{\delta x}\right)$$

Hence, $\dfrac{dy}{dx} = \dfrac{du}{dx} + \dfrac{dv}{dx}$

This can be extended to any number of terms.

$$\text{So } \frac{d}{dx}(x^2 + 7x - 5) = \frac{d}{dx}(x^2) + \frac{d}{dx}(7x) - \frac{d}{dx}(5) = 2x + 7$$

WORKED EXAMPLES

1 Differentiate **(a)** $3x^4 + 2x^3 - 8x$ **(b)** $x + \dfrac{3}{x}$ with respect to x,

(a) If $y = 3x^4 + 2x^3 - 8x$, then $\dfrac{dy}{dx} = 12x^3 + 6x^2 - 8$.

(b) If $y = x + \dfrac{3}{x} = x + 3x^{-1}$, then $\dfrac{dy}{dx} = 1 - 3x^{-2} = 1 - \dfrac{3}{x^2}$.

2 Find $f'(x)$ if **(a)** $f(x) = 3x^2 - 4x + 5$ **(b)** $f(x) = (x - 3)(x + 1)$ and   **(c)** $f(x) = \dfrac{x^3 - 3x}{2x^4}$.

(a) $f(x) = 3x^2 - 4x + 5$ $\Rightarrow$ $f'(x) = 6x - 4$

(b) $f(x) = (x - 3)(x + 1)$

Since this function consists of a product of two terms it must be rearranged into a polynomial **before** differentiating

$$f(x) = (x - 3)(x + 1) = x^2 - 2x - 3 \quad \Rightarrow \quad f'(x) = 2x - 2$$

(c) $f(x) = \dfrac{x^3 - 3x}{2x^4}$

Since $f(x)$ is a quotient, it must also be changed into a polynomial

$$f(x) = \frac{x^3 - 3x}{2x^4} = \frac{1}{2x} - \frac{3}{2x^3} = \tfrac{1}{2}x^{-1} - \tfrac{3}{2}x^{-3}$$

Hence, $f'(x) \quad = -\tfrac{1}{2}x^{-2} + \tfrac{9}{2}x^{-4} = -\dfrac{1}{2x^2} + \dfrac{9}{2x^4}$

3 Find the co-ordinates of the point on the curve $y = 3 - 2x - x^2$ at which the tangent is parallel to the x-axis.

If $y = 3 - 2x - x^2$, the gradient of the tangent $\dfrac{dy}{dx} = -2 - 2x$.

If the tangent is parallel to the x-axis its gradient is zero.

When $\dfrac{dy}{dx} = 0$, $\quad -2 - 2x = 0 \quad \Leftrightarrow \quad x = -1$.

Substituting $x = -1$ into the equation of the curve,

$$y = 3 - 2(-1) - (-1)^2 = 3 + 2 - 1 = 4$$

The co-ordinates of the point where the tangent is parallel to the x-axis are $(-1, 4)$.

Exercise 4.2

---

1 Differentiate the following functions with respect to x.

   (a) x^3 **(b)** $5x^4$ **(c)** 4 **(d)** $2x$ **(e)** x^{-2} **(f)** $\dfrac{2}{x}$ **(g)** $-\dfrac{3}{x^3}$

   (h) $2x - x^2$ **(i)** $4 - x$ **(j)** $2x^3 - 4x^2 + 7$

2 Find $\dfrac{dy}{dx}$ at a general point (x, y) for the following curves.

   (a) $y = 3x - 7$ **(b)** $y = x^2 + 7$ **(c)** $y = 2x^2 - 3x + 1$ **(d)** $y = 4x^3 - 6x^2$

   (e) $y = x + \dfrac{1}{x} + \dfrac{1}{x^2}$ **(f)** $y = \dfrac{2}{x} - \dfrac{3}{x^4}$

3 Differentiate the following functions to find $f'(x)$

   (a) $f(x) = 3 - 4x + 2x^2$ **(b)** $f(x) = \frac{1}{4}x^4 + \frac{1}{3}x^3 + \frac{1}{2}x^2$ **(c)** $f(x) = x(x - 2)$

   (d) $f(x) = x^2 - \dfrac{1}{x^2}$ **(e)** $f(x) = \dfrac{x^2 + 4x}{x}$

4 Rearrange the following functions into polynomials and hence find the gradient functions.

   (a) $x(2x - 1)$ **(b)** $\dfrac{1 + 4x}{x}$ **(c)** $(x - 1)(x + 4)$ **(d)** $\dfrac{x^3 + 4x}{x^2}$

5 Find $f'(x)$ and $f'(-1)$ for the following functions.

   (a) $f(x) = 3x^2 - 4x + 1$ **(b)** $f(x) = (x - 3)(2x + 1)$ **(c)** $f(x) = 2(x + 1)^2$

6 Find the values of $f'(1)$, $f'(2)$ and $f'(-2)$ if

   (a) $f(x) = \frac{2}{3}x^3 + 4x^2$ **(b)** $f(x) = \dfrac{1}{x} - \dfrac{3}{x^2}$ **(c)** $f(x) = \dfrac{(x + 2)^3}{x}$

7   Find the gradients of the following curves at the stated values of x.

   (a) $y = x^2 - 4x + 3$; $x = 2$ **(b)** $y = x^3 + x^2 - 2x$; $x = 0$ **(c)** $y = \dfrac{x^3 - 3}{x^2}$; $x = 1$

   (d) $y = 3x(x - 2)^3$; $x = 1$

8   Find the equations of the tangents to the following curves at the points given.

   (a) $y = x^2$; $x = 3$ **(d)** $y = \dfrac{1}{x^2}$; $x = -1$

   (b) $y = 3 - x^3$; $x = 1$ **(e)** $y = x(x - 1)(x + 2)$; $x = 0$

   (c) $y = x^2 - 3x + 2$; $x = -2$ **(f)** $y = 2 - \dfrac{1}{x}$; $x = -2$

9   Differentiate with respect to x,

   (a) $x^{\frac{1}{2}}$ **(b)** $3x^{\frac{1}{3}}$ **(c)** $x^{-\frac{1}{4}}$ **(d)** $-\dfrac{1}{\sqrt{x}}$ **(e)** $\sqrt[5]{x}$

   (f) $\sqrt{x}\,(x - 1)$ **(g)** $(1 - \sqrt{x})^2$ **(h)** $\dfrac{x + 2}{\sqrt{x}}$ **(i)** $(\sqrt{x} + \sqrt[3]{x})^2$

10   Find the co-ordinates of the points of the curve $y = \frac{1}{3}x^3 - 2x^2 - 11x + 7$, at which the gradient is 1.

11   Find the points on the following curves where the tangent is parallel to the x-axis.

   (a) $y = 1 - 4x - 2x^2$ **(b)** $y = 6x^3 - 2x + 1$

12   Find the points of intersection of the line $y = x + 5$ and the curve $y = x^2 - 3x$. Find the gradients of the tangents to the curve at these points and hence find the equations of these tangents.

13   Find the equations of the normals to the curve $y = x + \dfrac{1}{x}$, at the points where $x = \pm 2$.

14   Find the gradient of the normal to the curve $y = 6 - x - x^2$ at the point $(1, 4)$. Hence find the equation of the normal at this point and the co-ordinates of the point where this normal meets the curve again.

15   Find the equation of the tangent to the curve $y = x^3 + x^2 - 2x$ at the point where $x = -1$. Find the equation of the normal to the curve at the origin and hence find the co-ordinates of the point of intersection of these two straight lines.

16   Two variables, s and t, are related by the equation $s = t^2 - 8t + 7$. Differentiate s with respect to t to form $\dfrac{ds}{dt}$.

   Use the result to find the values of s and t when $\dfrac{ds}{dt}$ is zero.

17   Find the co-ordinates of the points where the curve $y = x^3 - 6x^2 + 9x$ cuts the x-axis. Find also the co-ordinates of the points on the curve at which the gradient of the tangent is zero. Hence sketch the curve.

The second derivative

In the previous sections you have seen how to form the derivative of a function to give $\dfrac{dy}{dx}$ or $f'(x)$.

For example, if $y = x^2 + x + \dfrac{3}{x}$, then $\dfrac{dy}{dx} = 2x + 1 - \dfrac{3}{x^2}$.

Clearly, the gradient function itself could be differentiated to obtain the **second derivative**. This is equivalent to finding $\dfrac{d}{dx}\left(\dfrac{dy}{dx}\right)$ and is usually written $\dfrac{d^2y}{dx^2}$ or $f''(x)$.

Thus, $\dfrac{d^2y}{dx^2} = \dfrac{d}{dx}(2x + 1 - 3x^{-2}) = 2 + 6x^{-3} = 2 + \dfrac{6}{x^3}$

Also, if $f(x) = x^3 + 3x^2 - 8x + 7$, then $f'(x) = 3x^2 + 6x - 8$

$$\text{and } f''(x) = 6x + 6$$

Higher derivatives

It is possible to continue to differentiate functions to produce the 3rd derivative, the 4th derivative and so on although these may well be zero.

Higher derivatives are denoted by $\dfrac{d^3y}{dx^3}$ or $f'''(x)$, etc.

So, if $f(x) = x^4 + 3x^3 - 7x^2 + 2x - 5$

$\Rightarrow \quad f'(x) = 4x^3 + 9x^2 - 14x + 2 \quad \Rightarrow \quad f''(x) = 12x^2 + 18x - 14$

$\Rightarrow \quad f'''(x) = 24x + 18, \quad$ and so on.

WORKED EXAMPLE

If $y = x^2 - 3x - 4$ find $\dfrac{dy}{dx}$ and $\dfrac{d^2y}{dx^2}$. Sketch the graph of the curve.

Differentiating with respect to x, $\quad \dfrac{dy}{dx} = 2x - 3 \quad$ and $\quad \dfrac{d^2y}{dx^2} = 2$

To sketch the curve, $y = x^2 - 3x - 4 = (x - 4)(x + 1)$.

Hence $y = 0 \quad \Leftrightarrow \quad x = 4$ or $x = -1$. The quadratic curve will cut the x-axis at $(4, 0)$ and $(-1, 0)$.

Also $x = 0 \quad \Rightarrow \quad y = -4$.

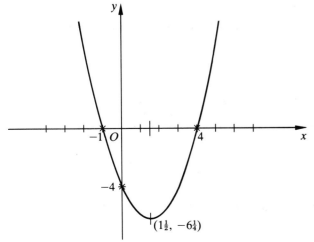

Since $\dfrac{d^2y}{dx^2} = 2$, which is positive, the rate of change of the gradient is positive which implies that the gradient is increasing.

The lowest point of the curve is $(1\frac{1}{2}, -6\frac{1}{4})$ which is the value of x for which the gradient is zero.

This gives a method for finding a turning point of a curve—where the gradient of the tangent to the curve is zero.

Maximum and minimum points

Consider a general curve $y = f(x)$, as shown in the next figure. By drawing tangents to the curve it is clear that the gradient of the curve at B, D and F is zero; it is positive between A and B, D and F and F and G, and negative between B and D.

The points B, D and F are called **stationary points** where the gradient of the tangent is zero.

At B, the curve rises to a greatest value before falling again and the gradient of the tangent changes from positive to negative as x increases through the point. B is called a **local maximum**.

At D the curve falls to a least value before rising again and the gradient of the tangent changes from negative to positive as x increases through the point. D is called a **local minimum**.

Points B and D are also called **turning points**.

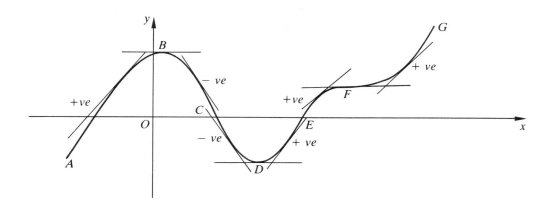

At F, the gradient of the tangent is zero but this is not a turning point since the curve does not change direction. Point F is called a **point of inflexion**. A point of inflexion is a point on a curve where the curve stops turning in one direction and begins to turn in the opposite direction, like an S-bend on a road.

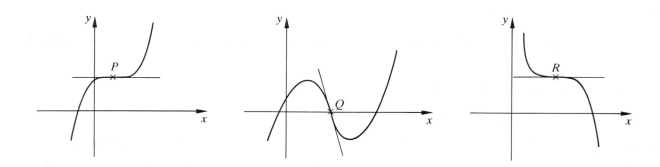

The points P, Q and R in the figure all show points of inflexion. Notice that, at a point of inflexion, the tangent crosses the curve and the gradient does not have to be zero, as at point Q.

In fact, where the curve is concave downwards (to the left of Q) the gradient is decreasing and $\dfrac{d}{dx}\left(\dfrac{dy}{dx}\right) = \dfrac{d^2y}{dx^2} < 0$.

Where the curve is concave upwards (to the right of Q) the gradient is increasing and $\dfrac{d^2y}{dx^2} > 0$.

Thus at the point of change, Q,　　$\dfrac{d^2y}{dx^2} = 0$ and $\dfrac{d^2y}{dx^2}$ changes sign, so $\dfrac{d}{dx}\left(\dfrac{d^2y}{dx^2}\right) = \dfrac{d^3y}{dx^3} \neq 0$.

A curve has a **maximum point** if $\dfrac{dy}{dx} = 0$ and $\dfrac{d^2y}{dx^2} < 0$

(the sign of the gradient changes from $+$ve to $-$ve).

A curve has a **minimum point** if $\dfrac{dy}{dx} = 0$ and $\dfrac{d^2y}{dx^2} > 0$

(the sign of the gradient changes from $-$ve to $+$ve).

A curve has a **point of inflexion** if $\dfrac{d^2y}{dx^2} = 0$ and $\dfrac{d^3y}{dx^3} \neq 0$

(the sign of the gradient does not change). A point of

inflexion may also have $\dfrac{dy}{dx} = 0$.

WORKED EXAMPLES

1 Find the turning point of the curve $y = x^2 - 4x + 3$ and sketch the curve.

The gradient of the tangent is $\dfrac{dy}{dx} = 2x - 4$.

A turning point exists if $\dfrac{dy}{dx} = 0$, i.e. $2x - 4 = 0 \iff x = 2$

Substitute $x = 2$ into the equation of the curve, $y = (2)^2 - 4(2) + 3 = -1$

The curve has a turning point at $(2, -1)$.

To determine the nature of the turning point, either consider the change in the gradient at $x = 2$ or consider the second differential.

For the first method draw a table to show the sign of the gradient.

x	$x < 2$	2	$x > 2$
$\dfrac{dy}{dx}$	$-$ve	0	$+$ve

i.e. as $\dfrac{dy}{dx} = 2x - 4$ it will be negative when x is just less than 2 but positive when x is just greater than 2.

Since the gradient changes from negative to positive and $\dfrac{dy}{dx} = 0$, the curve has a minimum point at $(2, -1)$.

Or, using the second differential,

$$\frac{dy}{dx} = 2x - 4 \implies \frac{d^2y}{dx^2} = 2 > 0$$

As $\dfrac{d^2y}{dx^2} > 0$ the curve has a minimum at $x = 2$.

Differentiation 1

Since $y = x^2 - 4x + 3$

$$= (x - 3)(x - 1)$$

When $y = 0 \Rightarrow x = 1$ or $x = 3$

When $x = 0$, $y = 3$.

The curve is sketched in the figure, right, where the minimum at $(2, -1)$ is clearly shown.

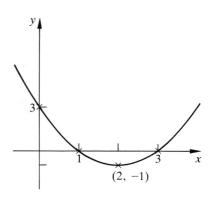

2 Find the stationary points of the curve $y = x^4 - 2x^3$ and sketch the curve.

Differentiating with respect to x,　$y = x^4 - 2x^3 \Rightarrow \dfrac{dy}{dx} = 4x^3 - 6x^2$

For stationary points, $\dfrac{dy}{dx} = 0$.

Hence, $4x^3 - 6x^2 = 0 \Leftrightarrow 2x^2(2x - 3) = 0$

$$\Leftrightarrow x = 0 \text{ or } x = \tfrac{3}{2}$$

Substituting into the equation,

If $x = 0$,　$y = 0$

If $x = \tfrac{3}{2}$,　$y = (\tfrac{3}{2})^4 - 2(\tfrac{3}{2})^3 = \tfrac{81}{16} - \tfrac{54}{8} = -\tfrac{27}{16}$

The curve has stationary points at $(0,0)$ and $(\tfrac{3}{2}, -\tfrac{27}{16})$

To determine the nature of these points, consider how the gradient changes at $x = 0$ and $x = \tfrac{3}{2}$ by constructing a table.

$$\frac{dy}{dx} = 4x^3 - 6x^2 = 2x^2(2x - 3)$$

x	$x < 0$	0	$x > 0$
$\dfrac{dy}{dx}$	$-$ve	0	$-$ve

x	$x < \tfrac{3}{2}$	$\tfrac{3}{2}$	$x > \tfrac{3}{2}$
$\dfrac{dy}{dx}$	$-$ve	0	$+$ve

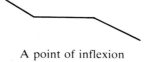

A point of inflexion

A minimum

Using the second differential method,

$$\frac{dy}{dx} = 4x^3 - 6x^2 \Rightarrow \frac{d^2y}{dx^2} = 12x^2 - 12x$$

At $x = \tfrac{3}{2}$,　$\dfrac{d^2y}{dx^2} = 12(\tfrac{3}{2})^2 - 12(\tfrac{3}{2}) = 27 - 18 > 0$　a minimum.

78

At $x = 0$, $\dfrac{d^2y}{dx^2} = 0$ and may give a point of inflexion.

Since $\dfrac{d^3y}{dx^3} = 24x - 12$, which is **not** zero for $x = 0$, this **is** a point of inflexion.

The curve has a minimum at $(\frac{3}{2}, -\frac{27}{16})$ and a point of inflexion at $(0, 0)$.

Since $y = x^4 - 2x^3$
$\qquad = x^3(x - 2)$

$y = 0 \ \Rightarrow \ x = 0 \ $ or $\ x = 2$

The curve is sketched in the figure with the minimum and point of inflexion shown.

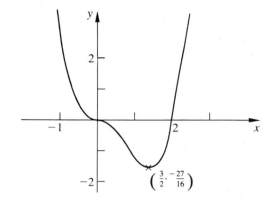

Notes

1   There is no need to use both methods to determine the nature of the stationary points. Choose whichever is easier for the particular question.

2   We have found only the point of inflexion for which $\dfrac{dy}{dx} = 0$. There is another point of inflexion, using $\dfrac{d^2y}{dx^2} = 0$ and $\dfrac{d^3y}{dx^3} \neq 0$.

    Now, $\quad \dfrac{d^2y}{dx^2} = 12x^2 - 12x = 12x(x - 1)$

    And $\quad \dfrac{d^2y}{dx^2} = 0$ when $x = 0$ or $x = 1$

    Since $\quad \dfrac{d^3y}{dx^3} = 24x - 12$, this is non-zero for both $x = 0$ and $x = 1$.

    There is a point of inflexion at $(1, -1)$ but at this point $\dfrac{dy}{dx} \neq 0$.

WORKED EXAMPLE

A solid cylinder is to be made so that it has a volume of $432\pi \ \text{cm}^3$. Find the minimum surface area.

The methods used for curves can be applied to any function, so long as only two variables are related.

Let the cylinder have a radius of r cm and a height of h cm.

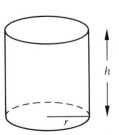

The total surface area A consists of the two ends and the curved face.

$$\therefore A = 2\pi r^2 + 2\pi rh \cdots\cdots (1)$$

Differentiation 1

A is now a function of two variables, r and h. Before differentiating to find the minimum value of A, either r or h must be replaced using the fact that the volume is $432\pi\,\text{cm}^3$.

Volume of a cylinder $= \pi r^2 h = 432\pi$

$$\Rightarrow\quad r^2 h = 432 \quad\Rightarrow\quad h = \frac{432}{r^2}\cdots\cdots(2)$$

Substituting (2) into (1),
$$A = 2\pi r^2 + 2\pi r\left(\frac{432}{r^2}\right)$$

$$\Leftrightarrow\quad A = 2\pi r^2 + \frac{864\pi}{r} = 2\pi r^2 + 864\pi r^{-1}\cdots\cdots(3)$$

Differentiating with respect to r,
$$\frac{dA}{dr} = 4\pi r - 864\pi r^{-2} = 4\pi r - \frac{864\pi}{r^2}\cdots\cdots(4)$$

For a minimum value, $\dfrac{dA}{dr} = 0$
$$\Rightarrow\quad 4\pi r - \frac{864\pi}{r^2} = 0 \quad\Leftrightarrow\quad 4\pi r^3 - 864\pi = 0$$
$$\Leftrightarrow\quad 4\pi(r^3 - 216) = 0$$
$$\Leftrightarrow\quad r^3 = 216 \quad\Leftrightarrow\quad r = \sqrt[3]{216} = 6$$

Differentiating equation (4) with respect to r gives
$$\frac{d^2 A}{dr^2} = 4\pi + 1728\pi r^{-3} = 4\pi + \frac{1728\pi}{r^3}$$

If $r = 6$, $\dfrac{d^2 A}{dr^2} > 0$ and hence A is a minimum.

Since $\dfrac{dA}{dr} = 0$ and $\dfrac{d^2 A}{dr^2} > 0$ when $r = 6$, the area is a minimum when the radius is $6\,\text{cm}$.

Substituting into equation (3) when $r = 6$,
$$A = 2\pi r^2 + \frac{864\pi}{r} = 2\pi(6)^2 + \frac{864\pi}{6}$$
$$= 72\pi + 144\pi = 216\pi\,\text{cm}^2$$

The minimum surface area is $216\pi\,\text{cm}^2$.

Exercise 4.3

(a) $x^2 + 3x - 2$ **(b)** $x^4 - 2x^3$ **(c)** $x^2 - \dfrac{1}{x}$ **(d)** $x^{\frac{1}{2}} - x^{\frac{3}{2}}$

2 Find $f''(2)$ if $f(x)$ equals,

(a) $x^3 + 3x^2 - 6x$ **(b)** $x^2(x^2 - 3)$ **(c)** $\dfrac{1}{x} + \dfrac{1}{x^2} + \dfrac{1}{x^3}$

3 Expand $f(x) = (x + 2)^4$ and find $f'''(x)$.

4 Find the co-ordinates of the turning points of the curves,

(a) $y = 2x^2 - 8x + 3$ (b) $y = 1 - x - x^2$

5 Find the stationary points of the following curves, stating the nature of these points.

(a) $y = x^2 - 2x$ (b) $y = 3x^4 - 4x^3 + 2$ (c) $y = x^3 - x^2$ (d) $y = 2x^3 - 9x^2 + 12x - 1$

(e) $y = 2x^4$ (f) $y = (x - 3)^2$ (g) $y = x^3(x - 1)$

6 Find the stationary points of the following curves for which $\dfrac{dy}{dx} = 0$ and sketch the curves.

(a) $y = x + \dfrac{1}{x}$ (b) $y = x^3 - 3x - 2$ (c) $y = x^2(x - 1)^2$ (d) $y = x(2 - x)^2$

7 Investigate the following curves to find any maximum or minimum points or points of inflexion for which $\dfrac{dy}{dx} = 0$.

(a) $y = x^2 + \dfrac{16}{x}$ (b) $y = 4x^3 - x^4$ (c) $y = x^3 + 3x^2 - 9x + 6$

8 By using the condition $\dfrac{d^2 y}{dx^2} = 0$, find whether the following curves possess a point of inflexion. Sketch the curves.

(a) $y = x(x - 1)(x + 1)$ (b) $y = x(x - 1)^2$ (c) $y = x^4 - 6x^2 + 10$

9 A rectangular piece of card 3 cm by 8 cm has squares of side x cm cut from each corner and the remainder folded to form a box of volume V cm^3. Show that $V = 24x - 22x^2 + 4x^3$ and find the maximum volume of the box.

10 A particle moves so that its displacement s metres at time t seconds is given by $s = 2t^3 - 15t^2 + 36t + 10$. Find its minimum velocity.

11 A man wants a rectangular lawn in his garden with an area of 1600 m^2. Find the dimensions of the lawn if the perimeter is to be a minimum.

12 A gardener wishes to fence a rectangular plot of land which has one side against the wall of the house. If 20 m of fencing are to be used, find the maximum area of the plot of land.

13 An open cylindrical can has an external surface area of 18π cm^2. Find its maximum volume.

14 A solid cylinder has a volume of 54π cm^3. Find the dimensions of the cylinder if the surface area is to be a minimum.

15 An open box in the shape of a cuboid has a square base and must have a volume of 256 cm^3. Find the dimensions of the box if the surface area is to be a minimum.

16 A box in the form of a cuboid has sides x cm, $(8 - 2x)$ cm and $(5 - 2x)$ cm. Find the maximum volume of the box.

17 A pentagon $ABCDE$ is formed by a rectangle $ABCE$ with an isosceles triangle ECD on the side CE. If $AB = 6x$ cm, $CD = 5x$ cm and the total perimeter of the pentagon is 48 cm find an expression for the area of the pentagon in terms of x. Find the value of x for which the area is a maximum.

5 Integration 1

Indefinite integration

The reverse of differentiation

In Chapter 4 we saw how to differentiate a function to obtain the gradient of the tangent. For example, if $y = 3x^2 + 5$,

then $\dfrac{dy}{dx} = 6x$. However, if $y = 3x^2 + 2$, then $\dfrac{dy}{dx}$ is still $6x$. In fact, if we differentiate any function of the form

$y = 3x^2 + c$, where c is a constant, the gradient of the curve will be $6x$.

The gradient function $\dfrac{dy}{dx} = 6x$, thus derives from an equation of the form $y = 3x^2 + c$. This equation represents a

family of curves as shown in the figure below.

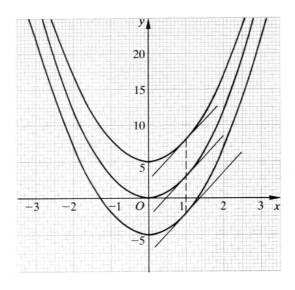

Notice that the gradient of the tangent at any value of x is the same for each member of the family.

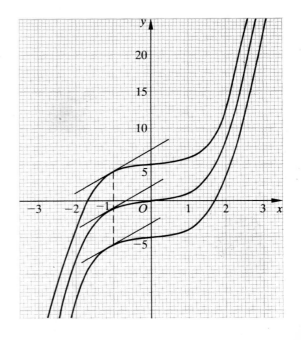

Similarly, if the gradient function is $\dfrac{dy}{dx} = 3x^2$ it is clear

that this could be obtained from $y = x^3 + c$ by differentiation.

The family of curves given by $y = x^3 + c$ is illustrated in the figure, right.

The process of obtaining an expression for the equation of a curve from its gradient function is called **integration.**

Thus, $y = 3x^2 + c$ is the integral of $\dfrac{dy}{dx} = 6x$ with respect to x,

and $y = x^3 + c$ is the integral of $\dfrac{dy}{dx} = 3x^2$ with respect to x.

The value of the constant, c, cannot be evaluated without further information being given. It is known as the **arbitrary constant** and the resulting function, y, is called the **indefinite integral.**

The general rule for integrating powers of x is found by reversing the rule for differentiation. Thus, to integrate, increase the power of x by 1 and divide by the new index.

If a and n are constants and

$$\frac{dy}{dx} = ax^n, \quad \text{then} \quad y = \frac{ax^{n+1}}{n+1} + c, \quad \text{unless} \quad n = -1.$$

The rule fails if $n = -1$, since there is no power of x whose differential is x^{-1}. This case will be met in Book 2.

WORKED EXAMPLES

1 Find y if $\dfrac{dy}{dx}$ is given by   **(a)** x^3 **(b)** $2x$ **(c)** $\dfrac{1}{x^4}$ **(d)** $2x^{\frac{1}{2}}$.

(a)   If $\dfrac{dy}{dx} = x^3$, then $y = \dfrac{x^4}{4} + c$

(b)   If $\dfrac{dy}{dx} = 2x$, then $y = \dfrac{2x^2}{2} + c \iff y = x^2 + c$

(c)   If $\dfrac{dy}{dx} = \dfrac{1}{x^4} = x^{-4}$, then $y = \dfrac{x^{-3}}{-3} + c \iff y = -\dfrac{1}{3x^3} + c$

(d)   If $\dfrac{dy}{dx} = 2x^{\frac{1}{2}}$, then $y = \dfrac{2x^{\frac{3}{2}}}{\frac{3}{2}} + c \iff y = \dfrac{4}{3}x^{\frac{3}{2}} + c$

2 Find $f(x)$ if $f'(x) = 5x^4$.

This problem is expressed in functional notation, but the same rules apply.

Since the gradient function $f'(x) = 5x^4$,

$$f(x) = \frac{5x^5}{5} + c$$

i.e. $f(x) = x^5 + c$

Sum of terms

A sum of terms can be integrated by applying the rule to each separate term.

If $\dfrac{dy}{dx} = 3x^2 - 8x + 7$, then $y = \dfrac{3x^3}{3} - \dfrac{8x^2}{2} + 7x + c$

$$\Leftrightarrow \quad y = x^3 - 4x^2 + 7x + c$$

The arbitrary constant can be determined if a set of conditions is given – for example, the co-ordinates of one point on the resulting curve.

WORKED EXAMPLES

1 If a curve has a constant gradient of 2 and passes through the point $(-1, 3)$, find the equation of the curve. Sketch the family of lines.

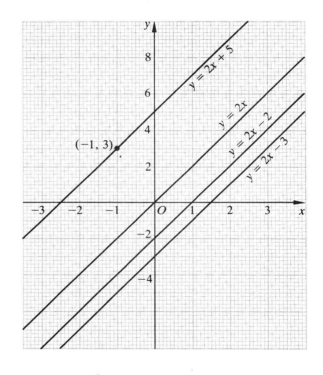

Since $\dfrac{dy}{dx} = 2$, the equation of the line is given by integration.

$$y = 2x + c$$

Since the line passes through the point $(-1, 3)$, these values must satisfy the equation $y = 2x + c$.

$\therefore$ Substituting,

$$3 = 2(-1) + c \quad \Leftrightarrow \quad c = 5$$

The particular solution is

$$y = 2x + 5$$

Some of the lines of the family $y = 2x + c$ are shown in the figure.

2 If $\dfrac{dA}{dr} = 8\pi r + 6\pi$ where A and r are the surface area and radius of a cylinder respectively, find the surface area when the radius is 2 cm, given that $A = 5\pi \text{ cm}^2$ when $r = 1$ cm.

Since $\dfrac{dA}{dr} = 8\pi r + 6\pi$, where π is a constant $\Rightarrow A = \dfrac{8\pi r^2}{2} + 6\pi r + c$

$$\Rightarrow \quad A = 4\pi r^2 + 6\pi r + c$$

Now $A = 5\pi$ where $r = 1$.

Hence, substituting, $\qquad\qquad 5\pi = 4\pi(1)^2 + 6\pi(1) + c \quad \Leftrightarrow \quad c = -5\pi$

The surface area A is given by the equation

$$A = 4\pi r^2 + 6\pi r - 5\pi$$

when $r = 2$, $\qquad\qquad\qquad A = 4\pi(2)^2 + 6\pi(2) - 5\pi = 23\pi$

The surface area is $23\pi \, \text{cm}^2$ when $r = 2 \, \text{cm}$.

Exercise 5.1

1 Integrate with respect to x,

    (a) 3 **(b)** $5x$ **(c)** $x^2 + 4x$ **(d)** $(x+1)^2$ **(e)** x^{-3}

2 Integrate with respect to t,

    (a) 4 **(b)** $-2t$ **(c)** $3t^2 - 7$ **(d)** t^{-2} **(e)** $t^{\frac{3}{2}}$

3 Sketch the family of curves whose gradient functions are,

    (a) -1 **(b)** $4x$ **(c)** $-x^{-2}$ **(d)** $4x^3$

4 Find the general solutions of the following.

    (a) $\dfrac{dy}{dx} = -\dfrac{5}{2}$ **(b)** $\dfrac{dy}{dx} = \sqrt{x}$ **(c)** $\dfrac{dy}{dx} = x(x-2)$

5 Find $f(x)$ if $f'(x)$ equals,

    (a) $x - 3$ **(b)** $4x^3 - 2x^2 + 3x - 5$ **(c)** $(x-1)(x-3)$

6 Integrate with respect to x,

    (a) $x^2 - 8x$ **(b)** $x^7 - 5x^4$ **(c)** $(x+3)^2$ **(d)** $\dfrac{x^2 + 2}{x^2}$ **(e)** $x - \dfrac{1}{x^2}$

    (f) $\dfrac{x^2 + 3x - 1}{x^4}$ **(g)** $\sqrt{x} - \dfrac{1}{\sqrt{x}}$ **(h)** $x^{\frac{1}{3}} + x^{-\frac{2}{3}}$ **(i)** $\dfrac{x^2 - x}{\sqrt{x}}$

    (j) $\left(1 + \dfrac{1}{x}\right)\left(1 - \dfrac{1}{x}\right)$ **(k)** $\sqrt[3]{x} + \sqrt{x^3}$ **(l)** $\left(2x - \dfrac{2}{x}\right)^2$

7 A curve has a gradient function of $2x + 3$ and passes through the point $(1, -7)$. Find the equation of the curve.

8 A curve passes through the point $(1, 3)$ and has a gradient function of $\dfrac{-4}{x^3}$. Find its equation and sketch the curve.

9 If two variables, s and t, are such that $\dfrac{ds}{dt} = 3t - \dfrac{1}{t^2}$, find s is terms of t given that $s = \frac{1}{2}$ when $t = 1$.

10 If $\dfrac{dv}{dt} = (t-1)(t+1)$, find v in terms of t given that $v = 10$ when $t = 3$.

11 The gradient of a curve at the point (x, y) is given by $3x^2 - 6x - 1$. If the curve passes through the point $(-1, 0)$, find the equation of the curve and the co-ordinates of the other points where it cuts the x-axis.

12 The gradient function of a curve is given by $2x - 1$. If the curve passes through the point $(0, -6)$, find the points where the curve cuts the x-axis and the equations of the tangents at these points.

13 If $\dfrac{ds}{dt} = 3t^2 - 4t + 5$, find an expression for s in terms of t given that $s = 9$ when $t = 1$.

14 The acceleration of a particle at time t seconds is given by $\dfrac{dy}{dt} = -4t$, where $v\,\mathrm{ms}^{-1}$ is the velocity. If $v = 3$ when $t = 0$, find an expression for the velocity at time t seconds.

15 If $\dfrac{d^2 y}{dx^2} = 24x - 6$, find $\dfrac{dy}{dx}$ and y in terms of x given that $y = 8$ and $\dfrac{dy}{dx} = 6$ when $x = 1$.

16 If $\dfrac{d^2 x}{dt^2} = 20t^3 - 2$, find $\dfrac{dx}{dt}$ and x in terms of t given that $\dfrac{dx}{dt} = 80$ and $x = 48$ when $t = 2$.

17 The gradient of a curve at the point (x, y) is given by $3x^2 - 4x - 1$. If the curve passes through the point $(-2, -12)$, find the equation of the curve and the co-ordinates of the points where it cuts the x-axis. Hence sketch the curve.

18 The gradient of a curve at the point (x, y) is given by $4x^3 + 6x^2 - 6x - 4$. If $y = 4$ when $x = 0$, find the equation of the curve.

Use the remainder theorem to find the co-ordinates of the turning points of the curve and hence sketch the curve. Hence, or otherwise, show that the equation of the curve can be written $y = (x - 1)^2 (x + 2)^2$.

Area under a curve

It is possible to use integration to evaluate the area enclosed by curves and lines.

The figure, right, shows part of the curve $y = f(x)$. The area between the curve, the x-axis and the lines $x = a$ and $x = b$ can be found by dividing it into strips which are approximately rectangular and adding the individual areas.

The accuracy of the result will depend upon the number of strips chosen. If a large number is taken, such that the width of each strip tends to zero, then the actual area will be found.

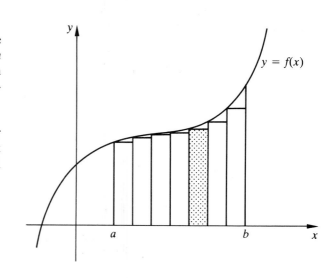

One such strip, or element, is shown in the figure, left. If $P(x, y)$ and $Q(x + \delta x, y + \delta y)$ are neighbouring points on the curve and the area $LMQP$ is δA, then δA lies between the area of the rectangle $LMSP$ and the rectangle $LMQT$.

$$\therefore y\delta x < \delta A < (y + \delta y)\delta x$$

Since $\delta x > 0$, this can be written

$$y < \frac{\delta A}{\delta x} < y + \delta y \qquad \dots\dots \text{ (1)}$$

As $\delta x \to 0$, $\delta y \to 0$ and $\dfrac{\delta A}{\delta x} \to \dfrac{dA}{dx}$

Thus, in the limit as $\delta x \to 0$, from **(1)**, $\dfrac{dA}{dx} = y$.

i.e. $\dfrac{dA}{dx} = f(x)$ and the area may be found by integrating $f(x)$ with respect to x.

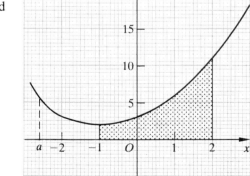

WORKED EXAMPLE

Find the area bounded by the curve $y = x^2 + 2x + 3$, the x-axis and the lines $x = -1$ and $x = 2$.

Since $\dfrac{dA}{dx} = y$

$$\frac{dA}{dx} = x^2 + 2x + 3$$

Integrating with respect to x,

$$A = \tfrac{1}{3}x^3 + x^2 + 3x + c$$

It is not possible to find the value of c, but the area between $x = -1$ and $x = 2$ can be determined by subtracting the areas formed from an arbitrary position, $x = a$.

For $x = 2$, $\qquad A_1 = \tfrac{1}{3}(2^3) + 2^2 + 3(2) + c$

$$= \frac{38}{3} + c$$

For $x = -1$, $\qquad A_2 = \tfrac{1}{3}(-1)^3 + (-1)^2 + 3(-1) + c$

$$= -\frac{7}{3} + c$$

The area between $x = -1$ and $x = 2$ is $A_1 - A_2$

$$\therefore \text{Area} = \left(\frac{38}{3} + c\right) - \left(-\frac{7}{3} + c\right) = \frac{45}{3} = 15 \text{ units}^2$$

Definite integration

Notice, in the Worked Example, that the constant c cancels and will always do so whatever values are used for x. This leads to a **special notation**, as follows.

If the area function obtained by integration is $F(x) + c$, then the area, $A = F(x) + c$.

Between the values of $x = a$ and $x = b$, we write

$$\text{Area} = A_b - A_a = [F(b) + c] - [F(a) + c]$$
$$= F(b) - F(a)$$

As the area has been found from a summation, the symbol $\int$, which is derived from the old English form of the letter S (the first letter of the word Sum), is used to denote the operation of integration.

The result for the area under a curve, earlier in this chapter, gave

$$\frac{dA}{dx} = y \quad \text{or} \quad \frac{dA}{dx} = f(x)$$

from which the area A can be found by integrating (i.e. summing between $x = a$ and $x = b$). We write,

$$A = \int_a^b y \, dx \quad \text{or} \quad A = \int_a^b f(x) \, dx$$
$$= \left[F(x) \right]_a^b = F(b) - F(a)$$

where $F(x)$ is the integral of $f(x)$ with respect to x.

Since this integral gives a numerical result and does not involve an arbitrary constant, it is known as a **definite integral**. The values of a and b are the limits of the integration.

The indefinite integrals we considered earlier in this chapter can also be written using the $\int$ notation, e.g.

$$\int (3x^2 + 2x + 7) \, dx = x^3 + x^2 + 7x + c$$

WORKED EXAMPLES

1   Find   (a) $\displaystyle\int (x - 3) \, dx$ (b) $\displaystyle\int_1^2 (x - 3) \, dx$ (c) $\displaystyle\int_{-3}^{-1} \left(x^2 + \frac{1}{x^2} \right) dx$

(a)   This is an indefinite integral.

$$\int (x - 3) \, dx = \tfrac{1}{2}x^2 - 3x + c$$

(b) $\displaystyle\int_1^2 (x - 3) \, dx = \left[\tfrac{1}{2}x^2 - 3x \right]_1^2 = \{\tfrac{1}{2}(2)^2 - 3(2)\} - \{\tfrac{1}{2}(1)^2 - 3(1)\}$

$$= -4 - (-2\tfrac{1}{2}) = -1\tfrac{1}{2}$$

(c) $\displaystyle\int_{-3}^{-1} \left(x^2 + \frac{1}{x^2} \right) dx = \int_{-3}^{-1} (x^2 + x^{-2}) \, dx = \left[\tfrac{1}{3}x^3 - x^{-1} \right]_{-3}^{-1}$

$$= \left[\tfrac{1}{3}x^3 - \frac{1}{x} \right]_{-3}^{-1}$$
$$= \{\tfrac{1}{3}(-1)^3 - (-1)\} - \{\tfrac{1}{3}(-3)^3 - (-\tfrac{1}{3})\}$$
$$= (-\tfrac{1}{3} + 1) - (-9 + \tfrac{1}{3}) = 9\tfrac{1}{3}$$

2 Find the area bounded by the curve $y = x^3$, the x-axis and the lines $x = 1$ and $x = 3$.

Since $\dfrac{dA}{dx} = y$,

$$A = \int_1^3 y\, dx = \int_1^3 x^3\, dx$$

$$= \left[\frac{x^4}{4} \right]_1^3$$

$$= \frac{1}{4}(3^4) - \frac{1}{4}(1)^4 = \frac{81}{4} - \frac{1}{4}$$

∴ The area required $= 20 \text{ units}^2$

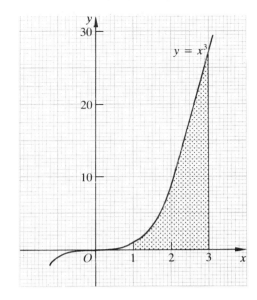

Investigation 1

(a) Sketch the curve $y = x(x^2 - 1)$ showing clearly the points where it meets the x-axis.

(b) Find the area between the curve and **(i)** $x = -1$ and $x = 0$, **(ii)** $x = 0$ and $x = 1$, and **(iii)** $x = -1$ and $x = 1$.

(c) Explain the significance of your answers to **(b)** parts **(ii)** and **(iii)**.

If $y = f(x)$ is a function that takes only negative values over an interval $x = a$ to $x = b$, then $y = -f(x)$ is a reflection of $y = f(x)$ in the x-axis, with the same area.

Thus, $$\int_a^b -f(x)\, dx = \left[-F(x) \right]_a^b = -\left[F(x) \right]_a^b = -\int_a^b f(x)\, dx$$

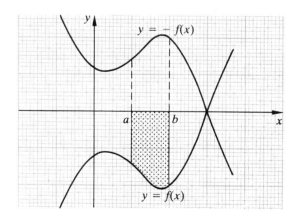

Therefore the integral evaluates the negative of the area bounded by $y = f(x)$, the x-axis and the lines $x = a$ and $x = b$.

Your answer to **(b)** part **(ii)** above should yield $-\frac{1}{4}$, although the actual area is $\frac{1}{4}$ unit^2.

It is therefore quite possible to obtain negative and zero answers to the integration. In **(b)** part **(iii)**, your result should be zero since the areas of the parts above and below the x-axis are numerically equal.

In general you should not integrate to find an area over a range $x = a$ to $x = b$, if the curve cuts the x-axis between these points.

Investigation 2

(a) Find $\displaystyle\int_{-1}^{2} \frac{1}{x^2}\,dx$.

(b) Sketch the curve of $y = \dfrac{1}{x^2}$ and decide whether your answer to (a) represents the area bounded by the curve, the x-axis and the lines $x = -1$ and $x = 2$.

In general, you should not integrate to find an area using a range of values of x in which the function becomes infinite. In this Investigation, the integral gives a value of $\frac{1}{2}$, but the area would be infinite.

> WORKED EXAMPLE

Find the area bounded by the curve $y = 3 - x - x^2$ and the line $y = 1$.

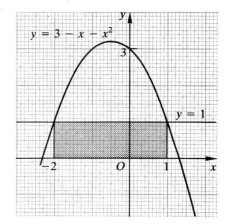

At the points of intersection of the line and the curve,

$$3 - x - x^2 = 1$$
$$\Leftrightarrow \quad x^2 + x - 2 = 0$$
$$\Leftrightarrow \quad (x-1)(x+2) = 0$$
$$\Leftrightarrow \quad x = 1 \quad \text{or} \quad -2$$

The points of intersection are $(-2, 1)$ and $(1, 1)$.

The area between the curve, the x-axis and the lines $x = -2$ and $x = 1$ is given by,

$$\text{Area} = \int_{-2}^{-1}(3 - x - x^2)\,dx = \left[3x - \tfrac{1}{2}x^2 - \tfrac{1}{3}x^3\right]_{-2}^{1}$$

$$\text{Area} = (3 - \tfrac{1}{2} - \tfrac{1}{3}) - (-6 - 2 + \tfrac{8}{3}) = 7\tfrac{1}{2} \text{ units}^2$$

The area of the shaded rectangle is $1 \times 3 = 3$ units^2

Hence, the required area $= 7\tfrac{1}{2} - 3 = 4\tfrac{1}{2}$ units^2

Exercise 5.2

1 Find (a) $\displaystyle\int(3 + 4x + 6x^2)\,dx$ (b) $\displaystyle\int\left(x - \frac{1}{x^2}\right)dx$ (c) $\displaystyle\int(1 + x)^2\,dx$ (d) $\displaystyle\int(x^{\frac{4}{3}} - x^{\frac{2}{3}})\,dx$

2 Evaluate (a) $\left[4x^3\right]_{1}^{2}$ (b) $\left[x + \dfrac{1}{x}\right]_{-1}^{1}$ (c) $\left[x^2 - 3x\right]_{0}^{3}$

3 Evaluate the definite integrals,

(a) $\displaystyle\int_{0}^{5}(1 + x^2)\,dx$ (b) $\displaystyle\int_{1}^{2} x(1 + x^3)\,dx$ (c) $\displaystyle\int_{-2}^{-1}\left(2 + \frac{1}{x^2}\right)dx$

(d) $\displaystyle\int_0^1 \sqrt{x}(x+2)\,dx$ **(e)** $\displaystyle\int_{-1}^2 (4+2x-2x^2)\,dx$ **(f)** $\displaystyle\int_1^4 \left(\sqrt{x}-\frac{1}{\sqrt{x}}\right)dx$

(g) $\displaystyle\int_1^2 (4x-x^2)\,dx$ **(h)** $\displaystyle\int_3^4 (x-1)^3\,dx$ **(i)** $\displaystyle\int_{-2}^2 (16-8x^2+x^4)\,dx$

4 Find the areas between the curves, the x-axis and the given lines.

  (a) $y=x^2$; $x=1$, $x=2$ **(b)** $y=2x^3$; $x=0$, $x=2$ **(c)** $y=\dfrac{1}{x^2}$; $x=3$, $x=5$

  (d) $y=x^2+2$; $x=-2$, $x=3$ **(e)** $y=x^3-x$; $x=-1$, $x=0$ **(f)** $y=x(x^2-1)$; $x=0$, $x=1$

  (g) $y=x(4-x)$; $x=4$, $x=6$ **(h)** $y=(x+1)(x-2)$; $x=-2$, $x=3$

5 Sketch the curve $y=x^2(x-3)$. Find the area bounded by the curve, the x-axis and **(a)** the lines $x=0$ and $x=2$, **(b)** the lines $x=3$ and $x=4$.

6 Sketch the curve $y=x^2-5x+4$ and find the area of the region cut off by the x-axis.

7 Sketch the curve $y=x(x-a)$, where a is a positive constant. If the area enclosed by the curve and the x-axis is $\dfrac{32}{81}$ units^2 find the value of a..

8 Find the areas of the regions enclosed by the following curves and lines.

  (a) $y=7+2x-x^2$, $y=4$ **(b)** $y=x^2$, $y=x$ **(c)** $y=(x-1)^2$, $y=1$

  (d) $y=x(4-x)$, $y=x$ **(e)** $y=x^2+x+3$, $y=5$ **(f)** $y=3\sqrt{x}$, $y=x+2$

Integration as the limit of a sum

In the previous section, the area under a curve was found by the definite integral $\displaystyle\int_a^b f(x)\,dx$.

It is possible to derive this result in an alternative way by considering the sum of the areas of the chosen elements as the number of elements tends to infinity.

   This process can be appreciated by considering a simple area under a straight line given by $y=x+1$ between $x=0$ and $x=5$.

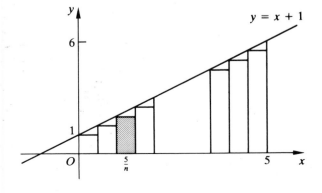

Divide the area between $x=0$ and $x=5$ into n strips (called elements) thus making the width of each element $\dfrac{5}{n}$.

Each element approximates to a rectangle of width $\dfrac{5}{n}$ and height given by the ordinate (y co-ordinate) of its left-hand edge.

The ordinates for each element are tabulated below for values of x.

Value of x	0	$\dfrac{5}{n}$	$2\left(\dfrac{5}{n}\right)$	$3\left(\dfrac{5}{n}\right)$	......	$(n-1)\dfrac{5}{n}$
Ordinate	1	$\dfrac{5}{n}+1$	$2\left(\dfrac{5}{n}\right)+1$	$3\left(\dfrac{5}{n}\right)+1$	......	$(n-1)\dfrac{5}{n}+1$

The area of the smallest rectangle $= 1 \times \dfrac{5}{n}$.

The area of the second rectangle $= \left(\dfrac{5}{n}+1\right)\dfrac{5}{n}$.

The area of the third rectangle $= \left[2\left(\dfrac{5}{n}\right)+1\right]\dfrac{5}{n}$, and so on.

The sum of the areas of all n rectangles is

$$1\left(\frac{5}{n}\right) + \left(\frac{5}{n}+1\right)\frac{5}{n} + \left[2\left(\frac{5}{n}\right)+1\right]\frac{5}{n} + \left[3\left(\frac{5}{n}\right)+1\right]\frac{5}{n} + \cdots + \left[(n-1)\frac{5}{n}+1\right]\frac{5}{n}$$

$$= \frac{5}{n}\left[1 + \left(\frac{5}{n}+1\right) + \left\{2\left(\frac{5}{n}\right)+1\right\} + \left\{3\left(\frac{5}{n}\right)+1\right\} + \cdots + \left\{(n-1)\frac{5}{n}+1\right\}\right]$$

The terms in the square bracket form an arithmetical progression (see Chapter 14) with a first term of 1 and a common difference of $\dfrac{5}{n}$.

As there are n terms, the sum $= \dfrac{n}{2}\left[2 + (n-1)\dfrac{5}{n}\right]$

Thus, the sum of the areas $= \dfrac{5}{n}\left[\dfrac{n}{2}\{2 + (n-1)\}\dfrac{5}{n}\right] = \dfrac{5}{2}\left\{2 + 5 - \dfrac{5}{n}\right\} = \dfrac{1}{2}\left(35 - \dfrac{25}{n}\right)$

Now as the number of elements is increased, $n \to \infty$ and the limit of the sum $= \dfrac{35}{2}$ since the term $\dfrac{25}{n} \to 0$. Check this result by finding the area of the trapezium.

To generalize this method, consider one such element of the curve $y = f(x)$ obtained by dividing the area under the curve into a large number of approximate rectangles of width δx.

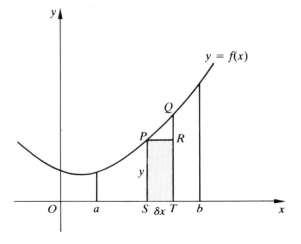

Let $P(x, y)$ and $Q(x + \delta x, y + \delta y)$ be neighbouring points on the curve such that $PQTS$ forms a typical element.

The area of the element is approximately equal to the area of the rectangle $PRTS$ i.e. $y\delta x$.

Hence, by adding the areas of all such rectangles between $x = a$ and $x = b$, an approximation for the total area is obtained. The Greek letter Σ is used to denote a sum.

$$\therefore \text{ Total area} \simeq \sum_{x=a}^{x=b} y\,\delta x$$

As a greater number of elements is taken, the width of each decreases and $\delta x \to 0$.

Thus, in the limit as $\delta x \to 0$

$$\text{Actual total area} = \lim_{\delta x \to 0} \sum_{x=a}^{x=b} y\,\delta x = \int_a^b y\,dx$$

Since the area was found as an integral at the beginning of the previous section, this defines the definite integral as the limit of a sum.

Choice of element

The above result shows that integration can be used to evaluate a quantity expressed as the limit of a sum in the form $y\,\delta x$ or $f(x)\,\delta x$. It is a general result which does not give any indication about how the rectangles should be formed.

It is possible to choose the element in different ways, although sometimes the resulting integral is difficult or impossible.

To find the area bounded by the curve $y = x^2$, the y-axis and the line $y = 4$, the element can be chosen parallel to the x-axis.

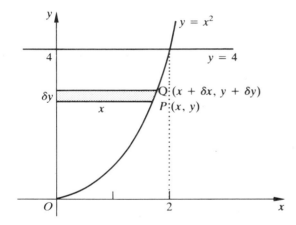

If $P(x, y)$ and $Q(x + \delta x, y + \delta y)$ are neighbouring points on the curve, the area of the element is $\simeq x\,\delta y$.

$$\text{The total area} \simeq \sum_{y=0}^{y=4} x\,\delta y$$

$$\therefore \text{Actual area} = \lim_{\delta y \to 0} \sum_{y=0}^{y=4} x\,\delta y = \int_0^4 x\,dy$$

Since $y = x^2$, $x = y^{\frac{1}{2}}$ and hence,

$$\text{Area} = \int_0^4 y^{\frac{1}{2}}\,dy, \text{ which can be evaluated.}$$

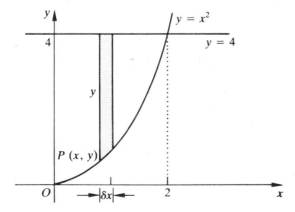

Alternatively, the element can be chosen parallel to the y-axis.

The area of the element $\simeq (4 - y)\,\delta x$

$$\therefore \text{Total area} \simeq \sum_{x=0}^{x=2} (4 - y)\,\delta x$$

$$\therefore \text{Actual area} = \lim_{\delta x \to 0} \sum_{x=0}^{x=2} (4 - y)\,\delta x$$

$$= \int_0^2 (4 - y)\,dx$$

Since $y = x^2$, $\quad \text{Actual area} = \int_0^2 (4 - x^2)\,dx$, $\quad$ which can be evaluated.

To find the area between the curve $y = x(4 - x)$, the y-axis and the line $y = 4$, it appears that a typical element can be chosen parallel to either axis.

If the element is parallel to the x-axis, its area is $\simeq x \, \delta y$.

Hence, total area $= \displaystyle\lim_{\delta y \to 0} \sum_{y=0}^{y=4} x \, \delta y$

$$= \int_0^4 x \, dy$$

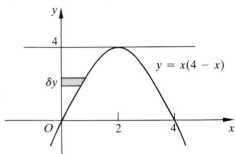

However, it is not possible to express x in terms of y from the equation and the integral cannot be evaluated.

The only option is to choose an element parallel to the y-axis, in which case we obtain

total area $= \displaystyle\int_0^2 (4 - y) \, dx$

$$= \int_0^2 (4 - 4x + x^2) \, dx$$

since $y = 4x - x^2$, and this can be evaluated.

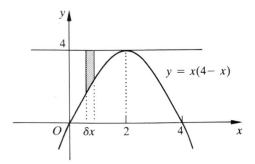

To find the area bounded by the curve $y = 3x - x^2$, the line $y = x$ and the x-axis, it is not possible to choose the element parallel to either axis and complete the solution.

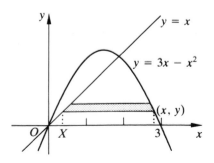

If it is parallel to the x-axis then the length of the element is $x - X$ where x and X are the x co-ordinates of points on the curve $y = 3x - x^2$ and the line $y = x$, respectively.

The total area gives $\displaystyle\int_0^3 (x - X) \, dy$

but again it is not possible to express x in terms of y from $y = 3x - x^2$ and so the integral cannot be evaluated.

If the element is chosen parallel to the y-axis (figure below, right) it is not a **typical** element, since its upper end lies on different graphs and its area would depend upon its position.

Hence, the element cannot be chosen parallel to the y-axis.

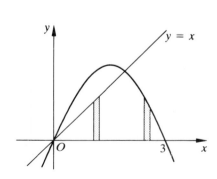

In this example (figure, right) we find the area between the curve, the *x*-axis and the lines $x = 2$ (the point of intersection) and $x = 3$ and the area of the shaded triangle.

The sum will give the total area.

Remember,

1 the element chosen must be typical over the whole area;
2 it must be possible to integrate the function obtained, preferably easily.

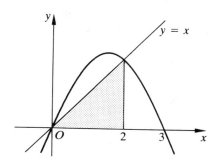

WORKED EXAMPLE

Find the area enclosed by the curve $y^2 = 4x$, the *y*-axis and the line $y = 2$.

This problem can be solved in three different ways.

Method 1

Choose an element of area *PQRS* parallel to the *x*-axis by taking points $P(x, y)$ and $Q(x + \delta x, y + \delta y)$ as neighbouring positions on the curve.

Area of element $\simeq x\,\delta y$

Total area $\simeq \sum\limits_{y=0}^{y=2} x\,\delta y$

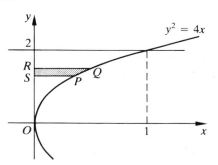

In the limit as $\delta y \to 0$,

Actual area $= \lim\limits_{\delta y \to 0} \sum\limits_{y=0}^{y=2} x\,\delta y = \int_0^2 x\,dy$

$$= \int_0^2 \frac{y^2}{4}\,dy = \left[\frac{y^3}{12}\right]_0^2 = \left(\frac{8}{12} - 0\right) = \frac{2}{3}\,\text{units}^2$$

Method 2

Choose an element of area *PQRS* parallel to the *y*-axis by taking points $P(x, y)$, and $Q(x + \delta x, y + \delta y)$ as neighbouring positions on the curve.

Area of element $\simeq (2 - y)\,\delta x$

Total area $\simeq \sum\limits_{x=0}^{x=1} (2 - y)\,\delta x$

Since $y = 2 \implies x = 1$.

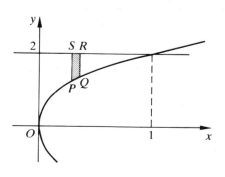

In the limit as $\delta x \to 0$,

Actual area $= \lim\limits_{\delta x \to 0} \sum\limits_{x=0}^{x=1} (2 - y)\,\delta x = \int_0^1 (2 - y)\,dx$

$$= \int_0^1 (2 - 2x^{\frac{1}{2}})\,dx = \left[2x - \frac{4}{3}x^{\frac{3}{2}}\right]_0^1 = \frac{2}{3}\,\text{units}^2$$

Method 3

Lastly, we can find the area under the curve and above the *x*-axis, and subtract this from the rectangle *OABC*.

Area of element $\simeq y\,\delta x$

Total area $\simeq \sum\limits_{x=0}^{x=1} y\,\delta x$

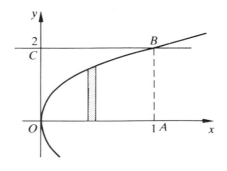

In the limit as $\delta x \to 0$,

Actual area $= \lim\limits_{\delta x \to 0} \sum\limits_{x=0}^{x=1} y\,\delta x$

$\qquad = \int_0^1 y\,dx = \int_0^1 2x^{\frac{1}{2}}\,dx$

$\qquad = \left[\dfrac{4}{3}x^{\frac{3}{2}}\right]_0^1 = \dfrac{4}{3}\,\text{units}^2$

The area of the rectangle $OABC = 2 \times 1 = 2\,\text{units}^2$ and hence the required area $= 2 - \dfrac{4}{3} = \dfrac{2}{3}\,\text{units}^2$.

You might notice that Method **2** and Method **3** are essentially the same.

WORKED EXAMPLE

Find the area enclosed by the curves $y = x^2 - 4x + 1$ and $y = 17 - x^2$.

First, find the points of intersection of the curves which occur when

$$x^2 - 4x + 1 = 17 - x^2 \quad \Leftrightarrow \quad 2x^2 - 4x - 16 = 0$$

$$\Leftrightarrow \quad x^2 - 2x - 8 = (x - 4)(x + 2) = 0$$

$$\Leftrightarrow \quad x = 4 \quad \text{or} \quad -2$$

The curves intersect at $(4, 1)$ and $(-2, 13)$. A sketch can be drawn since they are both quadratic curves.

Select $P(x, y)$ and $Q(x + \delta x, y + \delta y)$ as neighbouring points on the curve $y = x^2 - 4x + 1$ to form a typical element *PQRS*.

Let the other curve be written $Y = 17 - x^2$, so the points S and R have co-ordinates (x, Y) and $(x + \delta x, Y + \delta y)$, respectively.

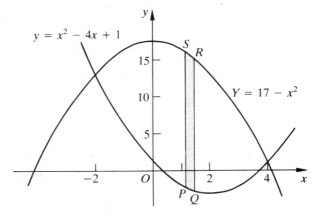

Note the choice of y and Y to distinguish the two ordinates of the curves.

i.e. $y = x^2 - 4x + 1$ and $Y = 17 - x^2$

Area of element $\simeq (Y - y)\,\delta x$

(Note that the difference $Y - y$ gives the length of the element, even in the position shown where y will be negative, which will actually produce the sum of the magnitudes of the ordinates as required.)

$$\text{Total area} \simeq \sum_{x=-2}^{x=4} (Y-y)\,\delta x$$

In the limit as $\delta x \to 0$

$$\text{Total area} = \lim_{\delta x \to 0} \sum_{x=-2}^{x=4} (Y-y)\,\delta x = \int_{-2}^{4} (Y-y)\,dx$$

$$= \int_{-2}^{4} (17-x^2) - (x^2 - 4x + 1)\,dx$$

$$= \int_{-2}^{4} (16 + 4x - 2x^2)\,dx = \left[16x + 2x^2 - \frac{2x^3}{3} \right]_{-2}^{4}$$

$$= \left(64 + 32 - \frac{128}{3} \right) - \left(-32 + 8 + \frac{16}{3} \right)$$

$$= 120 - 48 = 72\,\text{units}^2$$

Exercise 5.3

1 Find the areas between the curves, the x-axis and the given lines.

(a) $y = 2x^2 + 3$, $x = -1$, $x = 2$ (b) $y = x(3-x)$, $x = 0$, $x = 2$

(c) $y = \dfrac{1}{x^2}$, $x = 1$, $x = 4$

2 Find the areas enclosed by the following curves, the y-axis and the given lines.

(a) $x = 2y^2$, $y = 3$ (b) $y = x^2$, $y = 16$ (c) $y^2 = x + 4$, $x = 0$ (d) $y = x^3$, $y = 8$

3 Find the areas enclosed by the following curves and straight lines.

(a) $y^2 = 4x$, $x = 1$, $x = 4$ (b) $y^2 = x + 4$, $x = 5$
(c) $y = 2 + 5x - x^2$, $y = 0$, $x = 1$, $x = 3$

4 Sketch the curve $y = x(x-2)^2$ and find the area enclosed by the curve and the x-axis.

5 Sketch the curve $y = x(x+1)(x-2)$ and find the areas enclosed by the curve and the x-axis between (a) $x = -1$ and $x = 0$ (b) $x = 0$ and $x = 2$ and (c) $x = -1$ and $x = 2$.

6 Find the points of intersection of the curve $y = 3 + 2x - x^2$ and the line $y = 3 - x$ and sketch these on the same axes. Find the area enclosed by the curve and the straight line.

7 Find the areas of the finite region enclosed by the following curves and straight lines.

(a) $y = x(4-x)$, $y = x$ (b) $y = x^2 - 8x + 16$, $y = 4$ (c) $y = x^2 + 3$, $y = 3x + 7$
(d) $y^2 = 4x$, $y = x$

8 Find the areas enclosed by the following pairs of curves.

(a) $y^2 = 4x$ and $x^2 = 4y$ (b) $y = x^2 - 3x$ and $y = 5x - x^2$
(c) $y = x^2 - 3x - 5$ and $y = 1 + x - x^2$ (d) $y = 3x^2 + 4x + 3$ and $y = x^2 - 4x - 3$

9 P is a point on the curve $y = 2x - 3x^2$ whose x co-ordinate is a, where $0 < a < \frac{2}{3}$. If O is the origin and N is the foot of the perpendicular from P to the x-axis, find the area of triangle OPN in terms of a and the area between the curve and the line OP in terms of a. If these two areas are equal, find the value of a.

Volumes of revolution

We saw, earlier, that the sum of quantities of the form $f(u)\,\delta u$ can be evaluated as an integral. So far, the function $f(u)$ has been x, y, $x - X$ or $y - Y$, but it is possible to include other functions, such as y^2 or x^2.

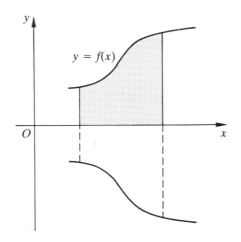

Consider the curve $y = f(x)$. If this curve is rotated about the x-axis, it will trace out a surface and enclose a volume, thus forming a solid.

This solid is called a **solid of revolution** or a **volume of revolution.**

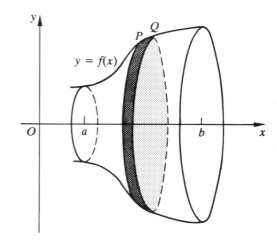

To find the volume of this solid, choose two points, $P(x, y)$ and $Q(x + \delta x, y + \delta y)$, on the curve so that, as the curve is rotated about the x-axis, an element of volume in the form of an approximate cylinder is formed.

The volumes of all such elements are summed between $x = a$ and $x = b$ to give an approximation to the total volume.

The actual volume will be found in the limit as $\delta x \to 0$.

The volume of a cylinder $= \pi r^2 h$.
In this case, the volume of the element is approximately equal to the volume of the cylinder of radius y and height δx.

$$\text{Volume of element} \simeq \pi y^2 \delta x$$

$$\text{Total volume} \simeq \sum_{x=a}^{x=b} \pi y^2 \, \delta x$$

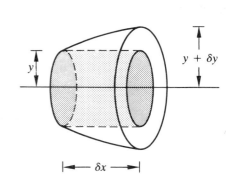

As a greater number of elements is taken, $\delta x \to 0$.

In the limit as $\delta x \to 0$

$$\text{Actual volume} = \lim_{\delta x \to 0} \sum_{x=a}^{x=b} \pi y^2 \, \delta x$$

As we saw in the case of areas, this limit of a sum can be written as an integral.

$$\therefore \text{Actual volume} = \int_a^b \pi y^2 \, dx \quad \text{or} \quad \pi \int_a^b y^2 \, dx$$

Note that π is a constant and thus can be removed from the integrand.

WORKED EXAMPLES

1 Find the volume generated by rotating the area enclosed by the curve $y = x^2 - 2x$ and the x-axis completely about the x-axis.

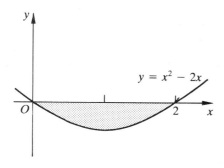

A sketch of the curve is always useful.

When $y = 0$,

$$x^2 - 2x = x(x - 2) = 0$$

$$\Leftrightarrow \quad x = 0 \quad \text{or} \quad x = 2$$

This quadratic curve cuts the x-axis at $(0, 0)$ and $(2, 0)$.

The area to be rotated is shown shaded in the sketch, left.

By rotating this shaded area about the x-axis through $360°$, a solid of revolution will be formed which will be similar to a rugby ball.

If $P(x, y)$ and $Q(x + \delta x, y + \delta y)$ are chosen as neighbouring points on the curve then, when the curve is rotated about the x-axis through $360°$, the arc PQ will generate an element in the form of a disc or approximate cylinder of radius y and height δx.

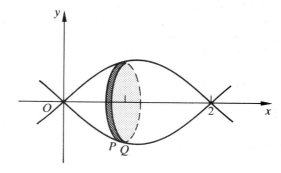

Volume of element $\simeq \pi y^2 \, \delta x$

Total volume $\simeq \displaystyle\sum_{x=0}^{x=2} \pi y^2 \, \delta x$

In the limit as $\delta x \to 0$

$$\text{Actual volume} = \lim_{\delta x \to 0} \sum_{x=0}^{x=2} \pi y^2 \, \delta x = \int_0^2 \pi y^2 \, dx$$

Now, $y = x^2 - 2x$

$$\therefore \text{Volume} = \pi \int_0^2 (x^2 - 2x)^2 \, dx$$

$$= \pi \int_0^2 (x^4 - 4x^3 + 4x^2) \, dx$$

$$= \pi \left[\frac{x^5}{5} - x^4 + \frac{4x^3}{3} \right]_0^2$$

$$= \pi \left[\left(\frac{32}{5} - 16 + \frac{32}{3} \right) - 0 \right]$$

$$= \frac{16\pi}{15} \text{ units}^3$$

Note that it is quite usual to leave the answer as a multiple of π.

2 Find the volume of the solid generated when the area enclosed by the curve $y^2 = x - 3$, the x-axis and the lines $x = 1$ and $y = 2$ is rotated completely about the line $x = 1$.

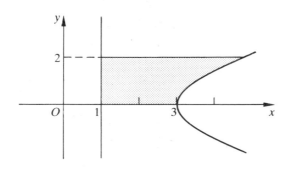

The equation of the curve $y^2 = x - 3$ gives

$$y = 0 \iff x = 3$$

The curve is symmetrical about the x-axis.

If $x < 3$, y cannot take real values. The curve and area to be rotated are shown in the sketch, left.

If $P(x, y)$ and $Q(x + \delta x, y + \delta y)$ are two neighbouring points on the curve, then the arc PQ and the shaded area will generate an element of volume in the form of an approximate cylinder of radius $x - 1$ and height δy when the curve is rotated about the line $x = 1$ through $360°$ (see figure, right).

$$\text{Volume of element} \simeq \pi(x - 1)^2 \, \delta y$$

$$\text{Total volume} \simeq \sum_{y=0}^{y=2} \pi(x - 1)^2 \, \delta y$$

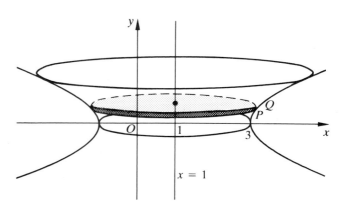

In the limit as $\delta y \to 0$

$$\text{Actual volume} = \lim_{\delta y \to 0} \sum_{y=0}^{y=2} \pi(x - 1)^2 \, \delta y = \int_0^2 \pi(x - 1)^2 \, dy$$

Now, $y^2 = x - 3 \iff x = y^2 + 3 \iff (x - 1) = y^2 + 2$

$$\text{Actual volume} = \pi \int_0^2 (y^2 + 2)^2 \, dy = \pi \int_0^2 (y^4 + 4y^2 + 4) \, dy$$

$$= \pi \left[\frac{y^5}{5} + \frac{4y^3}{3} + 4y \right]_0^2$$

$$= \pi \left[\left(\frac{32}{5} + \frac{32}{3} + 8 \right) - 0 \right] = \frac{376}{15} \pi \text{ units}^3$$

3 Find the volume of the solid generated when the area enclosed by the lines $y = x + 3$, $x = 3$, $y = 2$ and the y-axis is rotated about the x-axis.

The equation $y = x + 3$ gives a straight line of gradient 1 passing through the point $(0, 3)$.

The line and the area to be rotated are shown in the diagram, right.

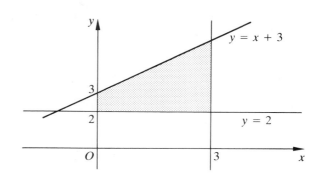

If $P(x, y)$ and $Q(x + \delta x, y + \delta y)$ are two neighbouring points on the curve, then the line PQ will generate an element of volume in the form of a disc with a hole through its centre, when the shaded area is rotated about the x-axis.

The cross-sectional area of the disc is the difference in the areas of two circles of radii y and 2 respectively.

i.e. $\pi y^2 - \pi(2)^2$ (**Not** $\pi(y - 2)^2$)

$\therefore$ Volume of element $\simeq \pi(y^2 - 4)\,\delta x$

$$\text{Total volume} \simeq \sum_{x=0}^{x=3} \pi(y^2 - 4)\,\delta x$$

In the limit as $\delta x \to 0$

$$\text{Actual volume} = \lim_{\delta x \to 0} \sum_{x=0}^{x=3} \pi(y^2 - 4)\,\delta x = \int_0^3 \pi(y^2 - 4)\,dx$$

Now $y = x + 3$,

$$\therefore \text{Actual volume} = \pi \int_0^3 \left[(x + 3)^2 - 4 \right] dx = \pi \int_0^3 (x^2 + 6x + 5)\,dx$$

$$= \pi \left[\frac{x^3}{3} + 3x^2 + 5x \right]_0^3$$

$$= \pi \left[\left(\frac{27}{3} + 27 + 15 \right) - 0 \right] = 51\pi \text{ units}^3$$

Exercise 5.4

1 Find the volumes of the solids generated when the areas enclosed by the following curves and lines are rotated completely about the x-axis.

(a) $y = 3x$, $y = 0$, $x = 1$ and $x = 3$

(b) $y = x^2$, $y = 0$, $x = 0$ and $x = 4$

(c) $y = x^2 + 2$, $y = 0$, $x = 0$ and $x = 2$

(d) $y = 4x - x^2$, $y = 0$, $x = 1$ and $x = 2$

(e) $y = x + \dfrac{1}{x}$, $y = 0$, $x = 1$ and $x = 3$

2 Find the volumes of the solids generated when the areas enclosed by the following curves and lines are rotated completely about the y-axis.

(a) $y = 2x + 1$, $x = 0$, $y = 2$

(b) $y = x^2 + 2$, $x = 0$, $y = 2$ and $y = 3$

(c) $y^2 = 4x$, $x = 0$, $y = 0$ and $y = 4$

(d) $y^2 = x + 4$, $x = 0$

(e) $y = \dfrac{1}{x}$, $x = 0$, $y = 1$ and $y = 2$

3 Find the volumes of the solids generated when each of the areas enclosed by the following curves and lines is rotated about the given line.

    (a) $y = 4x$, $x = 0$, $y = 8$;   about $x = 0$

    (b) $y = x^2$, $y = 4$,   about $y = 4$

    (c) $y^2 = 4x$, $y = 2$, $x = 0$;   about $y = 2$

    (d) $y^2 = x - 1$, $x = 2$;   about $x = 2$

    (e) $y = x(2 - x)$, $y = 2$, $x = 0$ and $x = 2$;   about $y = 2$

    (f) $y = x(2 - x)$, $y = 0$;   about $y = 2$

4 Find the points of intersection of the curve $y = x^2$ and the line $y = 4x$. Find the volume generated when the area enclosed between the curve and the line is rotated about the x-axis.

5 Sketch the curve $y^2 = x(x - 3)^2$. The area of the loop for which $x > 0$ and $y > 0$ is rotated about the x-axis. Find the volume of the solid generated.

6 The area bounded by the curve $y = x^2 - 4x + 7$ and the line $y = 7$ is rotated completely about the line $y = 7$. Find the volume of the solid of revolution.

7 Repeat question **6** by rotating that area about **(a)** the line $y = 8$ and **(b)** the x-axis.

8 The region enclosed by the curve $y = x + \dfrac{1}{x}$ and the lines $y = \frac{1}{2}x$, $x = 1$ and $x = 2$, is rotated about the x-axis. Find the volume of revolution.

9 Find the volume of the solid formed when the region enclosed by the curve $y = 2x^3$, the x-axis and the line $y = 3 - x$, is rotated completely about the x-axis.

10 The equation of a circle of radius r, centre at the origin, is $x^2 + y^2 = r^2$. Show that by considering the volume of the solid formed when the area bounded by that part of the curve above the x-axis and the x-axis is rotated about this axis, the volume of a sphere is $V = \frac{4}{3}\pi r^3$.

11 If the region bounded by the line $y = \dfrac{r}{h}x$, the x-axis and the line $x = h$, is rotated about the x-axis, show that the volume of a cone is $V = \frac{1}{3}\pi r^2 h$.

12 The area enclosed by the curve $y = x^2 - 8x + 15$ and the line $y = 3$, is rotated about the line $y = 3$. Find the volume of the solid formed.

6 Trigonometry 1

Sine, cosine and tangent of any angle

You probably met these three trigonometrical ratios when solving right-angled triangles.

Referring to triangle **(1)**,

$$\sin A^\circ = \frac{a}{b} = \frac{\text{Opposite}}{\text{Hypotenuse}}$$

$$\cos A^\circ = \frac{c}{b} = \frac{\text{Adjacent}}{\text{Hypotenuse}}$$

$$\tan A^\circ = \frac{a}{c} = \frac{\text{Opposite}}{\text{Adjacent}}$$

SIN OF
   HEAT

COS ALAN
   HAS

TAN ON
   ARMS

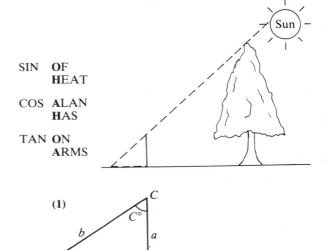

In every right-angled triangle with one angle 35° (the others are 90° and 55°), the ratio $\frac{a}{b}$ will always be the same whatever the size of the triangle. This ratio is defined as the sine of 35° (sin 35°) and is calculated to be equal to 0.574 (correct to 3 decimal places).

If b is given to be equal to 6, then the size of the triangle is specified and a and c can be calculated.

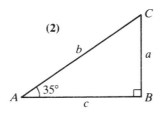

$$\frac{a}{b} = \frac{a}{6} = \sin 35^\circ \quad \Rightarrow \quad a = 6 \times \sin 35^\circ = 6 \times 0.574 = 3.441 \quad \text{(correct to 3 decimal places)}$$

$$\frac{c}{b} = \frac{c}{6} = \cos 35^\circ \quad \Rightarrow \quad c = 6 \times \cos 35^\circ = 6 \times 0.819 = 4.915$$

You can check that $\tan 35^\circ = \frac{a}{c} = \frac{6\sin 35^\circ}{6\cos 35^\circ} = \frac{3.441}{4.915} = 0.700$ and $\tan 35^\circ = \frac{\sin 35^\circ}{\cos 35^\circ}$

In general, $\boxed{\tan A^\circ = \dfrac{\sin A^\circ}{\cos A^\circ}}$

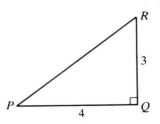

If a right-angled triangle is specified (figure, left) by the shorter sides being 3 and 4, the ratios can be used to solve the triangle i.e. find the remaining angles and side.

$$\tan P = \frac{3}{4} = 0.75, \text{ so } P = \arctan(0.75) \text{ or } \tan^{-1}(0.75)$$

$\arctan x$ and $\tan^{-1} x$ are both used to denote the angle whose tangent is x, i.e. the inverse function of $\tan x$.

$$P = \tan^{-1}(0.75) \quad \Rightarrow \quad P = 36.9° \quad \Rightarrow \quad R = 53.1°$$

You can check that using the sine or cosine will give the value of PR, but it is much easier to use Pythagoras' Theorem, to give

$$PR^2 = 3^2 + 4^2 \quad \Rightarrow \quad PR = \sqrt{3^2 + 4^2} = \sqrt{25} = 5$$

In fact, by now you should recognize a right-angled $3, 4, 5$ triangle, and it is used so often that you might learn that the other angles are $36.9°$ and $53.1°$.

Special triangles

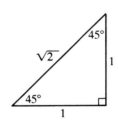

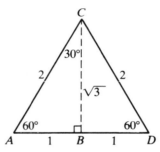

Pythagoras Theorem $\Rightarrow$

$$BC^2 + 1^2 = 2^2$$

$$\Rightarrow \quad BC^2 = 4 - 1 = 3$$

$$\Rightarrow \quad BC = \sqrt{3}$$

These ratios given accurately in terms of surds, are used so frequently that they are worth remembering.

$$\sin 60° = \frac{\sqrt{3}}{2} = \cos 30°$$

$$\cos 60° = \frac{1}{2} = \sin 30°$$

$$\tan 60° = \sqrt{3}; \quad \tan 30° = \frac{1}{\sqrt{3}}$$

Referring to triangles **(1)** and **(2)** earlier in the chapter,

$$\sin 35° = \frac{a}{b} = \cos 55°$$

In general, $\boxed{\sin A = \cos(90 - A)}$ and $\boxed{\cos A = \sin(90 - A)}$

$\tan 35° = \dfrac{a}{c}$ and $\tan 55° = \dfrac{c}{a} \Rightarrow \tan 55° = \dfrac{1}{\tan 35°}$, and in general $\boxed{\tan A = \dfrac{1}{\tan(90 - A)}}$

Using Pythagoras' Theorem for $\triangle ABC$ (triangle **(1)**) i.e. $a^2 + c^2 = b^2$

$$(\sin A)^2 + (\cos A)^2 = \left(\frac{a}{b}\right)^2 + \left(\frac{c}{b}\right)^2 = \frac{a^2}{b^2} + \frac{c^2}{b^2} = \frac{a^2 + c^2}{b^2} = \frac{b^2}{b^2} = 1$$

$(\sin A)^2 = (\sin A) \times (\sin A)$, but it is written more conveniently as $\sin^2 A$.

The Pythagorean identity is $\boxed{\sin^2 A + \cos^2 A \equiv 1}$ which is true for all values of A.

$\sin^{-1}(0.5)$ means the angle whose sine is 0.5 $\Rightarrow$ $\sin^{-1}(0.5) = 30°$.

$$\frac{1}{\sin 0.5} = \frac{1}{0.00872} = 114.6 \text{ which is \textbf{not} the same as } \sin^{-1}(0.5).$$

In fact, $\sin^{-1}(0.5)$ is an **angle** and $\dfrac{1}{\sin 0.5}$ is a **number**.

Perhaps the confusion in some students' minds led to our use of $\sin^{-1}(0.5) = \arcsin(0.5)$, and $\arcsin x$ to denote the inverse function of $\sin x$.

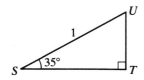

If the hypotenuse $SU = 1$, then $ST = \cos 35°$, $TU = \sin 35°$.

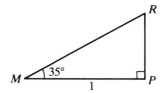

If $MP = 1$, then $RP = \tan 35°$.

Circular functions

Sine

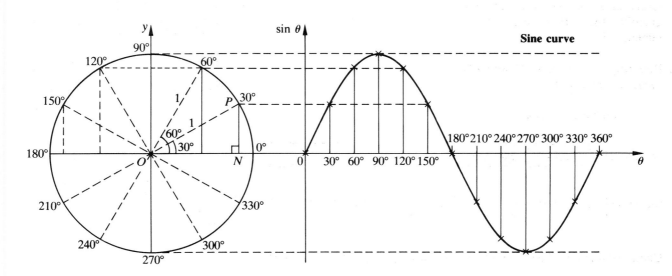

In the circle of radius 1 unit, $PN = \sin 30°$, so the graph of $\sin \theta$ can be plotted for values of θ from 0° to 360°.
   The co-ordinates of P are $(\cos 30°, \sin 30°)$ and $\sin 30°$ is the y co-ordinate of P i.e. PN.

As P traces out the circle, the y co-ordinate gives the value of $\sin \theta$ for any angle θ.

As θ increases from $0°$ to $90°$, the value of $\sin \theta$ increases from 0 to $+1$, giving $\sin 90° = +1$.

As θ increases from $90°$ to $180°$, the value of $\sin \theta$ decreases from $+1$ to 0, giving $\sin 180° = 0$.

From the diagram, $\sin 120° = \sin 60°$ and $\sin 150° = \sin 30°$.

For $180° < \theta < 360°$, the value of $\sin \theta$ is negative and $\sin 210° = \sin 330° = -\sin 30°$
$$\sin 240° = \sin 300° = -\sin 60°$$

Cosine

The value of $\cos 30°$ is the x co-ordinate of P, i.e. ON, so the values of $\cos \theta$ are the projections of OP onto the original x-axis.

The graph of $\cos \theta$ can best be drawn by rotating the circle through $90°$.

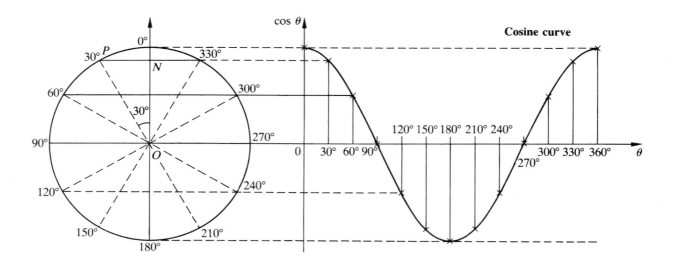

As θ increases from $0°$ to $90°$, $\cos \theta$ decreases from $+1$ to 0.

From the diagram, $\cos 330° = \cos 30°$ and $\cos 300° = \cos 60°$
$$\cos 120° = \cos 240° = -\cos 60°$$
$$\cos 150° = \cos 210° = -\cos 30°$$

Tangent

From the circle radius 1 unit,

In $\triangle OXT_1$, $OX = 1$ unit, $T_1\hat{O}X = 30° \Rightarrow XT_1 = \tan 30°$.
To plot $\tan 30°$, project OP_1 to T_1 and transfer the length XT_1 to the point T_1 on the graph.

Similarly P_2 represents $45°$ $(X\hat{O}P_2 = 45°)$. Produce OP_2 to T_2 so $XT_2 = \tan 45° = 1$ and locate T_2 on the graph. $P_3(\theta = 60°)$ gives $XT_3 = \tan 60°$ and hence T_3 on the graph.

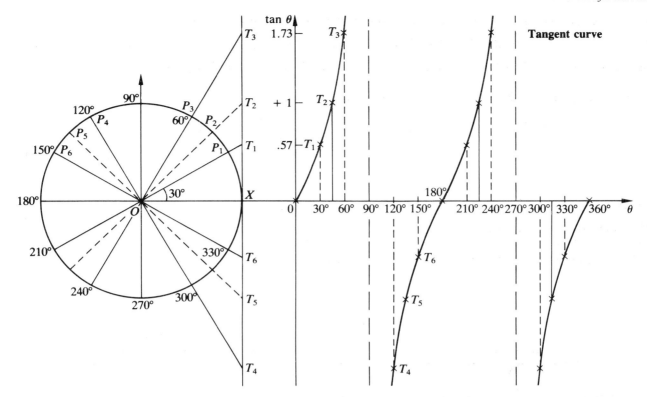

The value of the tangent of an angle is the length cut out by the radius OP on the vertical tangent to the circle at X.

When $\theta = 90°$ the value of the tangent is infinite so the tangent curve has an **asymptote** at $\theta = 90°$.

For $\theta = 120°$, $X\hat{O}P_4 = 120°$, so $\tan 120°$ is found by producing P_4O to T_4 $\Rightarrow$ $\tan 120°$ is negative.

Similarly, $\tan 135°(P_5)$ is found by locating T_5 and $\tan 150°$ by T_6; $\tan 180° = 0$.

$\tan 210° = \tan 30°$; $\tan 240° = \tan 60°$ and the graph repeats its values.
$\tan 330° = \tan 150° = -\tan 30°$; $\tan 300° = \tan 120° = -\tan 60°$.

Periodic functions

For $0° \leqslant \theta \leqslant 360°$, the sine curve completes one **cycle** of values. These values will be repeated for $360° \leqslant \theta \leqslant 720°$. The function $\sin \theta$ is an example of a **periodic function** whose values repeat after every $360°$. The **period** of $\sin \theta$ is $360°$.

Similarly, $\cos \theta$ is a periodic function whose period is also $360°$.
$\tan \theta$ is a periodic function whose period is $180°$.

Exercise 6.1

1 Use your calculator to find the values of the sine, cosine and tangent of the following angles and check that each agrees with values on the graphs.

(a) 30°   (b) 60°   (c) 45°   (d) 100°   (e) 200°

(f) 300°   (g) 120°   (h) 90°   (i) 360°

2 Using the accurate values of $\cos A$ and $\sin A$ in surd form, find the value of $\cos^2 A + \sin^2 A$ for
    (a) $A = 30°$,   **(b)** $A = 45°$ and   **(c)** $A = 60°$.

3 Compare the values of $2 \sin A$ and $\sin 2A$ for **(a)** $A = 30°$ **(b)** $A = 45°$ **(c)** $A = 60°$.

4 Work out **(a)** $\sin 30° + \sin 60°$ and **(b)** $\sin(30° + 60°)$ and comment.

5 Evaluate $2 \sin A \cos A$ for **(a)** $A = 30°$ **(b)** $A = 45°$ **(c)** $A = 60°$.

6 Find the value of $\cos^2 A - \sin^2 A$ for **(a)** $A = 30°$ **(b)** $A = 45°$ **(c)** $A = 60°$.

7 Find the value of $2 \cos^2 A - 1$ for **(a)** $A = 30°$ **(b)** $A = 45°$ **(c)** $A = 60°$.

8 Find the value of $1 - 2 \sin^2 A$ for **(a)** $A = 30°$ **(b)** $A = 45°$ **(c)** $A = 60°$.

Properties of the sine, cosine and tangent functions

The upper figure, right, shows the graph of $y = \sin x$ for values of x from $-360°$ to $+360°$.

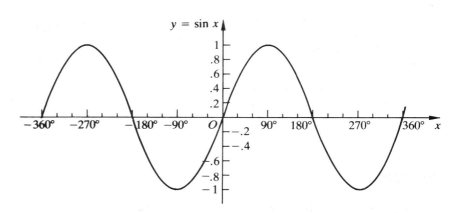

1 $\sin x$ is   periodic $\Rightarrow$ $\sin x$ repeats its values every $360°$.

Formally, $\boxed{\sin(x + 360°) = \sin x}$

2 $\sin x$ is an **odd** function; like $y = x^3$ or $y = x^5$ it has rotational symmetry of order 2 about the origin.

Formally, $\boxed{\sin(-x) = -\sin x}$

The lower figure, right, shows the graph of $y = \cos x$ for $-360° \leqslant x \leqslant +360°$.

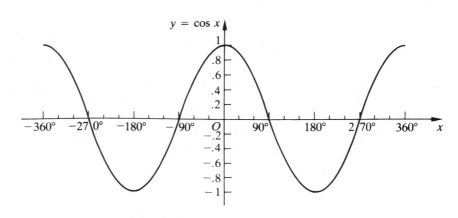

3 $\cos x$ is   periodic $\Rightarrow$ $\cos x$ repeats its values every $360°$.

Formally, $\boxed{\cos(x + 360°) = \cos x}$

4 $\cos x$ is an **even** function; like $y = x^2$ or $y = \dfrac{1}{x^2}$ it has the y-axis $(x = 0)$ as a line of symmetry.

Formally, $\boxed{\cos(-x) = \cos x}$

5 The sine curve is a translation of the cosine curve (and vice versa). $\cos 60° = \sin 150° = \sin(90° + 60°)$, and in general

$$\boxed{\cos x = \sin(x + 90°)}$$

Similarly, $\sin 150° = \cos 60° = \cos(150° - 90°)$, and in general $\boxed{\sin x = \cos(x - 90°)}$

The sine curve leads the cosine curve by 90°, or the cosine lags behind the sine by 90°, the two curves being 90° out of phase.

The figure, right, shows the graph of $y = \tan x$ for values of x from $-180°$ to $+180°$.

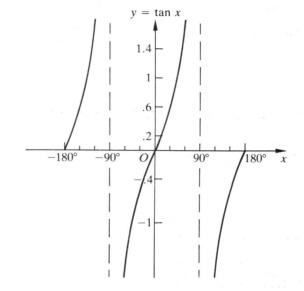

6 $\tan x$ is periodic of period 180°, repeating its values every 180°.

Formally, $\boxed{\tan(x + 180°) = \tan x}$

7 $y = \tan x$ is an **odd** function, like $\sin x$.

Formally, $\boxed{\tan(-x) = -\tan x}$

Equations

Trigonometrical equations can have many solutions.

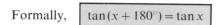

1 Solve $\sin x = 0.5$.

$\sin x = 0.5 \Rightarrow x = \sin^{-1}(0.5) = \arcsin(0.5) = 30°$ and this **principal value** is the value your calculator will provide. However, if you look at the figure for $y = \sin x$, there are other solutions, namely 150°, $-210°$ and $-330°$.

For $0° \leqslant x \leqslant 360°$, $\sin x = 0.5$ has two solutions, $x = 30°$ or $x = 150°$. The other solutions can be found by adding 360° to 30° or 150° and by subtracting 360° from 30° and 150°.

In general, $\sin x = 0.5 \Rightarrow x = 30° \pm 360n$ where $n = 1, 2, 3, \ldots$
or $x = 150° \pm 360n$ where $n = 1, 2, 3, \ldots$

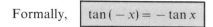

2 Solve $\cos x = 0.5$.

Your calculator gives $x = \cos^{-1}(0.5) = \arccos(0.5) = 60°$.
The figure for $y = \cos x$ will show you that x could also be equal to 300°.

In general, $\cos x = 0.5 \Rightarrow x = 60° + 360n$ (where $n \in \mathbb{Z}$) giving $x = 60°, 420°, -300°$ etc.
or $x = 300° + 360n$ $(n \in \mathbb{Z})$ giving $x = 300°, 660°, -60°$ etc.

Since $\cos x$ is an even function $x = 360n \pm 60°$ gives the complete set of solutions.

3 Solve $\tan x = 0.5$.

Your calculator gives $x = \tan^{-1}(0.5) = \arctan(0.5) = 26.6°$, but $x = 26.6° + 180° = 206.6°$ or $x = 26.6° - 180° = -153.4°$, since $\tan x$ is a periodic function of period $180°$.

In general, $\tan x = 0.5 \quad \Rightarrow \quad x = 180n + 26.6°$ where $n \in \mathbb{Z}$.

4 Solve $\sin^2 \theta = \sin \theta$.

$\sin^2 \theta = \sin \theta \quad \Rightarrow \quad \sin^2 \theta - \sin \theta = 0 \quad \Rightarrow \quad \sin \theta (\sin \theta - 1) = 0$

so $\sin \theta = 0 \quad \Rightarrow \quad \theta = 0, 180°, 360°$ etc.

or $\sin \theta - 1 = 0 \quad \Rightarrow \quad \sin \theta = 1 \quad \Rightarrow \quad \theta = 90°, 450°, -270°$ etc.

Note Take care not to cancel $\sin \theta$ from the original equation, otherwise you lose the solution $\sin \theta = 0$.

5 Solve $\sin^2 A + 3 \cos^2 A - 3 = 0$ for $0 \leqslant A \leqslant 360°$.

Remember that $\sin^2 A + \cos^2 A = 1 \quad \Rightarrow \quad \sin^2 A = 1 - \cos^2 A$, and substitute $1 - \cos^2 A$ for $\sin^2 A$.

$\sin^2 A + 3 \cos^2 A - 3 = 0 \quad \Rightarrow \quad 1 - \cos^2 A + 3 \cos^2 A - 3 = 0 \quad \Rightarrow \quad 2 \cos^2 A - 2 = 0 \quad \Rightarrow \quad \cos^2 A = 1$

$\Rightarrow \quad \cos A = \pm 1$

$\Rightarrow \quad A = 0°, 180°$ or $360°$.

Exercise 6.2

Solve the following equations for $0° \leqslant \theta \leqslant 360°$, giving if possible the general solution.

1 $\sin \theta = 0.8$	**4** $\tan \theta = 0.8$	**7** $\cos \theta = 0$
2 $\cos \theta = 0.8$	**5** $\sin \theta = 0.6$	**8** $\tan \theta = -1$
3 $\sin \theta = -0.7$	**6** $\tan \theta = 1.2$	**9** $\cos \theta = 1.2$

Use the symmetry of the graphs to simplify the following.

10 $\sin(180° - \theta)$	**12** $\cos(90° - \theta)$	**14** $\sin(90° + \theta)$
11 $\cos(180° - \theta)$	**13** $\sin(90° - \theta)$	**15** $\tan(180° + \theta)$

Solve for $0° \leqslant \theta \leqslant 360°$

16 $\sin \theta = \cos \theta$	**19** $\cos^2 \theta = \cos \theta$	**22** $5 \cos^2 \theta + 2 \sin^2 \theta + 5 \cos \theta = 4$
17 $\sin \theta = \tan \theta$	**20** $\sin^2 \theta = \cos^2 \theta$	**23** $1 + \sin^2 \theta = \cos \theta$
18 $\cos \theta = \tan \theta$	**21** $2 \cos^2 \theta + \sin^2 \theta = 2$	

Solving triangles

Sine and cosine rule

Right-angled triangles can be specified by giving,

1 one more angle and one of the sides (figures **(a)** and **(b)**, right)

2 two sides (and of course the right angle) (figure **(c)**)

and then all the angles and sides of the triangle can be calculated.

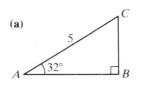

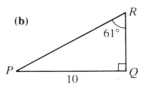

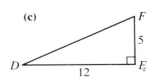

In $\triangle ABC$: $C = 180° - 90° - 32° = 58°$
$AB = 5\cos 32° = 4.24$
$BC = 5\sin 32° = 2.65$

In $\triangle PQR$: $P = 90° - 61° = 29°$
$QR = 10\tan 29° = 5.54$

$\dfrac{10}{PR} = \sin 61° \Rightarrow PR = \dfrac{10}{\sin 61°} = 11.43$

or by Pythagoras, $PR^2 = 10^2 + 5.54^2 = 100 + 30.69 = 130.69 \Rightarrow PR = \sqrt{130.69} = 11.43$

In $\triangle DEF$: $DF^2 = 12^2 + 5^2 = 169 \Rightarrow DF = 13$

$\tan D = \dfrac{5}{12} = 0.416 \Rightarrow D = 22.6° \Rightarrow F = 90° - 22.6° = 67.4°$

Can we solve triangles which do not contain a right angle?

How much information is needed to specify a triangle?

Investigation 1

How many sides and/or angles are needed to specify the shape and size of a triangle?

Case 1 Two angles and one side given (figure, right).

$I = 70°$. Draw HJ perpendicular ($\perp$) to GI

$HJ = 4\sin 50° = 3.064$

$G\hat{H}J = 90° - 50° = 40° \Rightarrow J\hat{H}I = 20° \Rightarrow \dfrac{HJ}{HI} = \cos 20°$

$\Rightarrow HI = \dfrac{HJ}{\cos 20°} = \dfrac{3.064}{\cos 20°} = 3.26$

$GJ = 4\cos 50° = 2.571$ and $JI = HJ\tan 20° = 3.064\tan 20° = 1.115 \Rightarrow GI = 2.571 + 1.115 = 3.686$

$HI = \dfrac{HJ}{\cos 20°} = \dfrac{4\sin 50°}{\sin 70°} = \dfrac{GH\sin 50°}{\sin 70°} \Rightarrow \dfrac{HI}{\sin 50°} = \dfrac{GH}{\sin 70°}$

By drawing $IK \perp GH$, we have $IK = GI\sin 50° = HI\sin 60°$

$\Rightarrow \dfrac{GI}{\sin 60°} = \dfrac{HI}{\sin 50°} = \dfrac{GH}{\sin 70°}$

(cont.)

This is an example of the **Sine Rule**

Note In $\triangle GHI$, the largest side (4) is opposite the largest angle (70°) and the smallest side (3.26) is opposite the smallest angle (50°).

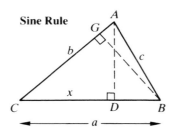

Sine Rule

In $\triangle ACD$, $AD = b \sin C$ (see figure, left)
In $\triangle ABD$, $AD = c \sin B$

$$\Rightarrow \quad AD = b \sin C = c \sin B \quad \Rightarrow \quad \frac{b}{\sin B} = \frac{c}{\sin C}$$

By drawing $BG \perp AC$,

$$BG = c \sin A = a \sin C \quad \Rightarrow \quad \frac{a}{\sin A} = \frac{c}{\sin C}$$

$$\Rightarrow \quad \boxed{\frac{a}{\sin A} = \frac{b}{\sin B} = \frac{c}{\sin C}} \quad \text{This is the \textbf{Sine Rule}}$$

If four quantities are known (in $\triangle GHI$, 3 angles and 1 side) the triangle can be solved completely (i.e. all sides and angles found).

Case 2 **One angle and two sides** are given (figure, right).

In $\triangle LMN$, the sine rule gives

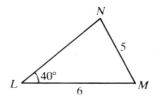

$$\frac{5}{\sin 40°} = \frac{6}{\sin N} = \frac{LN}{\sin M}$$

$$\Rightarrow \quad \sin N = \frac{6 \sin 40°}{5} = 0.7713 \quad \Rightarrow \quad N = 50.5° \text{ \textbf{or} } 129.5°$$

There are **two** values for N because there are **two triangles** that can be drawn with angle $L = 40°$, $LM = 6$ and $MN = 5$.

$$N = 50.5° \quad \Rightarrow \quad M = 180° - 40° - 50.5° = 89.5° \quad \Rightarrow \quad LN = \frac{5 \sin 89.5°}{\sin 40°} = \frac{4.99981}{\cdot 6428} = 7.78$$

$$N = 129.5° \quad \Rightarrow \quad M = 180° - 40° - 129.5° = 10.5° \quad \Rightarrow \quad LN = \frac{5 \sin 10.5°}{\sin 40°} = \frac{0.91117}{\cdot 6428} = 1.42$$

Case 3 If **one angle and two sides** are given (**neither side being opposite the angle**), the sine rule does not solve the triangle (figure, right).

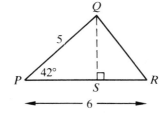

$$\frac{QR}{\sin 42°} = \frac{5}{\sin R} = \frac{6}{\sin Q} \quad \text{and whichever pair is selected,}$$
no progress can be made.

Draw $QS \perp PR \quad \Rightarrow \quad QS = 5 \sin 42° = 3.3457$
$$PS = 5 \cos 42° = 3.7157$$

$$\Rightarrow \quad SR = 6 - 3.7157 = 2.2843$$

(cont.)

In $\triangle QSR$, $\tan R = \dfrac{QS}{SR} = \dfrac{3.3457}{2.2843} = 1.4646$ $\Rightarrow$ $R = 55.7°$ $\Rightarrow$ $Q = 180° - 42° - 55.7° = 82.3°$.

By Pythagoras $QR^2 = 3.3457^2 + 2.2843^2 = 16.411$ $\Rightarrow$ $QR = 4.051$

The **Cosine Rule** gives a more direct solution. See the Sine Rule figure, with $CD = x$.

In $\triangle ACD$, Pythagoras $\Rightarrow$ $AD^2 + x^2 = b^2$ $\left.\vphantom{\begin{array}{c}a\\a\end{array}}\right\}$ $\Rightarrow$ $AD^2 = b^2 - x^2 = c^2 - (a - x)^2$
In $\triangle ABD$, Pythagoras $\Rightarrow$ $AD^2 + (a - x)^2 = c^2$

$\Rightarrow$ $b^2 - x^2 = c^2 - a^2 + 2ax - x^2$

Cancelling x^2 and rearranging $\Rightarrow$ $c^2 = a^2 + b^2 - 2ax$
From $\triangle ACD$, $x = b \cos C$ $\Rightarrow$ $c^2 = a^2 + b^2 - 2ab \cos C$
Using this in $\triangle PQR$, above, $\Rightarrow$ $QR^2 = 5^2 + 6^2 - 2 \times 5 \times 6 \times \cos 42° = 25 + 36 - 44.589 = 16.4113$

$\Rightarrow$ $QR = 4.051$

Now the Sine Rule will give $\dfrac{5}{\sin R} = \dfrac{6}{\sin Q} = \dfrac{4.051}{\sin 42°}$ $\Rightarrow$ $\sin R = \dfrac{5 \sin 42°}{4.051}$ $\Rightarrow$ $R = 55.7°$.

$R = 55.7°$ and not $180° - 55.7° = 124.3°$, as $\triangle PQR$ is acute. We can tell that all the angles are less than $90°$ since $6^2 < 5^2 + 4.05^2$.

Perhaps you did not know that $a^2 = b^2 + c^2$ (Pythagoras) $\Rightarrow$ angle $A = 90°$.
$a^2 < b^2 + c^2$ $\Rightarrow$ angle A is less than $90°$.
$a^2 > b^2 + c^2$ $\Rightarrow$ angle A is more than $90°$.

All these facts arise from the **Cosine Rule** in the form $\boxed{a^2 = b^2 + c^2 - 2bc \cos A}$

Pythagoras' theorem $\Rightarrow$ $a^2 = b^2 + c^2$ $\Rightarrow$ $\cos A = 0$ $\Rightarrow$ $A = 90°$
$A < 90°$ $\Rightarrow$ $\cos A > 0$ $\Rightarrow$ $a^2 < b^2 + c^2$ (from the Cosine Rule)
$90° < A < 180°$ $\Rightarrow$ $\cos A < 0$ $\Rightarrow$ $a^2 > b^2 + c^2$ (from the Cosine Rule)

Case 4 Given **three sides** of the triangle **and no angles** (figure, right).

The Sine Rule gives $\dfrac{7}{\sin X} = \dfrac{5}{\sin Y} = \dfrac{6}{\sin Z}$ from which no progress is made yet.

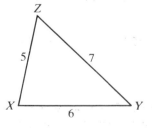

The Cosine Rule gives $7^2 = 6^2 + 5^2 - 2 \times 6 \times 5 \cos X$

$\Rightarrow$ $49 = 36 + 25 - 60 \cos x$

$\Rightarrow$ $60 \cos X = 61 - 49 = 12$

$\Rightarrow$ $\cos X = \dfrac{12}{60} = 0.2$ $\Rightarrow$ $X = 78.5°$ (Note $\cos x < 0$ $\Rightarrow$ $90° < x < 180°$)

Similarly $y^2 = x^2 + z^2 - 2xz \cos Y$ $\Rightarrow$ $\cos Y = \dfrac{x^2 + z^2 - y^2}{2xz} = \dfrac{7^2 + 6^2 - 5^2}{2 \times 7 \times 6} = \dfrac{60}{84} = .7142$ $\Rightarrow$ $Y = 44.4°$

Hence, $Z = 180° - 78.5° - 44.4° = 57.1°$, or it may be quicker to use the Sine Rule after finding $X = 78.5°$.

(end of Investigation 1)

Exercise 6.3

1 With the notation of $\triangle ABC$ in the figure, find all the angles and sides of the triangles.

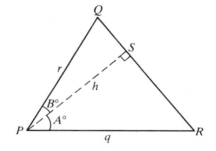

    (a) $B = 90°, A = 37°, a = 8\,\text{cm}$
    (b) $B = 90°, C = 27°, b = 7\,\text{cm}$
    (c) $B = 90°, C = 8\,\text{cm}, b = 17\,\text{cm}$
    (d) $A = 42°, B = 69°, a = 10\,\text{cm}$
    (e) $A = 59°, b = 8\,\text{cm}, c = 7\,\text{cm}$
    (f) $a = 8\,\text{m}, b = 9\,\text{m}, c = 10\,\text{m}$
    (g) $A = 50°, a = 6\,\text{m}, b = 7\,\text{m}$

2 Find the area of each triangle in question **1**.

3 A boat travels 6 km on a bearing of 140° and then 7 km on a bearing of 200°. Find the distance and bearing from its original position.

4 A yacht sails 2 km on a bearing 049°, then x km on a bearing 330° to finish 5 km from its starting position. Find x and the bearing of the start from the finishing position.

Addition formulae

$$\sin(A + B) \neq \sin A + \sin B.$$

With the notation in the figure, right, the area of

$$\triangle PQR = \tfrac{1}{2}qr\sin(A + B)$$

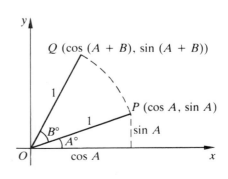

In $\triangle PRS$, $h = q\cos A$ and $RS = q\sin A$

In $\triangle PQS$, $h = r\cos B$ and $QS = r\sin B$

Area $\triangle PRS = \tfrac{1}{2}h \times RS = \tfrac{1}{2}r\cos B \times q\sin A$

Area $\triangle PQS = \tfrac{1}{2}h \times QS = \tfrac{1}{2}q\cos A \times r\sin B$

  being careful over our choice for h in each case.

$\Rightarrow$ Area $\triangle PQR$ = Area $\triangle PRS$ + Area $\triangle PQS$ $\Rightarrow$ $\tfrac{1}{2}qr\sin(A + B) = \tfrac{1}{2}qr\sin A\cos B + \tfrac{1}{2}qr\cos A\sin B$

Cancelling $\tfrac{1}{2}qr$ throughout $\Rightarrow$ $\boxed{\sin(A + B) = \sin A\cos B + \cos A\sin B}$

For example, $A = 60°$, $B = 30°$ $\Rightarrow$ $\sin A = \dfrac{\sqrt{3}}{2}$, $\cos A = \dfrac{1}{2}$, $\sin B = \dfrac{1}{2}$, $\cos B = \dfrac{\sqrt{3}}{2}$

$\Rightarrow$ $\sin A\cos B + \cos A\sin B = \dfrac{\sqrt{3}}{2} \times \dfrac{\sqrt{3}}{2} + \dfrac{1}{2} \times \dfrac{1}{2} = \dfrac{3}{4} + \dfrac{1}{4} = 1 = \sin(60° + 30°)$

The triangle method for deriving $\sin(A + B)$ is more complicated if either angle is larger than 90°, so for the general result we use a matrix method.

In the figure, right, OP has length 1 and P has co-ordinates $(\cos A, \sin A)$. P is rotated through angle $B°$ about O to Q, so that the co-ordinates of Q are $[\cos(A + B), \sin(A + B)]$.

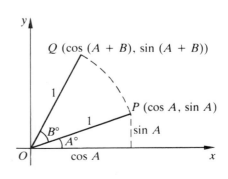

The matrix for a rotation of $B°$ about the origin is

$$\begin{pmatrix} \cos B & -\sin B \\ \sin B & \cos B \end{pmatrix}, \text{ so}$$

$$\begin{pmatrix} \cos(A+B) \\ \sin(A+B) \end{pmatrix} = \begin{pmatrix} \cos B & -\sin B \\ \sin B & \cos B \end{pmatrix}\begin{pmatrix} \cos A \\ \sin A \end{pmatrix} = \begin{pmatrix} \cos B \cos A - \sin B \sin A \\ \sin B \cos A + \cos B \sin A \end{pmatrix}$$

$\Rightarrow$ $\boxed{\cos(A+B) = \cos A \cos B - \sin A \sin B}$ .... (1)

$\boxed{\sin(A+B) = \sin A \cos B + \sin B \cos A}$ .... (2)

Replacing B by $-B$ in (2) gives $\sin(A+(-B)) = \sin A \cos(-B) + \cos A \sin(-B)$

Using $\cos(-B) = \cos B$
and $\sin(-B) = -\sin B$ $\Rightarrow$ $\boxed{\sin(A-B) = \sin A \cos B - \cos A \sin B}$ .... (3)

Doing the same in (1) $\Rightarrow$ $\boxed{\cos(A-B) = \cos A \cos B + \sin A \sin B}$ .... (4)

Putting $B=A$ in (2) $\Rightarrow$ $\sin(A+A) = \sin A \cos A + \sin A \cos A$ $\Rightarrow$ $\boxed{\sin 2A = 2 \sin A \cos A}$ .... (5)

Putting $B=A$ in (4) $\Rightarrow$ $\cos(A-A) = \cos A \cos A + \sin A \sin A$ $\Rightarrow$ $\boxed{\cos^2 A + \sin^2 A = 1}$ .... (6)

Putting $B=A$ in (1) $\Rightarrow$ $\cos(A+A) = \cos A \cos A - \sin A \sin A$ $\Rightarrow$ $\boxed{\cos 2A = \cos^2 A - \sin^2 A}$ .... (7)

From (6), $\cos^2 A = 1 - \sin^2 A$, so (7) can be written $\cos 2A = 1 - \sin^2 A - \sin^2 A$ $\Rightarrow$ $\boxed{\cos 2A = 1 - 2\sin^2 A}$ ... (8)

From (6), $\sin^2 A = 1 - \cos^2 A$, so (7) becomes $\cos 2A = \cos^2 A - (1 - \cos^2 A)$ $\Rightarrow$ $\boxed{\cos 2A = 2\cos^2 A - 1}$ ... (9)

These formulae are important identities and worth learning, but frequent use will make then familiar.

WORKED EXAMPLES

Solve (a) $\sin 2A = \sin A$ and (b) $\cos 2A = \cos A$, for $0° \leqslant A \leqslant 360°$.

(a)   Using $\sin 2A = 2\sin A \cos A$ gives $\sin 2A = \sin A$ $\Rightarrow$ $2\sin A \cos A = \sin A$

$\Rightarrow$ $2\sin A \cos A - \sin A = 0$ $\Rightarrow$ $\sin A(2\cos A - 1) = 0$ $\Rightarrow$ $\sin A = 0$ or $\cos A = \frac{1}{2}$.

$\sin A = 0$ $\Rightarrow$ $A = 0°, 180°$ or $360°$ and $\cos A = 0.5$ $\Rightarrow$ $A = 60°$ or $300°$.

Complete solution $A = 0°, 60°, 180°, 300°$ or $360°$.

(b)   Use $\cos 2A = 2\cos^2 A - 1$ so that $\cos 2A = \cos A$ $\Rightarrow$ $2\cos^2 A - 1 = \cos A$.

Rearranging $\Rightarrow$ $2\cos^2 A - \cos A - 1 = 0$, a quadratic equation for $\cos A$.

Factorize $\Rightarrow$ $(\cos A - 1)(2\cos A + 1) = 0$ $\Rightarrow$ $\cos A = 1$ or $\cos A = -\frac{1}{2}$.

$\cos A = 1$ $\Rightarrow$ $A = 0°$ or $360°$ and $\cos A = -0.5$ $\Rightarrow$ $A = 120°$ or $240°$

Complete solution $A = 0°, 120°, 240°$ or $360°$ for $0° \leqslant A \leqslant 360°$

General solution $A = 120n°$ where $n = \{0, \pm 1, \pm 2, ...\}$ i.e. $n \in \mathbb{Z}$

Exercise 6.4

1 Using the values of sin, cos and tan of 30°, 45°, 60°, find the values of sin 15°, cos 15°, tan 15°, sin 75°, cos 75° and tan 75°.

2 Given $\sin A = \dfrac{4}{5}$ and $\cos B = \dfrac{5}{13}$, find all possible values of $\cos(A - B)$ and $\sin(A + B)$.

3 Solve the following equations for $0° \leqslant A \leqslant 360°$ and give the general solution

    (a) $\sin 2A = \cos A$ **(b)** $\cos 2A = \sin A$ **(c)** $\cos 2A + \sin 2A = 0$
    (d) $\tan A = 2 \sin A$ **(e)** $\tan A = \sin 2A$ **(f)** $\tan A = \cos A$.

4 Draw the graph of $\sin \theta$ for $0° \leqslant \theta \leqslant 360°$. On the same graph plot values of $\sin^2 \theta$ for $\theta = 0°, 30°, 60°, 90°$ etc., and draw the graph of $\sin^2 \theta$. Comment!

5 Draw the graph of $\cos \theta$ for $0° \leqslant \theta \leqslant 360°$. On the same graph plot values of $\cos^2 \theta$ for $\theta = 0°, 30°, 60°, 90°$ etc., and draw the graph of $\cos^2 \theta$.

6 Expressing $\sin 3A$ as $\sin(2A + A)$ use formula **(2)** to expand $\sin(2A + A)$, and formulae for $\sin 2A$ and $\cos 2A$ to express $\sin 3A$ in terms of $\sin A$ only.

7 Repeat question **6** for $\cos 3A$ by expanding $\cos(2A + A)$ to express $\cos 3A$ in terms of $\cos A$ only.

8 Use formulae **(2)** and **(3)** to expand and simplify $\sin(A + B) + \sin(A - B)$.

9 Use formulae **(1)** and **(4)** to expand and simplify $\cos(A + B) + \cos(A - B)$.

10 Use formulae **(2)** and **(3)** to expand and simplify $\sin(A + B) - \sin(A - B)$.

11 Use formulae **(1)** and **(4)** to expand and simplify $\cos(A + B) - \cos(A - B)$.

Amplitude and frequency

Sine and Cosine waves occur frequently in electricity and electronics. You may know that the frequency of our electricity supply is 50 cycles per second.
   What exactly does this mean?

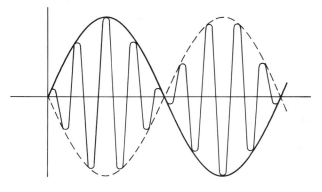

Investigation 2

For $0° \leqslant \theta \leqslant 360°$ the graph of $y = \sin \theta$ completes one revolution or one complete cycle of values. The figure, right, shows the graph of $y = \sin \theta$ completing one **cycle** of values.

The maximum value of $\sin \theta$ is $+1$ ($\theta = 90°$).
The maximum value of $\sin \theta$ is -1 ($\theta = 270°$).

What can you say about the graphs of $y_2 = 2 \sin \theta$, $y_3 = 3 \sin \theta$ etc.?

Investigate $y = A \sin \theta$ for $A = \frac{1}{2}$, $A = \frac{1}{3}$, $A = -1$, $A = -2$. This may be more easily achieved with a computer graph plotting program.

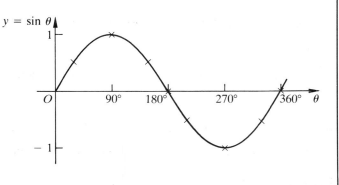

Investigation 3

Plot the graph of $Y_2 = \sin 2\theta$ for values of $0° \leqslant \theta \leqslant 360°$.
Examine the graph of $Y_3 = \sin 3\theta$, $Y_4 = \sin 4\theta$, $Y_{\frac{1}{2}} = \sin \frac{1}{2}\theta$.

Can you describe the graph of $y = 3\sin 2\theta$? Can you describe the graph of $y = 2\cos 3\theta$?

For the graph of $y = A\sin n\theta$, A is called the **amplitude** which gives the maximum value $+A$ and the minimum value $-A$.

n is related to the frequency of the function; $y = \sin 2\theta$ completes 2 cycles between $0°$ and $360°$ while $y = \sin \theta$ completes one cycle between $0°$ and $360°$.

Investigation 4

The figure shows the graphs of $y_1 = \cos\theta$, $y_2 = \sin\theta$ and $y_3 = \cos\theta + \sin\theta$ (dotted).

What are the periods of y_1, y_2 and y_3?

What are the amplitudes of y_1, y_2 and y_3?

y_3 looks like a cosine or sine curve. Can you express y_3 in the form $A\cos(\theta - \alpha)$ or $B\sin(\theta + \beta)$?

Do the same for $y_4 = \cos\theta + 2\sin\theta$, $y_5 = 2\cos\theta + \sin\theta$, $y_6 = 3\cos\theta + 4\sin\theta$.

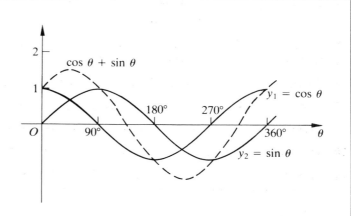

The function $y = a\cos\theta + b\sin\theta$

From our investigations it looks as if $y = a\cos\theta + b\sin\theta \equiv r\cos(\theta - \alpha) = s\sin(\theta + \beta)$, where r, α or s, β are related to the values of a and b.

Perhaps you guessed that $\cos\theta + \sin\theta = \sqrt{2}\cos(\theta - 45) = \sqrt{2}\sin(\theta + 45)$.
  We can check by using the formulae for $\cos(A - B)$ and $\sin(A + B)$.

$$\sqrt{2}\cos(\theta - 45) = \sqrt{2}(\cos\theta \cos 45 + \sin\theta \sin 45) = \sqrt{2}\cos\theta \times \frac{1}{\sqrt{2}} + \sqrt{2}\sin\theta \times \frac{1}{\sqrt{2}} = \cos\theta + \sin\theta.$$

$$\sqrt{2}\sin(\theta + 45) = \sqrt{2}(\sin\theta \cos 45 + \cos\theta \sin 45) = \sqrt{2}\sin\theta \times \frac{1}{\sqrt{2}} + \sqrt{2}\cos\theta \times \frac{1}{\sqrt{2}} = \cos\theta + \sin\theta.$$

If we want an equivalent form for $y_4 = \cos\theta + 2\sin\theta$, we put $y_4 = r\cos(\theta - \alpha)$.

  Then, $\cos\theta + 2\sin\theta \equiv r\cos\theta \cos\alpha + r\sin\theta \sin\alpha$ which must be true for all values of θ.

Requiring $r\sin\alpha = 2 \ldots$ **(1)**
and $\quad\quad r\cos\alpha = 1 \ldots$ **(2)**

117

Dividing **(1)** by **(2)** $\Rightarrow$ $\tan \alpha = 2$ $\Rightarrow$ $\alpha = \tan^{-1} 2 = 63.4°$.

Squaring **(1)** and **(2)** and adding $\Rightarrow$ $r^2 \sin^2 \alpha + r^2 \cos^2 \alpha = 2^2 + 1^2 = 5$

$$\Rightarrow \quad r^2 (\sin^2 \alpha + \cos^2 \alpha) = 5$$

$$\Rightarrow \quad r^2 = 5 \quad (\text{since } \sin^2 \alpha + \cos^2 \alpha = 1)$$

So $y_4 = \cos \theta + 2 \sin \theta = r \cos (\theta - \alpha) = \sqrt{5} \cos (\theta - 63.4°)$.

In general, $\boxed{y = a \cos \theta + b \sin \theta = r \cos (\theta - \alpha)}$ where $r = \sqrt{a^2 + b^2}$ and $\alpha = \tan^{-1} \dfrac{b}{a}$

Express $y = 3 \cos \theta + 4 \sin \theta$ in the form $y = s \sin (\theta + \beta)$.

We require, $3 \cos \theta + 4 \sin \theta \equiv s \sin (\theta + \beta) \equiv s \sin \theta \cos \beta + s \cos \theta \sin \beta$, for all θ.

$\Rightarrow \quad \begin{array}{l} s \cos \beta = 4 \\ s \sin \beta = 3 \end{array} \Rightarrow \quad \tan \beta = \frac{3}{4} \Rightarrow \beta = 36.9°$ and $s^2 \cos^2 \beta - s^2 \sin^2 \beta = 4^2 + 3^2$

$$\Rightarrow \quad s^2 = 25 \quad \Rightarrow \quad s = 5$$

So, $3 \cos \theta + 4 \sin \theta = 5 \sin (\theta + 36.9°)$

Exercise 6.5

1 Express $y = \cos \theta + 2 \sin \theta$ in the form $y = s \sin (\theta + \beta)$.

2 Express $y = 3 \cos \theta + 4 \sin \theta$ in the form $y = r \cos (\theta - \alpha)$.

3 Express y in both forms **(a)** $y = r \cos (\theta - \alpha)$ and **(b)** $y = s \sin (\theta + \beta)$
    (i) $y = 3 \cos \theta + \sin \theta$ **(ii)** $y = 6 \cos \theta + 8 \sin \theta$ **(iii)** $y = 5 \cos \theta + 12 \sin \theta$

Solving equations of this form $a \cos \theta + b \sin \theta = p$

WORKED EXAMPLE

Solve $\cos \theta + 2 \sin \theta = 0.5$ for $0 \leqslant \theta \leqslant 360°$.

Express $\cos \theta + 2 \sin \theta$ in the form $r \cos (\theta - \alpha)$ $\Rightarrow$ $r^2 = 1^2 + 2^2 = 5$

$$\Rightarrow \quad r = \sqrt{5} \text{ and } \alpha = \tan^{-1} 2 = 63.4° \text{ from working above.}$$

$\cos \theta + 2 \sin \theta = \frac{1}{2}$ $\Rightarrow$ $\sqrt{5} \cos (\theta - 63.4°) = 0.5$ $\Rightarrow$ $\cos (\theta - 63.4°) = \dfrac{0.5}{\sqrt{5}} = 0.2236$

$$\Rightarrow \quad \theta - 63.4° = 77.1° \quad \text{or} \quad 360° - 77.1°$$

$$\Rightarrow \quad \theta = 140.5° \quad \text{or} \quad 360° - 77.1° + 63.4° = 346.3°$$

Exercise 6.6

Solve the following equations for $0° \leqslant \theta \leqslant 360°$.

1 $\cos\theta + \sin\theta = p$, where **(a)** $p = 1$ **(b)** $p = \sqrt{2}$ **(c)** $p = 2$ **(d)** $p = \dfrac{1}{\sqrt{2}}$.

2 $\cos\theta + 2\sin\theta = q$, where **(a)** $q = 0$ **(b)** $q = 1$ **(c)** $q = \sqrt{5}$ **(d)** $q = 1.5$

3 $4\cos\theta + 3\sin\theta = m$, where **(a)** $m = 1$ **(b)** $m = 2$ **(c)** $m = 3$ **(d)** $m = 4$

4 $\sin 2\theta + \cos 2\theta = n$, where **(a)** $n = 0$ **(b)** $n = 1$ **(c)** $n = 1.5$ **(d)** $h = 2$

Alternative solution of a $\cos\theta + \sin\theta = p$;   t formulae

Some students are confused by the methods for changing $a\cos\theta + b\sin\theta$ into $r\cos(\theta - \alpha)$ or $s\sin(\theta + \beta)$, not knowing which to use when either will do. Here is a more direct method, but one which requires more formulae.

$$\sin 2A = 2\sin A \cos A = \frac{2\sin A \cos A}{\sin^2 A + \cos^2 A} \quad \text{since } \sin^2 A + \cos^2 A = 1$$

Dividing top and bottom by $\cos^2 A$ $\Rightarrow$ $\sin 2A = \dfrac{\dfrac{2\sin A \cos A}{\cos^2 A}}{\dfrac{\sin^2 A + \cos^2 A}{\cos^2 A}} = \dfrac{\dfrac{2\sin A \cos A}{\cos A \cos A}}{\dfrac{\sin^2 A}{\cos^2 A} + \dfrac{\cos^2 A}{\cos^2 A}} = \dfrac{2\tan A}{\tan^2 A + 1}$

Similarly, $\cos 2A = \cos^2 A - \sin^2 A = \dfrac{\cos^2 A - \sin^2 A}{\cos^2 A + \sin^2 A} = \dfrac{\dfrac{\cos^2 A}{\cos^2 A} - \dfrac{\sin^2 A}{\cos^2 A}}{\dfrac{\cos^2 A}{\cos^2 A} + \dfrac{\sin^2 A}{\cos^2 A}} = \dfrac{1 - \tan^2 A}{1 + \tan^2 A}$

So, $\sin\theta = \dfrac{2\tan(\theta/2)}{1 + \tan^2(\theta/2)} = \dfrac{2t}{1 + t^2}$ and $\cos\theta = \dfrac{1 - \tan^2(\theta/2)}{1 + \tan^2(\theta/2)} = \dfrac{1 - t^2}{1 + t^2}$ where $t = \tan\dfrac{\theta}{2}$.

WORKED EXAMPLE

Solve $\cos\theta + 2\sin\theta = 0.5$ for $0° \leqslant \theta \leqslant 360°$.

Substitute $\cos\theta = \dfrac{1 - t^2}{1 + t^2}$ and $\sin\theta = \dfrac{2t}{1 + t^2}$ into $\cos\theta + 2\sin\theta = 1$

$\Rightarrow \dfrac{1 - t^2}{1 + t^2} + \dfrac{2 \times 2t}{1 + t^2} = 0.5 \Rightarrow 1 - t^2 + 4t = \tfrac{1}{2}(1 + t^2) \Rightarrow 2 - 2t^2 + 8t = 1 + t^2$

$\Rightarrow 3t^2 - 8t - 1 = 0 \Rightarrow t = \dfrac{8 \pm \sqrt{64 - 4(3)(-1)}}{6} = \dfrac{8 \pm \sqrt{76}}{6} = \dfrac{8 \pm 8.718}{6} = 2.786 \text{ or } -0.1197$

$$\Rightarrow \quad t = \tan\frac{\theta}{2} = 2.786 \quad \Rightarrow \quad \frac{\theta}{2} = \tan^{-1} 2.786 = 70.26° \text{ or } 180° + 70.26° \quad \Rightarrow \quad \theta = 140.5° \text{ or } 360° + 140.5°$$

$$\text{or } t = \tan\frac{\theta}{2} = -0.1197 \quad \Rightarrow \quad \frac{\theta}{2} = -6.8° \text{ or } 180° - 6.8° \quad \Rightarrow \quad \theta = -13.6° \text{ or } 360° - 13.6° = 346.3°$$

Solutions are $\theta = 140.5°$ or $346.3°$.

Note There is a t formula for $\tan 2A$ which can be found by dividing $\sin 2A$ by $\cos 2A$.

$$\tan 2A = \frac{\sin 2A}{\cos 2A} = \frac{\dfrac{2\tan A}{1 + \tan^2 A}}{\dfrac{1 - \tan^2 A}{1 + \tan^2 A}} = \frac{2\tan A}{1 - \tan^2 A} \quad \Rightarrow \quad \tan\theta = \frac{2t}{1 - t^2} \text{ where } t = \tan\frac{\theta}{2}.$$

Exercise 6.7

Solve the equations from Exercise 6.6 by using the t formulae.

7 Vectors

Introduction

Have you ever been lost when hill walking or come to a road junction and found the signpost missing or broken?

You could consult a map, use a compass or ask someone to guide you. What you need are two pieces of information: the distance to your destination and the direction in which you should travel. It is of little help to be told that the town is 4 km away, unless you are also told that it is, say, north of your present position.

Some physical quantities, such as the displacement above, require both magnitude and direction to specify them completely. These are called **vector quantities**.

On the other hand some quantities, such as time and temperature, do not need to be associated with a direction and are described completely by their magnitude. These are **scalar quantities**.

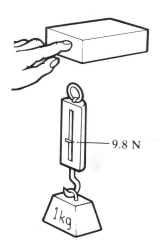

Some vectors	Some scalars
Force	Mass
Weight	Temperature
Velocity	Distance
Acceleration	Speed
Displacement	Time
Momentum	Kinetic Energy
Impulse	Volume
	Density

Mass 20 kg

It is important to use the correct terms. For example, speed describes the magnitude of the velocity. Speed is a scalar quantity and velocity is a vector quantity requiring direction as well as magnitude.

121

Representation of a vector

A vector can be represented by a line segment whose length is proportional to the magnitude of the vector and whose direction is the same as that of the vector (see figure).

The ends of the line segment are often specified and the vector written by giving two letters e.g. $\overrightarrow{AB}$ or **AB**.

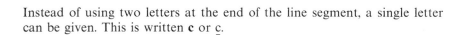

The order of the letters is important since they indicate the sense of the vector.
$\overrightarrow{AB} \neq \overrightarrow{PQ}$ although they are parallel and equal in length. However $\overrightarrow{AB}$ does equal $\overrightarrow{QP}$.

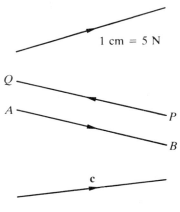

1 cm = 5 N

Instead of using two letters at the end of the line segment, a single letter can be given. This is written **c** or $\underline{c}$.

Alternatively, a column vector which defines the components parallel to a set of axes can be used. For example, $\mathbf{b} = \begin{pmatrix} 5 \\ 3 \end{pmatrix}$

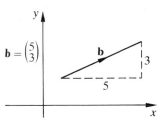

The magnitude of a vector

The magnitude (or modulus) of a vector is the length of the corresponding line segment.
It is written $|\overrightarrow{AB}|$, $|\mathbf{PQ}|$ or $|\mathbf{c}|$ and must be positive or zero (in which case it has an indeterminate direction and is written **0**).

The algebra of vectors

When you first used letters for numbers, or handled matrices, it was necessary to formulate rules by which they could be combined. We need an algebra (a set of rules of operation) to handle vectors.

Equality

Two vectors are equal if they have the same length, the same direction and are in the same sense.

$\overrightarrow{AB} = \overrightarrow{PQ}$ implies

(i) $|\overrightarrow{AB}| = |\overrightarrow{PQ}|$,
(ii) AB is parallel to PQ,
(iii) A to B is in the same sense as P to Q.

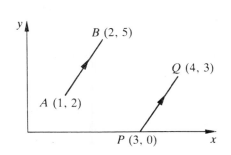

Addition of vectors

Consider a computer graph plotter which draws a line from $A(1,2)$ to $B(5,6)$ followed by a second line to $C(6,10)$.

Each of these lines represents a vector (a displacement) and it is clear that one single displacement of the pen along the line AC would leave the pen in the same final position at C.

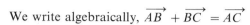

This is how addition is defined. It is the single vector which will achieve the same result as two (or more) separate vectors.

We write algebraically, $\overrightarrow{AB} + \overrightarrow{BC} = \overrightarrow{AC}$

For example, if $\overrightarrow{AB} = \begin{pmatrix} 4 \\ 4 \end{pmatrix}$ and $\overrightarrow{BC} = \begin{pmatrix} 4 \\ 1 \end{pmatrix}$ then $\overrightarrow{AC} = \begin{pmatrix} 4 \\ 4 \end{pmatrix} + \begin{pmatrix} 4 \\ 1 \end{pmatrix} = \begin{pmatrix} 8 \\ 5 \end{pmatrix}$

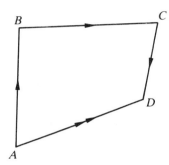

This principle can be extended to any number of vectors.

$$\overrightarrow{AB} + \overrightarrow{BC} + \overrightarrow{CD} = \overrightarrow{AD}$$

i.e. $\overrightarrow{AD}$ is the single vector to achieve the same result as the application of all three vectors.

A special case arises in the sum of $\overrightarrow{XY}$ and $\overrightarrow{YX}$. Clearly, for displacements, a move from X to Y followed by a move from Y to X leaves a particle where it started.

$$\text{i.e. } \overrightarrow{XY} + \overrightarrow{YX} = 0 \quad \text{or} \quad \overrightarrow{YX} = -\overrightarrow{XY}$$

So $\overrightarrow{YX}$ is a vector equal in magnitude and parallel to $\overrightarrow{XY}$ but in the opposite sense.

The properties of vector addition

Vector addition is both **commutative** and **associative**.

Commutative $\quad \overrightarrow{AB} + \overrightarrow{BC} = \overrightarrow{BC} + \overrightarrow{AB}$

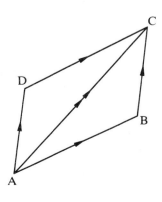

Consider the parallelogram $ABCD$. Since opposite sides are equal and parallel,

$$\overrightarrow{AB} = \overrightarrow{DC} \quad \text{and} \quad \overrightarrow{AD} = \overrightarrow{BC}$$

Hence, using $\triangle ABC$,

$$\overrightarrow{AB} + \overrightarrow{BC} = \overrightarrow{AC}$$

and using $\triangle ADC$,

$$\overrightarrow{BC} + \overrightarrow{AB} = \overrightarrow{AD} + \overrightarrow{DC} = \overrightarrow{AC}$$

Hence, $\quad \overrightarrow{AB} + \overrightarrow{BC} = \overrightarrow{BC} + \overrightarrow{AB}$

i.e. the order of addition is unimportant.

Vectors

Associative $(\overrightarrow{AB} + \overrightarrow{BC}) + \overrightarrow{CD} = \overrightarrow{AB} + (\overrightarrow{BC} + \overrightarrow{CD})$

Consider the left-hand side using diagram **(a)** in which $\overrightarrow{AB}$, $\overrightarrow{BC}$ and $\overrightarrow{CD}$ are three vectors.

$$(\overrightarrow{AB} + \overrightarrow{BC}) + \overrightarrow{CD} = \overrightarrow{AC} + \overrightarrow{CD}$$
$$= \overrightarrow{AD}$$

Consider the right-hand side, using diagram **(b)**. This time combine $\overrightarrow{BC}$ and $\overrightarrow{CD}$ first.

$$\overrightarrow{AB} + (\overrightarrow{BC} + \overrightarrow{CD}) = \overrightarrow{AB} + \overrightarrow{BD}$$
$$= \overrightarrow{AD}$$

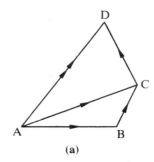

(a)

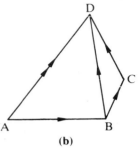

Since both sides are equivalent to the vector $\overrightarrow{AD}$,

$$(\overrightarrow{AB} + \overrightarrow{BC}) + \overrightarrow{CD} = \overrightarrow{AB} + (\overrightarrow{BC} + \overrightarrow{CD})$$

i.e. if three or more vectors are added, it is possible to group them for addition in different ways so long as the overall order is unchanged.

(b)

> WORKED EXAMPLES

---

1 Find the resultant of the displacements $\overrightarrow{OS}$, $\overrightarrow{TV}$ and $\overrightarrow{WX}$ where $O\,(0, 0)$, $S\,(2, -2)$, $T\,(3, 1)$, $V\,(6, 2)$, $W\,(1, 3)$ and $X\,(1, 5)$ are points on a graph.

In diagram **(a)** following, the actual vectors $\overrightarrow{OS}$, $\overrightarrow{TV}$ and $\overrightarrow{WX}$ are shown represented by line segments. To find the resultant, replace $\overrightarrow{TV}$ and $\overrightarrow{WX}$ by equal vectors $\overrightarrow{SA}$ and $\overrightarrow{AB}$ respectively as in diagram **(b)**.

Clearly,

$$\overrightarrow{OS} + \overrightarrow{SA} + \overrightarrow{AB} = \overrightarrow{OB}$$

i.e. $\begin{pmatrix} 2 \\ -2 \end{pmatrix} + \begin{pmatrix} 3 \\ 1 \end{pmatrix} + \begin{pmatrix} 0 \\ 2 \end{pmatrix} = \begin{pmatrix} 5 \\ 1 \end{pmatrix}$

The resultant is a vector $\overrightarrow{OB} = \begin{pmatrix} 5 \\ 1 \end{pmatrix}$

or we say that the resultant is a vector of magnitude

$$\sqrt{5^2 + 1^2} = \sqrt{26}$$

at $\tan^{-1}\left(\dfrac{1}{5}\right)$ to the x-axis.

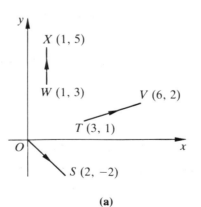

(a)

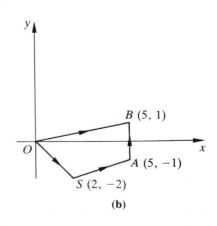

(b)

2 If *ABCDE* is a pentagon, show that $\overrightarrow{AE} + \overrightarrow{ED} + \overrightarrow{DC} = \overrightarrow{AB} - \overrightarrow{CB}$

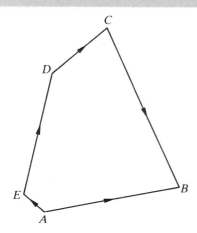

Consider the left-hand side, $\overrightarrow{AE} + \overrightarrow{ED} + \overrightarrow{DC}$, and find equivalent vectors in order to rearrange it to match the right-hand side.

$$\overrightarrow{AE} + \overrightarrow{ED} + \overrightarrow{DC} = (\overrightarrow{AB} + \overrightarrow{BC} + \overrightarrow{CD} + \overrightarrow{DE}) + \overrightarrow{ED} + \overrightarrow{DC}$$
$$= \overrightarrow{AB} + \overrightarrow{BC} + (\overrightarrow{CD} + \overrightarrow{DC}) + (\overrightarrow{DE} + \overrightarrow{ED})$$
$$= \overrightarrow{AB} + \overrightarrow{BC}$$

(since $\overrightarrow{CD} + \overrightarrow{DC} = 0$ and $\overrightarrow{DE} + \overrightarrow{ED} = 0$)

$$\therefore \ \overrightarrow{AE} + \overrightarrow{ED} + \overrightarrow{DC} = \overrightarrow{AB} - \overrightarrow{CB}$$

3 A marble is rolled across a table in a railway carriage at $2\,\text{ms}^{-1}$ at an angle of $60°$ to the direction of travel of the train. If the train is travelling at $6\,\text{ms}^{-1}$, find the actual velocity of the marble.

Since the particle is subject to two simultaneous velocities, we need to find the resultant of two vectors.

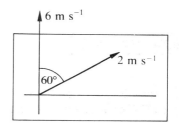

The vector triangle is drawn with $\overrightarrow{AB}$ and $\overrightarrow{BC}$ representing the velocities of the marble and the train, respectively, placed end to end.
    The law of vector addition gives

$$\overrightarrow{AB} + \overrightarrow{BC} = \overrightarrow{AC}$$

Hence, $\overrightarrow{AC}$ represents the resultant velocity of the particle and can be found by solving triangle *ABC* using the cosine rule and sine rule.

Since the angle between the given velocities is $60°$, angle $ABC = 120°$.

Using the cosine rule in $\triangle ABC$,

$$AC^2 = AB^2 + BC^2 - (2.AB.BC \cos A\hat{B}C)$$
$$= 2^2 + 6^2 - (2 \times 2 \times 6 \cos 120°)$$
$$= 4 + 36 + 12 = 52 \quad \text{(since } \cos 120° = -\cos 60° = -\tfrac{1}{2})$$
$$\therefore \ AC = \sqrt{52} = 7.21$$

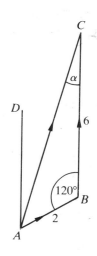

The magnitude of the velocity is $7.21\,\text{ms}^{-1}$.

To find the direction, use the sine rule in $\triangle ABC$. Let angle $ACB = \alpha$.

$$\frac{AB}{\sin \alpha} = \frac{AC}{\sin A\hat{B}C} \quad \Rightarrow \quad \frac{2}{\sin \alpha} = \frac{7.21}{\sin 120°}$$

$$\Rightarrow \quad 7.21 \sin \alpha = 2 \sin 120° \quad \Leftrightarrow \quad \sin \alpha = \frac{2 \sin 120°}{7.21} = 0.2402$$

Thus $\alpha = 13.9°$ or $13°\,54'$.

Since $C\hat{A}D = \alpha$, the direction of the marble is $13°\,54'$ to the direction of the train. The velocity of the marble is thus $7.2\,\text{ms}^{-1}$ at $13°\,54'$ to the direction of the train.

Exercise 7.1

1 In a regular hexagon $ABCDEF$, $\overrightarrow{AB} = \mathbf{a}$, $\overrightarrow{AE} = \mathbf{b}$ and $\overrightarrow{BC} = \mathbf{c}$.

Find expressions for **(i)** $\overrightarrow{EF}$ **(ii)** $\overrightarrow{AC}$ **(iii)** $\overrightarrow{AD}$ **(iv)** $\overrightarrow{EC}$ and **(v)** $\overrightarrow{DF}$

in terms of $\mathbf{a}, \mathbf{b}$ and $\mathbf{c}$.

2 Using the figure, right, state whether the following are true or false.

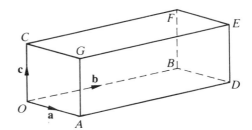

(i) $\overrightarrow{OA} = \overrightarrow{BD}$ **(ii)** $\overrightarrow{OC} = \overrightarrow{GA}$

(iii) $\overrightarrow{AE} = \overrightarrow{OF}$ **(iv)** $|\overrightarrow{DE}| = |\overrightarrow{GA}|$

(v) $\overrightarrow{OA} + \overrightarrow{AD} = \overrightarrow{OB} + \overrightarrow{BD}$

(vi) $\overrightarrow{OA} + \overrightarrow{AD} + \overrightarrow{DE} = \overrightarrow{CF} + \overrightarrow{FE} + \overrightarrow{ED}$

3 Use the figure from question **2** to find expressions for the following vectors in terms of $\mathbf{a}, \mathbf{b}$ and $\mathbf{c}$.

(i) $\overrightarrow{OD}$ **(ii)** $\overrightarrow{OG}$ **(iii)** $\overrightarrow{OE}$ **(iv)** $\overrightarrow{DF}$ **(v)** $\overrightarrow{DC}$ **(vi)** $\overrightarrow{GB}$ **(vii)** $\overrightarrow{AF}$

4 $PQRS$ is a quadrilateral and T is any point on QR.

Find a vector equivalent to $\overrightarrow{PQ} + \overrightarrow{QR} + \overrightarrow{RS} - \overrightarrow{PT}$.

5 O is any point inside a triangle ABC.

Show that $\overrightarrow{OA} - \overrightarrow{OB} + \overrightarrow{OC} = \overrightarrow{BA} + \overrightarrow{OB} + \overrightarrow{BC}$.

6 $ABCD$ is a rectangle with a triangle ABE drawn so that E is outside the rectangle. Show that

(a) $\overrightarrow{EC} - \overrightarrow{ED} + \overrightarrow{AB} + \overrightarrow{CD} = \overrightarrow{EB} - \overrightarrow{EA}$

and **(b)** $\overrightarrow{EC} + \overrightarrow{ED} - \overrightarrow{EB} - \overrightarrow{EA} = \overrightarrow{BC} + \overrightarrow{AD}$

7 Find the resultant of displacements $\overrightarrow{OA}$, $\overrightarrow{AB}$ and $\overrightarrow{BC}$ where $A(5,2)$, $B(8,7)$ and $C(9,-4)$ are points on a Cartesian graph. Find also $\overrightarrow{BA} + \overrightarrow{AC} + \overrightarrow{CO}$ and $\overrightarrow{AO} + \overrightarrow{BC} + \overrightarrow{AC}$.

8 Find, by scale drawing, the resultant of a displacement $\overrightarrow{AB}$ where $|\overrightarrow{AB}| = 10$ m in a direction N 60° E and a displacement $\overrightarrow{BC}$ where $|\overrightarrow{BC}| = 6$ m in a direction due south.

9 Find the resultant of two displacements given by 8.2 cm N 63° E and 4.7 cm S 27° E by **(a)** scale drawing and **(b)** calculation.

10 Find by calculation the resultant of the following vectors.

(a) a velocity of 3 m s⁻¹, N 30° W and a velocity of 4 m s⁻¹, due west.

(b) accelerations of 5 m s⁻², N 77° W and 3 m s⁻², S 33° W.

11 Find the displacement which when added to a displacement of 2 km north-east gives a resultant of 5 km south-east.

12 Find the displacement which, when added to a displacement of 7 m S 50° W, gives a resultant of 10 m S 60° E.

Further operations on vectors

Scalar multiples

A scalar multiple of a vector **a** can be defined using the addition property

i.e. $\mathbf{a} + \mathbf{a} + \mathbf{a} = 3\mathbf{a}$

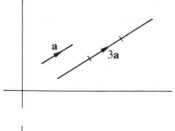

Thus a vector 3**a** would be a vector parallel to **a** and in the same sense with three times its magnitude.

However $-\mathbf{a} - \mathbf{a} - \mathbf{a} = -3\mathbf{a}$

The vector $-3\mathbf{a}$ is a vector parallel to **a**, but in the opposite sense, with three times its magnitude.

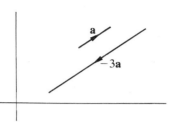

Similarly, since $\frac{1}{2}\mathbf{a} + \frac{1}{2}\mathbf{a} = \mathbf{a}$, the vector $\frac{1}{2}\mathbf{a}$ has the same direction and sense as **a**, but half its magnitude.

i.e. division of a vector **a** by a scalar t is defined as the scalar multiplication $\left(\dfrac{1}{t}\right)\mathbf{a}$.

In general, if λ is a scalar, then $\lambda\mathbf{a}$ is a vector parallel to **a** with a magnitude $|\lambda|$ times the magnitude of **a**, in the same sense if $\lambda > 0$ and in opposite sense if $\lambda < 0$.

Properties of scalar multiplication

If λ and μ are two scalar quantities, then

1 $\lambda(\mu\mathbf{a})\ \ \ = \mu(\lambda\mathbf{a})$

2 $(\lambda + \mu)\mathbf{a} = \lambda\mathbf{a} + \mu\mathbf{a}$

and **3** $\lambda(\mathbf{a} + \mathbf{b}) = \lambda\mathbf{a} + \lambda\mathbf{b}$

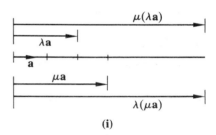

(i)

1 Multiplication by real numbers is associative and hence $\lambda\mu = \mu\lambda$. Diagram **(i)** shows the case for $\lambda = 2$ and $\mu = 3$.

2 Using the distributive property of real numbers,

$$(\lambda + \mu)\mathbf{a} = \lambda\mathbf{a} + \mu\mathbf{a}$$

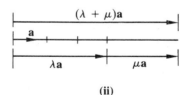

(ii)

The case for $\lambda = 3$ and $\mu = 2$ is shown in diagram **(ii)**.

i.e. $5\mathbf{a} = 3\mathbf{a} + 2\mathbf{a}$

3 Using diagram **(iii)** if $\lambda > 0$ and **(iv)** if $\lambda < 0$,

let $\overrightarrow{PQ} = \mathbf{a}$ and $\overrightarrow{QR} = \mathbf{b}$.

By vector addition, $\overrightarrow{PR} = \mathbf{a} + \mathbf{b}$.

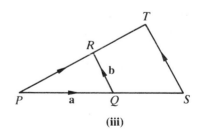

(iii)

If $\overrightarrow{PS} = \lambda\mathbf{a}$ and $\overrightarrow{ST} = \lambda\mathbf{b}$, then

$$\frac{|\overrightarrow{PS}|}{|\overrightarrow{PQ}|} = \frac{|\overrightarrow{ST}|}{|\overrightarrow{QR}|} = \lambda$$

But triangles whose corresponding sides are in the same ratio are similar. Hence triangles PQR and PST are similar and P, R and T are collinear. Hence, $\overrightarrow{PT} = \lambda\overrightarrow{PR}$

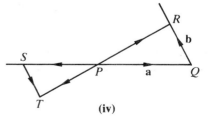

i.e. $\lambda\mathbf{a} + \lambda\mathbf{b} = \lambda(\mathbf{a} + \mathbf{b})$

(iv)

WORKED EXAMPLE

In a triangle ABC, the mid-points of BC, AC and AB are D, E and F respectively. If $\overrightarrow{AB}$ = **a** and $\overrightarrow{AC}$ = **b**, express
(i) $\overrightarrow{AF}$, **(ii)** $\overrightarrow{FE}$, **(iii)** $\overrightarrow{AD}$ and **(iv)** $\overrightarrow{FC}$ in terms of **a** and **b**.

(i) $\overrightarrow{AF} = \frac{1}{2}\overrightarrow{AB} = \frac{1}{2}\mathbf{a}$

(ii) $\overrightarrow{FE} = \overrightarrow{AE} + \overrightarrow{FA}$
$\quad = \overrightarrow{AE} - \overrightarrow{AF} = \frac{1}{2}\mathbf{b} - \frac{1}{2}\mathbf{a}$
$\quad = \frac{1}{2}(\mathbf{b} - \mathbf{a})$

(iii) $\overrightarrow{AD} = \overrightarrow{AB} + \overrightarrow{BD}$
$\quad = \overrightarrow{AB} + \frac{1}{2}\overrightarrow{BC}$
$\quad = \mathbf{a} + \frac{1}{2}(\overrightarrow{BA} + \overrightarrow{AC})$
$\quad = \mathbf{a} + \frac{1}{2}(-\mathbf{a} + \mathbf{b}) = \frac{1}{2}(\mathbf{a} + \mathbf{b})$

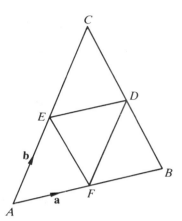

Alternatively, since by the mid-point theorem $\overrightarrow{FD} = \overrightarrow{AE} = \frac{1}{2}\mathbf{b}$,
$$\overrightarrow{AD} = \overrightarrow{AF} + \overrightarrow{FD} = \frac{1}{2}\mathbf{a} + \frac{1}{2}\mathbf{b} = \frac{1}{2}(\mathbf{a} + \mathbf{b})$$

(iv) $\overrightarrow{FC} = \overrightarrow{FA} + \overrightarrow{AC} = -\frac{1}{2}\mathbf{a} + \mathbf{b} = \mathbf{b} - \frac{1}{2}\mathbf{a}$

Subtraction of vectors

Subtraction of vectors is achieved by combining scalar
multiplication by -1 with addition (see figure, right).

So, $\mathbf{a} - \mathbf{b} = \mathbf{a} + (-\mathbf{b})$

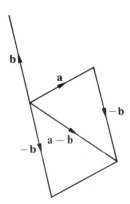

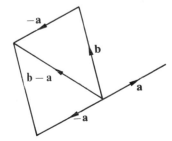

Refer to the figure, left.

Notice that $\mathbf{b} - \mathbf{a} = \mathbf{b} + (-\mathbf{a})$.

Hence $\mathbf{b} - \mathbf{a}$ will be a vector with the same magnitude as $\mathbf{a} - \mathbf{b}$ and parallel
to it, but in the opposite sense.

A geometrical property

If **a** and **b** are two vectors represented by $\overrightarrow{OA}$ and $\overrightarrow{OB}$, respectively, then **a** + **b** is represented by $\overrightarrow{OC}$ and **a** − **b** by $\overrightarrow{OE}$.

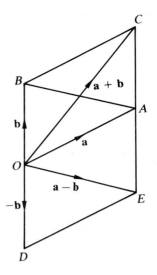

Note The length of OE is equal to the length of BA.

Hence, $|\overrightarrow{BA}| = |\mathbf{a} - \mathbf{b}|$.

Since OAC is a triangle, the sum of two sides is greater than the third side.

$$\therefore |\overrightarrow{OC}| < |\overrightarrow{OA}| + |\overrightarrow{AC}|$$

$$\Rightarrow \quad |\mathbf{a} + \mathbf{b}| < |\mathbf{a}| + |\mathbf{b}|$$

An equality applies when **a** and **b** are parallel, as the diagram then reduces to a straight line.

Hence, $\quad |\mathbf{a} + \mathbf{b}| \leqslant |\mathbf{a}| + |\mathbf{b}|$

WORKED EXAMPLE

The diagonals of a parallelogram $ABCD$ meet at O. If E is the mid-point of OD, show that $\overrightarrow{AD} + \overrightarrow{AB} - \overrightarrow{AE} = \overrightarrow{EC}$.

Consider the left-hand side of the result.

$$\overrightarrow{AD} + \overrightarrow{AB} - \overrightarrow{AE} = \overrightarrow{AD} + \overrightarrow{AB} + \overrightarrow{EA}$$
$$= (\overrightarrow{AO} + \overrightarrow{OD}) + (\overrightarrow{AO} + \overrightarrow{OB}) + \overrightarrow{EA}$$
$$= 2\overrightarrow{AO} + \overrightarrow{OD} + \overrightarrow{OB} + \overrightarrow{EA}$$
$$= 2\overrightarrow{AO} + \overrightarrow{EA}$$

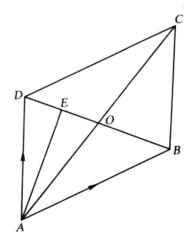

$(\overrightarrow{OD} = -\overrightarrow{OB}$, as O is the mid-point of $\overrightarrow{BD}$)

$$\therefore \overrightarrow{AD} + \overrightarrow{AB} - \overrightarrow{AE} = \overrightarrow{AC} + \overrightarrow{EA}$$
$$= \overrightarrow{EC}$$

Alternatively, since $ABCD$ is a parallelogram, $\overrightarrow{AB} = \overrightarrow{DC}$

Hence, $\quad \overrightarrow{AD} + \overrightarrow{AB} - \overrightarrow{AE} = \overrightarrow{AD} + \overrightarrow{DC} - \overrightarrow{AE}$
$$= \overrightarrow{AC} - \overrightarrow{AE}$$
$$= \overrightarrow{AC} + \overrightarrow{EA}$$
$$= \overrightarrow{EC}$$

129

The ratio theorem

If a point P divides a line segment AB internally in the ratio $\lambda:\mu$ and O is any point not on AB, then

$$\overrightarrow{OP} = \frac{\mu\mathbf{a} + \lambda\mathbf{b}}{\mu + \lambda} \quad \text{where } \overrightarrow{OA} = \mathbf{a} \quad \text{and} \quad \overrightarrow{OB} = \mathbf{b}$$

$$\overrightarrow{AB} = \overrightarrow{AO} + \overrightarrow{OB} = -\overrightarrow{OA} + \overrightarrow{OB}$$
$$= \mathbf{b} - \mathbf{a}$$

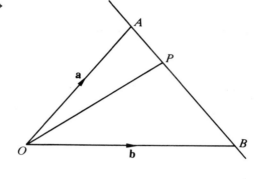

Since $|\overrightarrow{AP}|:|\overrightarrow{PB}| = \lambda:\mu$, it follows that
$$|\overrightarrow{AP}|:|\overrightarrow{AB}| = \lambda:\mu + \lambda$$

or $\overrightarrow{AP} = \dfrac{\lambda}{\mu + \lambda}\overrightarrow{AB}$

$$= \frac{\lambda}{\mu + \lambda}(\mathbf{b} - \mathbf{a})$$

Now, $\overrightarrow{OP} = \overrightarrow{OA} + \overrightarrow{AP} = \mathbf{a} + \overrightarrow{AP} = \mathbf{a} + \dfrac{\lambda}{\mu + \lambda}(\mathbf{b} - \mathbf{a}) = \dfrac{(\mu + \lambda)\mathbf{a} + \lambda(\mathbf{b} - \mathbf{a})}{\mu + \lambda}$

$$\therefore \overrightarrow{OP} = \frac{\mu\mathbf{a} + \lambda\mathbf{b}}{(\mu + \lambda)}$$

This is known as the **ratio theorem**.

So, if P divides AB in the ratio of 4:3, then $\overrightarrow{OP} = \dfrac{3\mathbf{a} + 4\mathbf{b}}{7}$.

A special case arises when $\lambda = \mu$. i.e. P is the mid-point of AB and $\overrightarrow{OP} = \frac{1}{2}(\mathbf{a} + \mathbf{b})$

WORKED EXAMPLE

> L, M and N are the mid-points of the sides OB, OA and AB of a triangle OAB. If ON and MB intersect at G such that $OG:ON = \lambda:1$ and $MG:MB = \mu:1$, find expressions for $\overrightarrow{AG}$ in terms of **(i)** λ and **(ii)** μ.
> Hence show that $\lambda = \frac{2}{3}$ and $\mu = \frac{1}{3}$, and show that A, G and L are collinear.

(i) Since $OG:ON = \lambda:1 \Rightarrow \overrightarrow{OG} = \lambda\overrightarrow{ON}$

Let $\overrightarrow{OA} = \mathbf{a}$ and $\overrightarrow{OB} = \mathbf{b}$.

Since N is the mid-point of AB, $\overrightarrow{ON} = \frac{1}{2}(\mathbf{a} + \mathbf{b})$ using the ratio theorem.

$$\therefore OG = \frac{\lambda}{2}(\mathbf{a} + \mathbf{b})$$

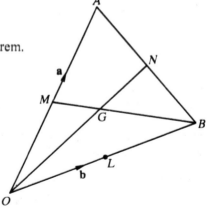

Now, $\overrightarrow{AG} = \overrightarrow{AO} + \overrightarrow{OG}$

$$= -\mathbf{a} + \frac{\lambda}{2}(\mathbf{a} + \mathbf{b})$$

$$\therefore \overrightarrow{AG} = (\tfrac{1}{2}\lambda - 1)\mathbf{a} + \frac{\lambda}{2}\mathbf{b} \cdots\cdots (1)$$

(ii) Since $MG:MB = \mu:1$

$\Rightarrow\quad MG:GB = \mu:1-\mu$

Using the ratio theorem where $\overrightarrow{OM} = \frac{1}{2}\mathbf{a}$ and $\overrightarrow{OB} = \mathbf{b}$,

$$\overrightarrow{OG} = \frac{(1-\mu)\frac{1}{2}\mathbf{a} + \mu\mathbf{b}}{1} \quad\Rightarrow\quad \overrightarrow{OG} = \frac{1}{2}(1-\mu)\mathbf{a} + \mu\mathbf{b}$$

Now, $\overrightarrow{AG} = \overrightarrow{AO} + \overrightarrow{OG} = -\mathbf{a} + \frac{1}{2}(1-\mu)\mathbf{a} + \mu\mathbf{b}$

$$= -\frac{1}{2}(1+\mu)\mathbf{a} + \mu\mathbf{b} \cdots\cdots (2)$$

In the expressions **(1)** and **(2)**, for $\overrightarrow{AG}$ to be identical, the components of $\mathbf{a}$ and $\mathbf{b}$ must be equal.

$$\therefore\ (\tfrac{1}{2}\lambda - 1) = -\tfrac{1}{2}(1+\mu) \quad\Rightarrow\quad \lambda - 2 = -1 - \mu \quad\Rightarrow\quad \lambda + \mu = 1 \cdots\cdots (3)$$

$$\text{and}\quad \frac{\lambda}{2} = \mu \qquad\qquad \Rightarrow\quad \lambda = 2\mu \cdots\cdots (4)$$

Solving **(3)** and **(4)** by substitution, gives $3\mu = 1 \Rightarrow \mu = \frac{1}{3}$. Hence, $\lambda = \frac{2}{3}$.

Now, using **(1)** with $\lambda = \frac{2}{3}$, gives $\overrightarrow{AG} = -\frac{2}{3}\mathbf{a} + \frac{1}{3}\mathbf{b} = \frac{1}{3}(-2\mathbf{a} + \mathbf{b})$

Also, $\overrightarrow{AL} = \overrightarrow{AO} + \overrightarrow{OL} = -\mathbf{a} + \frac{1}{2}\mathbf{b} = \frac{1}{2}(-2\mathbf{a} + \mathbf{b})$

Thus since $\overrightarrow{AL}$ is a multiple of $\overrightarrow{AG}$, the points A, G and L are collinear.

Note that this shows that the three **medians** of a triangle are concurrent and divide each other in the ratio of 2:1. The point G is called the centroid of the triangle.

Exercise 7.2

---

1 If three displacements are denoted by $\mathbf{a} = \begin{pmatrix} 1 \\ 2 \end{pmatrix}$, $\mathbf{b} = \begin{pmatrix} -3 \\ 4 \end{pmatrix}$ and $\mathbf{c} = \begin{pmatrix} 5 \\ 5 \end{pmatrix}$ draw diagrams to represent the following vectors.

    (i) $3\mathbf{a}$ **(ii)** $2\mathbf{a} + \mathbf{b}$ **(iii)** $\mathbf{b} - \mathbf{c}$ **(iv)** $\mathbf{a} - 2\mathbf{b} + \mathbf{c}$ **(v)** $\mathbf{a} + \frac{1}{2}\mathbf{b}$

2 X is a point on the side BC of a triangle ABC such that $CX:XB = 3:1$. If $\overrightarrow{AB} = \mathbf{a}$ and $\overrightarrow{AC} = \mathbf{c}$ find an expression for $\overrightarrow{AX}$ in terms of $\mathbf{a}$ and $\mathbf{c}$.

3 The diagram, right, shows a tetrahedron $OABC$ in which $\overrightarrow{OA} = \mathbf{a}$, $\overrightarrow{OB} = \mathbf{b}$ and $\overrightarrow{OC} = \mathbf{c}$. If W, X, Y and Z are the mid-points of the sides OA, AB, OB and OC, respectively, find expressions for the following vectors in terms of $\mathbf{a}, \mathbf{b}$ and $\mathbf{c}$.

    (i) $\overrightarrow{AC}$ **(ii)** $\overrightarrow{OX}$ **(iii)** $\overrightarrow{XY}$

    (iv) $\overrightarrow{XZ}$ **(v)** $\overrightarrow{CW}$

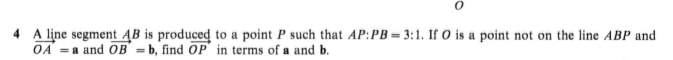

4 A line segment AB is produced to a point P such that $AP:PB = 3:1$. If O is a point not on the line ABP and $\overrightarrow{OA} = \mathbf{a}$ and $\overrightarrow{OB} = \mathbf{b}$, find $\overrightarrow{OP}$ in terms of $\mathbf{a}$ and $\mathbf{b}$.

5 $ABCD$ is a parallelogram with A and C as opposite vertices. If E and F are points on the diagonal AC such that $AE = FC$, prove that $BEDF$ is a parallelogram.

6 $ABCDEF$ is a regular hexagon in which AD and BE intersect at O. Show that $\overrightarrow{AB} + \overrightarrow{AC} + \overrightarrow{AD} + \overrightarrow{AE} + \overrightarrow{AF} = 6\overrightarrow{AO}$.

7 *ABCDE* is a pentagon and *O* is any point in the plane of the pentagon. Show that

$$\overrightarrow{AB} + \overrightarrow{OB} + \overrightarrow{CD} + \overrightarrow{ED} + \overrightarrow{EA} = 2(\overrightarrow{OB} + \overrightarrow{CO} + \overrightarrow{OD} + \overrightarrow{EO})$$

8 Using vector methods, show that the line joining the mid-points of two sides of a triangle is parallel to the third side and equal to half its length.

Components of a vector

It is possible to express a given vector as the sum of two or more non-parallel vectors, uniquely.

In two dimensions

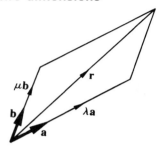

If **a** and **b** are two non-parallel vectors, then $\lambda\mathbf{a}$ and $\mu\mathbf{b}$ will be vectors in the direction of **a** and **b**, respectively.

Since these directions are not parallel it is always possible to form the parallelogram, and hence the single vector **r** is equivalent to the sum of $\lambda\mathbf{a}$ and $\mu\mathbf{b}$ (see figure, left).

$$\therefore \mathbf{r} = \lambda\mathbf{a} + \mu\mathbf{b}$$

We say that **r** has been resolved into two component vectors $\lambda\mathbf{a}$ and $\mu\mathbf{b}$ in the directions of non-parallel vectors **a** and **b**.

In three dimensions

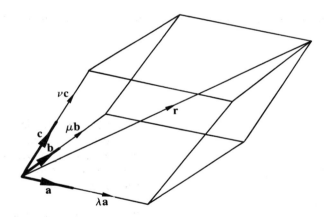

It is necessary to express **r** in terms of three component vectors parallel to three given non-coplanar vectors **a**, **b** and **c**.

The diagram, left, shows a general vector **r** expressed as the sum of multiples of the base vectors **a**, **b** and **c**.

i.e. $\mathbf{r} = \lambda\mathbf{a} + \mu\mathbf{b} + v\mathbf{c}$

where λ, μ and v are scalar multiples.

In both dimensions the expressions for **r** are unique and only one set of values exist for λ, μ and v for a given vector **r**.

Consider the two dimensional case. Assume that **r** can be expressed as $\mathbf{r} = \lambda_1\mathbf{a} + \mu_1\mathbf{b}$ or $\mathbf{r} = \lambda_2\mathbf{a} + \mu_2\mathbf{b}$.

If this is true, $\lambda_1\mathbf{a} + \mu_1\mathbf{b} = \lambda_2\mathbf{a} + \mu_2\mathbf{b}$

$\Rightarrow$ $(\lambda_1 - \lambda_2)\mathbf{a} = (\mu_2 - \mu_1)\mathbf{b}$

Now $(\lambda_1 - \lambda_2)\mathbf{a}$ and $(\mu_2 - \mu_1)\mathbf{b}$ are multiples of **a** and **b** respectively, and are thus parallel to **a** and **b**. However, **a** and **b** have different directions and so,

$$(\lambda_1 - \lambda_2)\mathbf{a} \neq (\mu_2 - \mu_1)\mathbf{b} \qquad \text{unless both vectors are zero.}$$

Hence $\lambda_1 = \lambda_2$ and $\mu_1 = \mu_2$ and the expression for **r** is unique.

Investigation 1

By assuming that $\mathbf{r} = \lambda_1 \mathbf{a} + \mu_1 \mathbf{b} + \nu_1 \mathbf{c}$ or $\mathbf{r} = \lambda_2 \mathbf{a} + \mu_2 \mathbf{b} + \nu_2 \mathbf{c}$, show that the above method can also be used to prove that $\mathbf{r}$ is uniquely expressed in terms of component vectors in three dimensions.

The i, j, k *notation*

It is usual to choose the base vectors $\mathbf{a}, \mathbf{b}$ and $\mathbf{c}$ in mutually perpendicular directions. For this we use the Cartesian axes x, y and z arranged so that they form a **right-handed set**. i.e. imagine a screw being turned so that it progresses along the z-axis by a rotation that moves the x-axis towards the y-axis. The usual diagrams are as in the figure.

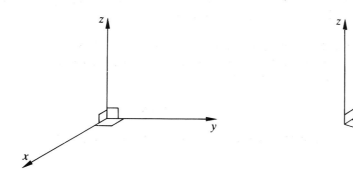

We also choose the base vectors $\mathbf{a}, \mathbf{b}$ and $\mathbf{c}$ to have a magnitude of 1, and in this case they are denoted by $\mathbf{i}, \mathbf{j}$ and $\mathbf{k}$, along the x, y and z axes, respectively.

A **unit vector** is a vector whose magnitude is 1.

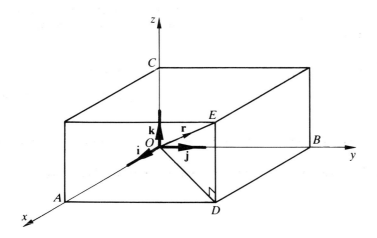

The parallelepiped in the figure for the three dimensional case, now becomes a cuboid with $\mathbf{i}, \mathbf{j}$ and $\mathbf{k}$ as the unit vectors along the x, y and z axes respectively.

If $\mathbf{r} = 3\mathbf{i} + 4\mathbf{j} + 2\mathbf{k}$, then
$$\overrightarrow{OA} = 3\mathbf{i}, \quad \overrightarrow{AD} = \overrightarrow{OB} = 4\mathbf{j}$$
and $\overrightarrow{DE} = \overrightarrow{OC} = 2\mathbf{k}$.

Since $\overrightarrow{OE} = \overrightarrow{OA} + \overrightarrow{AD} + DE$

$$\mathbf{r} = 3\mathbf{i} + 4\mathbf{j} + 2\mathbf{k}.$$

In general, we write $\mathbf{r} = x\mathbf{i} + y\mathbf{j} + z\mathbf{k}$, where x, y and z are the scalar multiples of the unit sectors $\mathbf{i}, \mathbf{j}$ and $\mathbf{k}$ required to form the vector $\mathbf{r} = \overrightarrow{OA} + \overrightarrow{AD} + \overrightarrow{DE}$.

The **magnitude of a vector** is the length of the line segment representing it, i.e. $|\overrightarrow{OE}|$

Using the theorem of Pythagoras, $OE^2 = OD^2 + DE^2$

$$= (OA^2 + AD^2) + DE^2$$

$$= x^2 + y^2 + z^2$$

The magnitude of the vector $\mathbf{r} = x\mathbf{i} + y\mathbf{j} + z\mathbf{k}$ is $|\mathbf{r}| = \sqrt{x^2 + y^2 + z^2}$

So the magnitude of $\mathbf{r} = 3\mathbf{i} + 4\mathbf{j} + 2\mathbf{k}$ is $|\mathbf{r}| = \sqrt{3^2 + 4^2 + 2^2} = \sqrt{29}$

The magnitude of $\mathbf{r} = 3\mathbf{i} + \mathbf{j} - 2\mathbf{k}$ is $|\mathbf{r}| = \sqrt{3^2 + 1^2 + (-2)^2} = \sqrt{14}$

WORKED EXAMPLE ▶

Find the unit vector in the direction of the vector $\mathbf{r} = 2\mathbf{i} + 3\mathbf{j} + 2\sqrt{3}\mathbf{k}$.

Now $|\mathbf{r}| = \sqrt{2^2 + 3^2 + (2\sqrt{3})^2} = \sqrt{4 + 9 + 12} = \sqrt{25} = 5$.

Since this vector has a magnitude of 5 units, clearly the vector $\frac{1}{5}\mathbf{r}$ will have unit magnitude.

The unit vector in the direction of $\mathbf{r} = \frac{1}{5}(2\mathbf{i} + 3\mathbf{j} + 2\sqrt{3}\mathbf{k})$.

In general, to find the unit vector in the direction of a given vector $\mathbf{a}$, divide by its magnitude. The unit vector in the direction of $\mathbf{a}$, is $\dfrac{\mathbf{a}}{|\mathbf{a}|}$.

When a vector is expressed in this form, the components give the cosines of the angles that the vector makes with each of the x, y and z axes.

If $\mathbf{r} = 4\mathbf{i} + 3\mathbf{j} + 9\mathbf{k}$, then the magnitude of $\mathbf{r}$ is given by $|\mathbf{r}| = \sqrt{4^2 + 3^2 + 9^2} = \sqrt{106}$.

The unit vector along OR is $\dfrac{4}{\sqrt{106}}\mathbf{i} + \dfrac{3}{\sqrt{106}}\mathbf{j} + \dfrac{9}{\sqrt{106}}\mathbf{k}$.

Since $\triangle ORL$ is right-angled,

$$\cos\theta = \frac{OL}{OR} = \frac{4}{\sqrt{106}}$$

Similarly, $\triangle$s OMR and ORN are right-angled,

$$\therefore \cos\phi = \frac{OM}{OR} = \frac{3}{\sqrt{106}} \quad \text{and} \quad \cos\psi = \frac{ON}{OR} = \frac{9}{\sqrt{106}}$$

The values $\dfrac{4}{\sqrt{106}}, \dfrac{3}{\sqrt{106}}, \dfrac{9}{\sqrt{106}}$ defined the direction of $\mathbf{r}$

and are called the **direction cosines**.

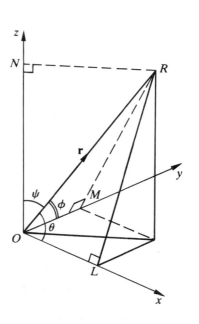

Hence, $\cos\theta = \dfrac{4}{\sqrt{106}} = 0.3885 \quad\Rightarrow\quad \theta = 67.14°$ or $67°\,8'$

$\cos\phi = \dfrac{3}{\sqrt{106}} = 0.2914 \quad\Rightarrow\quad \phi = 73.06°$ or $73°\,3'$

and $\cos\psi = \dfrac{9}{\sqrt{106}} = 0.8742 \quad\Rightarrow\quad \psi = 29.05°$ or $29°\,3'$

The angles made with the x, y and z axes are $67°\,8'$, $73°\,3'$ and $29°\,3'$ respectively.

Clearly if a vector is multiplied by a scalar it changes its magnitude but not its direction. Thus any multiple of the direction cosines (called the **direction ratios**) will also define the direction of the vector.
   The direction ratios in the Example could be $4, 3, 9$ or $2, 1\frac{1}{2}, 4\frac{1}{2}$ or $8, 6, 18$ or even $-4, -3, -9$.

Exercise 7.3

1 Find the magnitude of the following vectors.

   (a) $5\mathbf{i} - 12\mathbf{j}$ (b) $\mathbf{i} - 4\mathbf{k}$ (c) $7\mathbf{i} + 24\mathbf{j}$ (d) $2\mathbf{i} + \mathbf{j} + \mathbf{k}$ (e) $3\mathbf{j} + 4\mathbf{k}$

   (f) $2\mathbf{i} + 2\mathbf{j} + \mathbf{k}$ (g) $6\mathbf{i} + 3\mathbf{j} + 2\mathbf{k}$ (h) $2\mathbf{i} - \mathbf{j} + \mathbf{k}$ (i) $4\mathbf{i} + 3\mathbf{j} - 2\mathbf{k}$

2 Find the angles that the following vectors make with each of the mutually prependicular axes.

   (a) $3\mathbf{i} + 4\mathbf{j}$ (b) $2\mathbf{i} - 3\mathbf{j} + 6\mathbf{k}$ (c) $-3\mathbf{i} + 2\mathbf{j} - \mathbf{k}$ (d) $\mathbf{i} - 3\mathbf{j} - \sqrt{6}\mathbf{k}$

3 If $\mathbf{a} = \mathbf{i} + \mathbf{j} + \mathbf{k}$, $\mathbf{b} = -2\mathbf{i} - \mathbf{j} + 3\mathbf{k}$ and $\mathbf{c} = 3\mathbf{i} + 2\mathbf{j} - \mathbf{k}$, find the magnitude of the vectors,

   (i) $\mathbf{a} + \mathbf{b}$ (ii) $2\mathbf{b} + \mathbf{c}$ (iii) $2\mathbf{a} + \mathbf{b} + \mathbf{c}$ (iv) $\mathbf{a} + \mathbf{b} - \mathbf{c}$

4 If $\mathbf{a} = \mathbf{i} - 2\mathbf{j} + \mathbf{k}$, $\mathbf{b} = \mathbf{i} + 3\mathbf{j} - \mathbf{k}$ and $\mathbf{c} = 2\mathbf{i} - 4\mathbf{j} + 3\mathbf{k}$, find the unit vectors in the directions of the following.

   (i) $\mathbf{a} + \mathbf{b}$ (ii) $2\mathbf{a} - \mathbf{c}$ (iii) $\mathbf{a} + \mathbf{b} + \mathbf{c}$ (iv) $3\mathbf{a} - 2\mathbf{b} + \mathbf{c}$

5 Find the direction cosines of the vector $2\mathbf{i} - \mathbf{j} + 4\mathbf{k}$ and hence find the angles that the vector makes with each of the axes.

6 If $\overrightarrow{OA} = \mathbf{x}$, $\overrightarrow{OC} = \mathbf{y}$ and $\overrightarrow{OD} = \mathbf{z}$, use the parallelepiped in the diagram to express each of the following vectors in terms of components in the directions of $\mathbf{x}, \mathbf{y}$ and $\mathbf{z}$.

   (a) $\overrightarrow{OB}$ (b) $\overrightarrow{FE}$ (c) $\overrightarrow{AC}$ (d) $\overrightarrow{CF}$

   (e) $\overrightarrow{AF}$ (f) $\overrightarrow{OF}$ (g) $\overrightarrow{AG}$

7 Use the diagram to find the following vectors in terms of $\mathbf{i}, \mathbf{j}$ and $\mathbf{k}$ if
   $\mathbf{x} = 3\mathbf{i} + \mathbf{j} + 2\mathbf{k}$, $\mathbf{y} = \mathbf{i} + 2\mathbf{j} + 4\mathbf{k}$ and $\mathbf{z} = 2\mathbf{i} - \mathbf{j} + 5\mathbf{k}$.

   (a) $\overrightarrow{OE}$ (b) $\overrightarrow{GC}$ (c) $\overrightarrow{BG}$ (d) $\overrightarrow{OF}$ (e) $\overrightarrow{DF}$

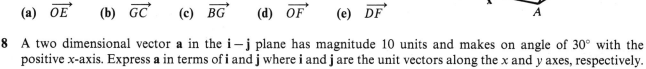

8 A two dimensional vector $\mathbf{a}$ in the $\mathbf{i} - \mathbf{j}$ plane has magnitude 10 units and makes on angle of $30°$ with the positive x-axis. Express $\mathbf{a}$ in terms of $\mathbf{i}$ and $\mathbf{j}$ where $\mathbf{i}$ and $\mathbf{j}$ are the unit vectors along the x and y axes, respectively.

9 If $\mathbf{i}$ and $\mathbf{j}$ are unit vectors along the x and y axes, respectively, and $A(3, 4), B(-1, 6), C(-4, -2)$ and $D(1, -4)$ are points in the x-y plane, find the vectors **(a)** $\overrightarrow{OA}$ **(b)** $\overrightarrow{AB}$ **(c)** $\overrightarrow{BC}$ and **(d)** $\overrightarrow{CD}$ in terms of $\mathbf{i}$ and $\mathbf{j}$. Verify that $\overrightarrow{OA} + \overrightarrow{AB} + \overrightarrow{BC} + \overrightarrow{CD} = \overrightarrow{OD}$.

Position vectors

So far, the position of the vectors in space has not been of importance and only the magnitude, direction and sense distinguished one from another.

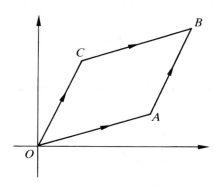

So, $\overrightarrow{AB} = \overrightarrow{OC}$, since they are represented by the opposite sides of a parallelogram.

Such vectors are called **free vectors**.

However, it is useful to be able to define the position of a point in space by a vector displacement relative to a fixed origin.

In the diagram, right, $\overrightarrow{OP}$ gives the displacement required to move from the origin O to point P.

$\overrightarrow{OP}$ is called the **position vector** of P relative to O.

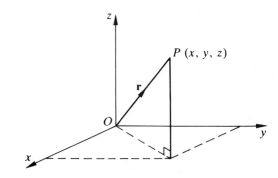

If $P(x, y, z)$ is a general point, then $\overrightarrow{OP} = \mathbf{r} = x\mathbf{i} + y\mathbf{j} + z\mathbf{k}$ is the position vector of P relative to O.

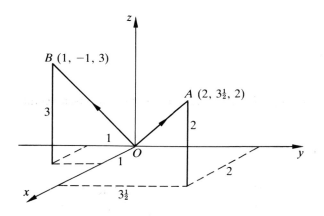

The position vector of $A(2, 3\frac{1}{2}, 2)$ relative to O is

$$\mathbf{r} = 2\mathbf{i} + 3\tfrac{1}{2}\mathbf{j} + 2\mathbf{k}$$

The position vector of $B(1, -1, 3)$ relative to O is

$$\mathbf{r} = \mathbf{i} - \mathbf{j} + 3\mathbf{k}$$

The position vector of A relative to $B = \overrightarrow{BA} = \overrightarrow{OA} - \overrightarrow{OB}$

$$= (2\mathbf{i} + 3\tfrac{1}{2}\mathbf{j} + 2\mathbf{k}) - (\mathbf{i} - \mathbf{j} + 3\mathbf{k})$$

$$\mathbf{r} = \mathbf{i} + 4\tfrac{1}{2}\mathbf{j} - \mathbf{k}$$

WORKED EXAMPLE

The line joining the point $A(1, 3)$ to the point $B(6, 7)$ is divided internally in the ratio of $2:3$ by a point X. Find the position vector of X, **(a)** relative to the origin and **(b)** relative to $C(4, 1)$.

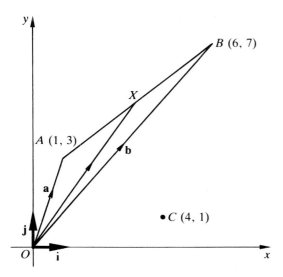

Let $\mathbf{i}$ and $\mathbf{j}$ be unit vectors along the x and y axes, respectively.

If the position vectors of A and B relative to O are $\mathbf{a}$ and $\mathbf{b}$, respectively,

$$\text{then,} \quad \mathbf{a} = \mathbf{i} + 3\mathbf{j}$$
$$\text{and} \quad \mathbf{b} = 6\mathbf{i} + 7\mathbf{j}$$

Using the ratio theorem (earlier in this chapter), with $\lambda = 2$ and $\mu = 3$.

$$\overrightarrow{OX} = \frac{\mu\mathbf{a} + \lambda\mathbf{b}}{\mu + \lambda}$$

The position vector of X relative to O is $\overrightarrow{OX} = \dfrac{3(\mathbf{i} + 3\mathbf{j}) + 2(6\mathbf{i} + 7\mathbf{j})}{5}$

$$= 3\mathbf{i} + \frac{23}{5}\mathbf{j}$$

The position vector of X relative to C is $\overrightarrow{CX} = \overrightarrow{OX} - \overrightarrow{OC}$

$$\therefore \overrightarrow{CX} = \left(3\mathbf{i} + \frac{23}{5}\mathbf{j}\right) - (4\mathbf{i} + \mathbf{j}) = -\mathbf{i} + \frac{18}{5}\mathbf{j}$$

Since the position vector of X relative to O is $3\mathbf{i} + \dfrac{23}{5}\mathbf{j}$, X is the point $\left(3, \dfrac{23}{5}\right)$. This can be checked by a non-vector method.

Draw the lines ALN parallel to the x-axis and BN and XL parallel to the y-axis to form similar triangles AXL and ABN.

Hence $\dfrac{AX}{AB} = \dfrac{2}{5} = \dfrac{AL}{AN} \quad \Rightarrow \quad AL = \dfrac{2}{5}AN$

$$\therefore AL = \frac{2}{5}AN = \frac{2}{5}(6 - 1) = 2$$

Similarly $LX = \dfrac{2}{5}NB = \dfrac{2}{5}(7 - 3) = \dfrac{8}{5}$

The co-ordinates of X are thus $(1 + AL, 3 + LX) = \left(3, 3 + \dfrac{8}{5}\right) = \left(3, \dfrac{23}{5}\right)$

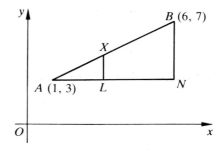

Exercise 7.4

1 The position vectors of points A and B relative to the origin are, $\mathbf{a} = \mathbf{i} + 2\mathbf{j} - 3\mathbf{k}$ and $\mathbf{b} = -3\mathbf{i} + 4\mathbf{j} + 2\mathbf{k}$, respectively. Find the position vector of the mid-point of AB relative to the origin.

2 In a rectangle $OABC$, X and Y are the mid-points of the sides AB and BC. If the position vectors of A and C with respect to O are $\mathbf{a}$ and $\mathbf{c}$, respectively, find the position vector of the mid-point of XY.

3 The points A, B and C have position vectors $2\mathbf{i} - \mathbf{j} + 3\mathbf{k}$, $4\mathbf{i} + 3\mathbf{j} - 2\mathbf{k}$ and $5\mathbf{i} + 5\mathbf{j} - \frac{9}{2}\mathbf{k}$ respectively with respect to the origin. Show that A, B and C are collinear.

4 The points A, B and C have position vectors $\mathbf{i} + \mathbf{j} + \mathbf{k}$, $3\mathbf{i} - 2\mathbf{j} + 2\mathbf{k}$ and $2\mathbf{i} + 4\mathbf{j} - \mathbf{k}$ respectively. By finding the magnitudes of $\overrightarrow{AB}$ and $\overrightarrow{AC}$, show that triangle ABC is isosceles.

5 A regular hexagon $OABCDE$ has a side of $2\,\text{cm}$. If O is the origin and unit vectors $\mathbf{i}$ and $\mathbf{j}$ are taken along $\overrightarrow{OA}$ and $\overrightarrow{OD}$, respectively, find the position vectors of **(a)** A, B and C with respect to O **(b)** X, the mid-point of BC and **(c)** Y, the point that divides AE in the ratio of $1:3$.

6 The position vectors of A and B with respect to O are $3\mathbf{i} + 2\mathbf{j} - \mathbf{k}$ and $\mathbf{i} - 4\mathbf{j} + \mathbf{k}$ respectively. Find the position vector of a point P on the line joining A and B such that $AP:PB = \lambda:1$.

7 If $A(1, 2, 0)$, $B(3, 4, 1)$ and $D(2, 1, -1)$ are three vertices of a parallelogram $ABCD$, write down the position vectors of A, B and D with respect to the origin. Find the position vector of X, the point of intersection of the diagonals and hence find the co-ordinates of C.

8 If $A(5, 12)$ and $B(3, -4)$ are two points in the x–y plane, write down the position vectors of A and B relative to the origin O in terms of $\mathbf{i}$ and $\mathbf{j}$, the unit vectors along the x and y axes, respectively. Find the unit vectors along the direction $\overrightarrow{OA}$ and $\overrightarrow{OB}$ and hence determine the y co-ordinate of a point C on the line bisecting the angle between OA and OB, if the x co-ordinate of C is 8.

9 If $A(1, 2, 1)$ and $B(-3, -1, 2)$ are two points, write down the vector $\overrightarrow{AB}$ in terms of $\mathbf{i}, \mathbf{j}$ and $\mathbf{k}$, the unit vectors along the x, y and z axes, respectively. Write down the position vector of a general point of the line and find the co-ordinates of the points on the line which are $\sqrt{14}$ units from the origin.

8 Forces and equilibrium

Introduction

Is this book that you are reading moving?
What forces act on it?
What is a force?

What happens if you lift the book to the side of the table (or desk) and let it go?

No doubt you have heard of the force on gravity, but what is it? What causes it?
How large is it? In what direction does it act?

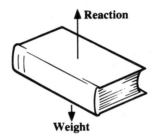
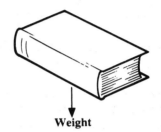

If we tilt the table top with a book, pencil, ruler and protractor lying on it, what happens? Do it!

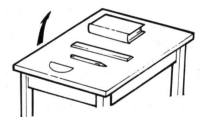

At a particular angle one of the objects will move.
Which one? Why? What forces are now acting?

Investigation 1

See figures, right.

Lay your pencil against the side of the book.
Will the pencil stay in a leaning position?
What forces are acting on it?

Lean your pencil against the book so that its point of contact is nearer the lower end than the upper end.
What happens? Why?

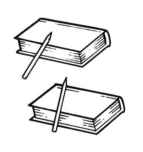

Forces

If an object is stationary under the action of the forces acting on it (like the book on the table) it is said to be in **equilibrium**. If the forces do not balance (e.g. the book is not supported) the object will more. The study of motion (**Dynamics**) occurs in later chapters. Here we will concentrate on **Statics**, the study of the forces acting when objects are in equilibrium.

Try to identify the forces acting on the objects in the following situations.

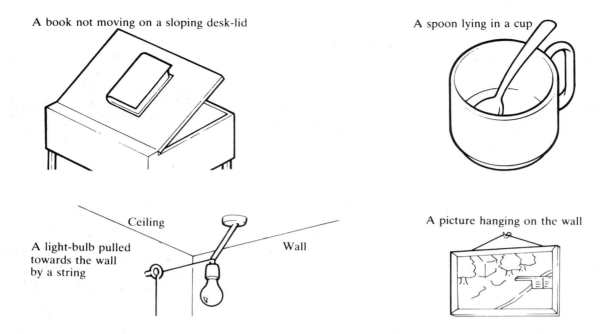

A book not moving on a sloping desk-lid

A spoon lying in a cup

A light-bulb pulled towards the wall by a string

Ceiling

Wall

A picture hanging on the wall

If you hold this book away from the table it will fall to the floor because the **force of gravity** acting on it (its weight) acts vertically downwards on it. If you support the book with your hand, so that the book is still, then you are providing a supporting force equal and opposite to the weight of the book. When the book lies on the table, there is a reaction force from the table supporting the book (see figure, below).

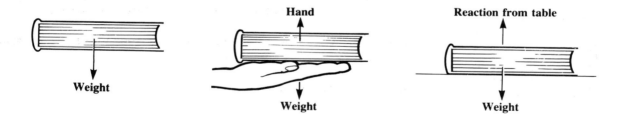

Weight

Hand

Weight

Reaction from table

Weight

If **I push** the book horizontally with my finger, the book moves in the direction of my push. Forces have magnitude and direction and are vectors. There is some resistance to my pushing on the book. This is **friction** between the book and the table which opposes the direction of motion. We now have four forces acting on the book, represented in direction by the arrows in the figure below, left.

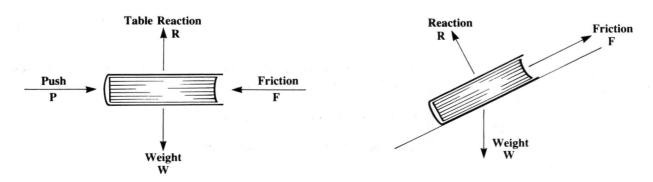

Table Reaction
R

Push
P

Friction
F

Weight
W

Reaction
R

Friction
F

Weight
W

When the book is lying still on the sloping desk-lid (previous figure, right) its tendency is to move downwards, so the friction force opposes this and acts up the line of greatest slope. There is a contact reaction force which will balance **F** and **W**; so **F + R + W = 0**

When my pencil is leaning against the book and not moving (figure, right) the forces on the pencil are as follows.

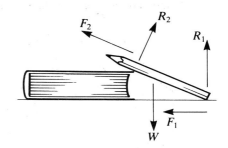

(a) Its weight acts vertically downwards.

(b) There is friction and a reaction on the end touching the table.

(c) Possibly friction F_2, and certainly reaction R_2, at the contact point of the pencil and book.

When the point of contact of the pencil with the book is nearer the lower end than the upper end, the pencil may not be in equilibrium. The pencil either slides down (to reach the position in the previous figure) or tips up to lie on top of the book. (The latter situation involves the turning effect (**moment**) of the forces and will be analysed in Book 2).

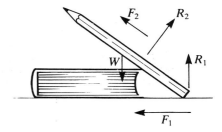

Some of the objects, such as the book and the light bulb, can be regarded as **particles** with the mass concentrated in one point (**centre of mass**) and other objects, such as the pencil and the spoon (although smaller in size!), have to be regarded as **rigid bodies** where the forces act at different points of the body but do not affect the shape of the body (hence, **rigid**).

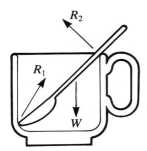

With the spoon in the cup (figure, left), the spoon rests against the inside at the base and the rim, so there must be contact reaction forces on the spoon.
The spoon presses on the cup so the cup presses on the spoon.

Would it make any difference if the base of the spoon were resting against the centre of the base of the cup?

Would there be friction at each contact?

In the figure, right, the horizontal string pulls the light-bulb towards the wall

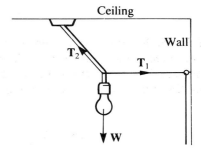

This pull is conveyed by the tension in the taut (tight) string, T_1.

T_1 and **W** are balanced by the tension T_2 in the supporting flex from the bulb to the ceiling.

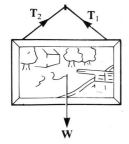

The figure, left, shows the picture with the weight balanced by the tensions T_1 and T_2 in the supporting string.

Can you say anything about T_1 and T_2?

What force acts on the supporting pin?

Exercise 8.1

Draw clearly-labelled diagrams showing the forces acting on the bodies underlined in the following situations. The bodies may not be in equilibrium.

1 A striplight of length 2 metres supported by two vertical chains 50 cm from each end.

2 A ladder leaning against a smooth (no friction) wall standing on rough ground.

3 The bob of a pendulum at its extreme point of displacement, where it is momentarily at rest.

4 Two bricks resting one on top of the other on a flat surface (draw forces on each brick separately).

5 A brick (a) resting on a flat smooth surface,

(b) resting on a flat rough surface,

(c) resting on a rough surface inclined at 25° to the horizontal,

(d) pulled up a rough surface inclined at 25° to the horizontal.

6 A rake resting against a garden roller.

7 A spoon of length 25 cm resting inside a smooth fixed hemispherical bowl of diameter 20 cm.

8 A spoon of length 10 cm resting inside a smooth fixed hemispherical bowl of diameter 20 cm.

9 An owner dragging a reluctant dog along the ground.

10 Masses of 3 kg and 5 kg hanging on a light string (whose weight is negligible compared with the masses) threaded over a smooth pulley (the bearing is frictionless).

11 A car moving up a hill.

12 A child on a pogo stick (a) at full compression (b) at no compression i.e. jumping.

13 The arrow of a bow (a) just before firing (b) at the moment of firing.

14 A ball falling vertically towards the ground (a) just before impact (b) at the moment of impact (c) just after impact.

15 A step ladder with no restraining bar standing (a) on rough ground (b) on a smooth surface.

16 A marble rolling in a hemispherical bowl.

17 A packing case resting (angled) against a step (figure, right).

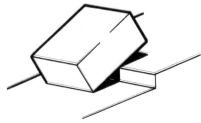

18 Two masses of 3 kg and 5 kg fixed at the points of trisection of a 3 metre string tied to two points, at the same level, 2 metres apart.

19 Two children sitting on the ends of a plank across an oil-barrel forming a seesaw.

Mass and weight

How much do you weigh?
  Eleven stones (in old British units)
  or 70 kg (in SI units)

How much do you weigh on the moon? The same?

Discuss the meaning of mass and weight with your teacher.

Strictly speaking, you should not talk about a **weight** of 70 kg.

  70 kg (or 11 stones) is a **mass**.

My mass is the same whether I am on earth or on the moon, but my weight is less on the moon (about one-sixth) because of the force of attraction (weight depends on the mass of the moon or the earth). The acceleration of the moon's gravitational field is less than that of the earth's.

This will be dealt with more fully in Chapter 15 on Newton's Laws. It is sufficient now to know that, on earth, an object of **mass** 70 kg has a **weight** of $70 \times 9.8 \simeq 700$ Newtons.
  Mass is measured in kilograms (kg) and **weight** in Newtons (N).

Forces in equilibrium

The upper figure, right, shows an object of weight 10 Newtons (10 N) supported by two strings of equal length. The weight must be balanced by the combined force of $\mathbf{T}_1$ and $\mathbf{T}_2$ upwards.

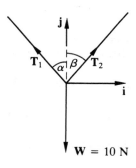

The three forces $\mathbf{T}_1, \mathbf{T}_2$ and $\mathbf{W}$ balance one other.

Forces are vectors (having magnitude and direction) and must be added together like vectors.

$\mathbf{T}_1, \mathbf{T}_2$ and $\mathbf{W}$ balance $\Rightarrow \mathbf{T}_1 + \mathbf{T}_2 + \mathbf{W} = 0$ and a scale-drawing of the triangle of forces will form a closed triangle (lower figure, right).

Using vectors $\mathbf{i}$ and $\mathbf{j}$ as shown in the upper figure,

$\mathbf{W} = -10\mathbf{j} \quad \mathbf{T}_2 = T_2 \sin\beta\,\mathbf{i} + T_2 \cos\beta\,\mathbf{j}$ where $|\mathbf{T}_2| = T_2$

$\qquad \mathbf{T}_1 = -T_1 \sin\alpha\,\mathbf{i} + T_1 \cos\alpha\,\mathbf{j}$

$\mathbf{T}_1 + \mathbf{T}_2 + \mathbf{W} = 0 \quad \Rightarrow \quad (-T_1 \sin\alpha + T_2 \sin\beta)\mathbf{i} + (T_1 \cos\alpha + T_2 \cos\beta - 10)\mathbf{j} = 0$

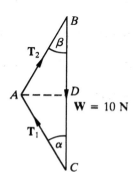

Horizontally $\rightarrow \quad -T_1 \sin\alpha + T_2 \sin\beta = 0 \quad \Rightarrow \quad T_1 \sin\alpha = T_2 \sin\beta$ .... **(1)**
$\qquad\qquad\qquad \Rightarrow$ the components of T_1 and T_2 in the horizontal direction are equal

Vertically $\uparrow \quad T_1 \cos\alpha + T_2 \cos\beta = 10$ .... **(2)**
$\qquad\qquad \Rightarrow$ the components of T_1 and T_2 in the vertical direction together balance the weight.

We can deduce this from $\triangle ABC$ (in the lower figure). $AD = T_1 \sin \alpha = T_2 \sin \beta$

$$10 = BC = BD + DC = T_2 \cos \beta + T_1 \cos \alpha$$

Finding the components in the **i** direction is called **resolving horizontally.**

Finding the components in the **j** direction is called **resolving vertically**.

If the supporting strings are the same length, then $\alpha = \beta$ (**W** is vertically downwards), and from **(1)** $T_1 = T_2 = T$.
From **(2)** $2T \cos \alpha = 10$ which will specify T if the angle is known.

The following example will illustrate different methods of solution.

> **WORKED EXAMPLE**

> Two strings, AB of length 8 cm and AC of length 6 cm, are attached to two points B and C on the same horizontal level so that $BC = 10$ cm. The strings support an object of weight 20 N attached at A. Find the tensions in the two strings.

A diagram of the problem is shown, right.

Notice that $\triangle ABC$ is right-angled

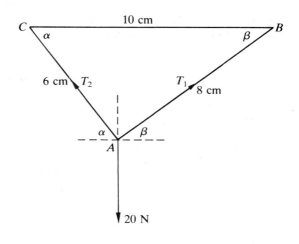

$$\tan \alpha = \frac{4}{3} \quad \Rightarrow \quad \alpha = 53.1° \quad \Rightarrow \quad \beta = 36.9°$$

$$\Rightarrow \cos \alpha = \sin \beta = \frac{3}{5}; \ \cos \beta = \sin \alpha = \frac{4}{5}$$

Considering horizontal components at A
($\leftarrow$ Resolving horizontally)

$$T_1 \cos \beta = T_2 \cos \alpha \quad \Rightarrow \quad T_1 \cdot \frac{4}{5} = T_2 \cdot \frac{3}{5} \quad \cdots\cdots\cdots \ (1)$$

$$\Rightarrow \quad 4T_1 = 3T_2$$

$\uparrow$ Resolving vertically,

$$T_1 \sin \beta + T_2 \sin \alpha = 20$$

$$\Rightarrow \quad T_1 \cdot \frac{3}{5} + T_2 \cdot \frac{4}{5} = 20 \quad \cdots\cdots\cdots \ (2)$$

$$\Rightarrow \quad 3T_1 + 4T_2 = 20 \times 5 \quad \text{and using } T_1 = \frac{3}{4}T_2 \ \Rightarrow \ 3 \cdot \frac{3}{4}T_2 + 4T_2 = 100$$

$$\Rightarrow \quad 9T_2 + 16T_2 = 400 \quad \Rightarrow \quad T_2 = \frac{400}{25} = 16 \text{ N}$$

$$T_1 = \frac{3}{4}T_2 = \frac{3}{4} \times 16 = 12 \text{ N}$$

These results are more quickly achieved from the triangle of forces (see figure, right). Since $\triangle PQR$ is right-angled.

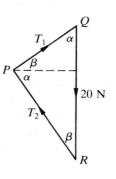

$$T_2 = 20 \cos \beta = 20 \times \frac{4}{5} = 16 \text{ N}$$

$$T_1 = 20 \cos \alpha = 20 \times \frac{3}{5} = 12 \text{ N}$$

This is equivalent to resolving (considering components) in the directions of the forces $\mathbf{T}_1$ and $\mathbf{T}_2$.

More simply still, $\triangle PQR$ has sides in the ratios 6, 8 and 10 which represent completely the forces T_1, T_2 and 20 N, so by simple proportion, $T_1 = 12\,\text{N}$ and $T_2 = 16\,\text{N}$.

Components and resultants of forces

We have seen the need for (a) breaking down a force into components and
(b) finding the resultant of a number of forces.

Components

A force $\mathbf{F} = 6\mathbf{i} + 8\mathbf{j}$, in the diagram, has two components, 6 in the $\mathbf{i}$ direction (represented by OX) and 8 in the $\mathbf{j}$ direction (represented by OY and XP).

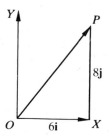

The resultant force has magnitude 10 N (represented by OP) and acts in the direction making on angle of $\tan^{-1}\frac{4}{3} = 53.1°$ with OX.

$\mathbf{F}$ can be resolved into components in any direction, but more usually we needs to resolve (break down) $\mathbf{F}$ into its components in two directions which are at right angles to each other.

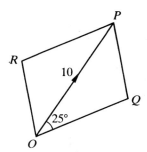

For example (see figure, left), OP can be resolved into two components, OQ (making an angle of $25°$ with OP) and OR (perpendicular to OQ).

$$OQ = 10\cos 25° = 9.06\,\text{N}$$

$$OR = QP = 10\sin 25° = 4.23\,\text{N}$$

Remember that, although the second component is represented in size and direction by QP, it does act through O and so is represented completely by OR.

Resultants of forces

To add forces we use a polygon of forces.

WORKED EXAMPLE

Find the resultant of the following forces all acting at the origin. $\mathbf{F}_1 = 4\,\text{N}$ acting along the x-axis, $\mathbf{F}_2 = 2\,\text{N}$ along the y-axis and $\mathbf{F}_3 = 3\,\text{N}$ at an angle of $40°$ with OX.

Diagram **(a)** shows the forces all acting through O.

Diagram **(b)** shows a polygon of forces with OX representing $\mathbf{F}_1$, XT representing $\mathbf{F}_3$, and TS representing $\mathbf{F}_2$ (in magnitude and direction, not position). $\mathbf{R}$, the resultant force, is represented by OS.

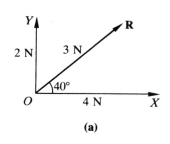

(a)

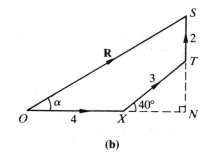

(b)

Forces and equilibrium

$\mathbf{R} = ON\mathbf{i} + NS\mathbf{j}$ where **i** and **j** are unit vectors in the x and y directions.

$ON = OX + XN = 4 + 3\cos 40° = 4 + 2.30 = 6.30$

$NS = NT + TS = 3\sin 40° + 2 = 3.93$

So, $\mathbf{R} = 6.30\mathbf{i} + 3.93\mathbf{j}$

To find the magnitude and direction of the resultant force **R**,

$OS^2 = 6.30^2 + 3.93^2 = 39.69 + 15.44 = 55.13$ $\Rightarrow$ $OS = 7.43\,\text{N}$

$\tan \alpha = \dfrac{SN}{NO} = \dfrac{3.93}{6.30} = 0.6238$ $\Rightarrow$ $\alpha = 32.0°$ so the resultant is a force of 7.43 N acting at an angle of 32° with OX.

This method is equivalent to resolving $\mathbf{F}_3$ into its components in the **i** and **j** directions and adding the total components in the **i** and **j** directions.

$\mathbf{F}_1 = 4\mathbf{i}$ $\mathbf{F}_2 = 2\mathbf{j}$ $\mathbf{F}_3 = 3\cos 40\mathbf{i} + 3\sin 40\mathbf{j} = 2.3\mathbf{i} + 1.93\mathbf{j}$

$\mathbf{R} = \mathbf{F}_1 + \mathbf{F}_2 + \mathbf{F}_3 = 6.3\mathbf{i} + 3.93\mathbf{j}$ hence $|\mathbf{R}| = 7.43\,\text{N}$ and $\alpha = 32°$.

The method of finding components by **resolving** is very important in solving problems in **dynamics** (motion) and **statics** (equilibrium) where we have to find the effect of different forces acting in different directions.

The following Example will give practice in resolving forces.

WORKED EXAMPLE

An object of weight 60 N is suspended from two strings inclined at 25° and 35° to the vertical. Find the tensions in the strings.

Method 1   With the notation in the diagram.

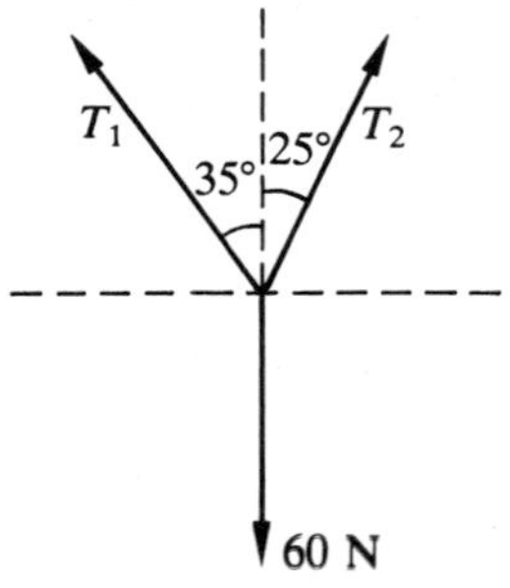

Resolving vertically   ↑ $T_1 \cos 35° + T_2 \cos 25° = 60$ .... **(1)**

Resolving horizontally   ← $T_1 \sin 35° = T_2 \sin 25°$

$\Rightarrow$ $T_1 = \dfrac{T_2 \sin 25°}{\sin 35°}$

Substitute for T_1 in **(1)**,

$T_2 \dfrac{\sin 25°}{\sin 35°} \cdot \cos 35° + T_2 \cos 25° = 60$

$\Rightarrow$ $T_2 (\sin 25° \cdot \cos 35° + \cos 25° \cdot \sin 35°) = 60 \sin 35°$ $\Rightarrow$ $T_2 \sin(25° + 35°) = 60 \sin 35°$

$\Rightarrow$ $T_2 = \dfrac{60 \sin 35°}{\sin 60°} \simeq 39.7\,\text{N}$

$\Rightarrow$ $T_1 = \dfrac{39.7 \sin 25°}{\sin 35°} \simeq 29.3\,\text{N}$

146

Method 2   From the triangle of forces in the diagram.

Using the sine rule in $\triangle ABD$

$$\frac{T_2}{\sin 35°} = \frac{T_1}{\sin 25°} = \frac{60}{\sin 120°}$$

$\Rightarrow \quad T_2 = \dfrac{60\sin 35°}{\sin 120°} = \dfrac{60\sin 35°}{\sin 60°} \simeq 39.7\,\text{N}$

$\Rightarrow \quad T_1 = \dfrac{60\sin 25°}{\sin 120°} = \dfrac{60\sin 25°}{\sin 60°} \simeq 29.3\,\text{N}$

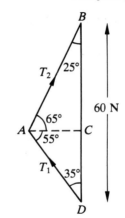

Method 3   In the diagram, right, draw $AB \perp T_2$

Resolve along AB, which eliminates T_2.

↖ $T_1 \cos 30° = 60 \cos 65°$

$\Rightarrow \quad T_1 = \dfrac{60\cos 65°}{\cos 30°} = \dfrac{60\sin 25°}{\sin 60°} \simeq 29.3\,\text{N}$

Resolve along CD ($\perp T_1$) which eliminates T_1.

↗ $T_2 \cos 30° = 60 \cos 55°$

$\Rightarrow \quad T_2 = \dfrac{60\cos 55°}{\cos 30°} = \dfrac{60\sin 35°}{\sin 60°} \simeq 39.7\,\text{N}$

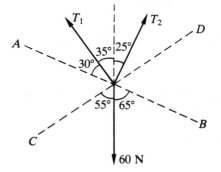

Perhaps, at this stage, Method 2 is the easiest to understand and apply. This application of the sine rule is known as **Lami's theorem**, and it states that when three concurrent forces balance, the size of each force is proportional to the sine of the angle between the other two forces. Thus,

$$\frac{60}{\sin 60°} = \frac{T_1}{\sin 155°} = \frac{T_2}{\sin 145°} \quad \Rightarrow \quad \frac{60}{\sin 60°} = \frac{T_1}{\sin 25°} = \frac{T_2}{\sin 35°} \quad \text{as in Method 2.}$$

For solving problems involving equilibrium at a point,

(a)   scale drawing of the polygon of forces,

(b)   calculation from the polygon of forces,

(c)   resolving,

(d)   Lami's Theorem (for three forces).

Exercise 8.2

1   Find the resultant of the following forces acting at a point,

(a) $\mathbf{F}_1 = 6\mathbf{i}$, $\mathbf{F}_2 = 4\mathbf{j}$, $\mathbf{F}_3 = 2\mathbf{i} + 3\mathbf{j}$, $\mathbf{F}_4 = \mathbf{i} - 5\mathbf{j}$. Draw these forces nose to tail on squared paper to check your result. Express the resultant in vector form and by giving its magnitude and direction.

(b) $\mathbf{F}_1 = 7\,\text{N}$ making an angle of 25° with OX (in the first quadrant).
$\mathbf{F}_2 = 8\,\text{N}$ in the first quadrant making an angle of 30° with the y-axis.
$\mathbf{F}_3 = 9\,\text{N}$ in the second quadrant making an angle of 20° with the y-axis.

(c) $\mathbf{F}_1 = 5\,\text{N}$ on a bearing 027°, $\mathbf{F}_2 = 6\,\text{N}$ on a bearing 100°, $\mathbf{F}_3 = 7\,\text{N}$ on a bearing 318°.

(d) $\mathbf{F}_1$ has magnitude 6 N, $\mathbf{F}_2$ has magnitude 5 N and the angle between $\mathbf{F}_1$ and $\mathbf{F}_2$ is
(i)   90°   (ii)   60°   (iii)   0°   (iv)   180°   (v)   120°

2   A ship travels 30 km on a bearing 062° at 10 km h⁻¹, followed by 20 km on a bearing 036° at 15 km h⁻¹. Find **(a)** its distance and **(b)** its bearing from the starting point; **(c)** how long it takes; and **(d)** its average speed; for the whole journey.

3   A small object of weight 10 N is hanging by a string. A force *P* pulls the object to the side so that the string makes an angle of 40° to the vertical. Find *P* and the tension in the string.

4   A picture of weight 20 N is supported by a string of length 30 cm with ends tied to two supporting pins fixed to the back of the picture 26 cm apart. The string is hooked over a nail in the wall and the picture is level, with the string symmetrical. Find the tension in the string.

5   Diagram **(a)** shows forces which are in equilibrium.
    Find *P* and *Q* by

    (a)   resolving horizontally and vertically (Method 1),

    (b)   drawing a triangle of forces (Method 2),

    (c)   resolving along *P* and *Q* (Method 3),

    (d)   Lami's Theorem.

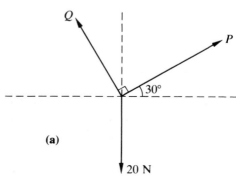

(a)

6   Diagram **(b)** shows forces in equilibrium. Find *R* and *P*.

7   In diagram **(c)** the forces are in equilibrium. Find *R* and *P*.

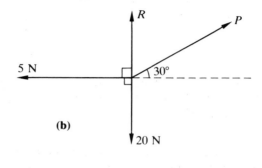

(b)

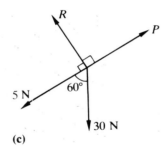

(c)

8   In diagram **(d)**, *BC* is horizontal. Find T_1, T_2, T_3 and α.

9   From diagram **(e)**, Find T_1, T_2, T_3 and α.

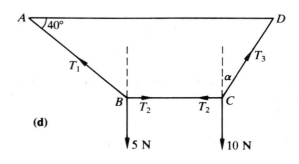

(d)

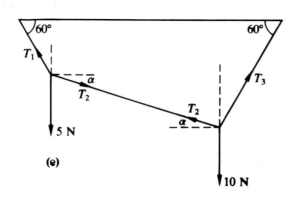

(e)

9 Differentiation 2

Composite functions

You will recall that, in Chapter 1, a function was defined as a mapping of one set of values onto another set of values. For example, the function f defined on a set of real numbers, may map values of x onto values of y and would be written $f: x \rightarrow y$ or $y = f(x)$.

If $f(x): x \rightarrow x^3$, the mapping for the interval $\{x: -1 \leqslant x \leqslant 3\}$ can be shown on a diagram, as in the figure, right.

If $g(x): x \rightarrow 2x + 5$, then the mapping for the interval $\{x: -3 \leqslant x \leqslant 3\}$ is as shown in the figure, far right.

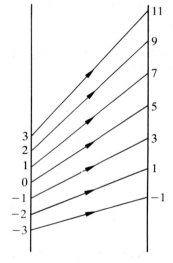

If these two mappings are combined, a **composite function** is created. This will be either $fg(x)$ or $gf(x)$, depending upon which of the function, f or g, operates upon x first.

$$\text{Now,} \quad fg(x) = f(2x + 5) = (2x + 5)^3$$
$$\text{and} \quad gf(x) = g(x^3) \quad = 2x^3 + 5$$

Since $fg(x) = f[g(x)]$ represents the function f applied to the result of the function g operating on x, the composite function is also known as a **function of a function**. This can be written using different letters to describe the two mappings.

If $f(u): u \rightarrow u^3$ and $g(x): x \rightarrow 2x + 5$, then $y = fg(x) = f(u)$ where $u = g(x) = 2x + 5$.

This is particularly useful when differentiating a composite function.

Differentiation of a function of a function

Consider the composite function $y = (x + 1)^3$.

Expand by multiplying the brackets, $\quad y = (x + 1)(x + 1)^2 = (x + 1)(x^2 + 2x + 1)$
$$\therefore \ y = x^3 + 3x^2 + 3x + 1$$

Differentiation 2

Differentiating with respect to x

$$\frac{dy}{dx} = 3x^2 + 6x + 3 = 3(x^2 + 2x + 1)$$

$$\Rightarrow \quad \frac{dy}{dx} = 3(x+1)^2$$

Thus it would appear that $\frac{d}{dx}(x+1)^3 = 3(x+1)^2$, and that the result can be obtained using the standard rule for differentiation. But this is not true in all cases.

If $\quad y = (x^2+1)^3 = (x^2+1)(x^2+1)^2 = (x^2+1)(x^4+2x^2+1)$

$$\Rightarrow \quad y = x^6 + 3x^4 + 3x^2 + 1$$

Differentiating with respect to x,

$$\frac{dy}{dx} = 6x^5 + 12x^3 + 6x = 6x(x^4 + 2x^2 + 1) = 6x(x^2+1)^2$$

Investigation 1

Expand each of the following functions. Differentiate your results with respect to x and simplify your answers to include the original function inside the bracket.

(a) $y = (2x+1)^2$ (d) $y = (x^2-5x)^3$ (f) $y = (1-x)^3$

(b) $y = (2x+5)^3$ (e) $y = \left(x - \frac{1}{x}\right)^4$ (g) $y = (x^3+1)^3$

(c) $y = (x^2+3)^2$ (h) $y = (x^3-x)^2$

Do you notice a method by which the differentials can be obtained?

General rule

To obtain a **general rule** for differentiating composite functions of the form $y = f[g(x)]$, let $u = g(x)$, which gives $y = f(u)$.

By taking an increment of δx in x, increments of δu and δy will be caused in u and y respectively.

Now, since δx, δu and δy are just small quantities, it must be true that

$$\frac{\delta y}{\delta x} = \frac{\delta y}{\delta u} \times \frac{\delta u}{\delta x}$$

In the limit as $\delta x \to 0$, δu and δy both tend to zero and

$$\lim_{\delta x \to 0}\left(\frac{\delta y}{\delta x}\right) = \lim_{\substack{\delta x \to 0 \\ \delta u \to 0}}\left[\frac{\delta y}{\delta u} \times \frac{\delta u}{\delta x}\right]$$

$$= \left[\lim_{\delta u \to 0}\frac{\delta y}{\delta u}\right] \times \left[\lim_{\delta x \to 0}\frac{\delta u}{\delta x}\right] \quad \text{(see Chapter 4)}$$

Hence, $\quad \boxed{\dfrac{dy}{dx} = \dfrac{dy}{du} \times \dfrac{du}{dx}}$

This formula is applied when differentiating a composite function (function of a function) and is often called the **chain rule**.

You should now be able to check your answers in Investigation 1. For instance part **(b)**, where $y = (2x + 5)^3$.

Let $u = 2x + 5$ which gives $y = u^3$.

Now, $\dfrac{du}{dx} = 2$ and $\dfrac{dy}{du} = 3u^2$

Using the chain rule, $\dfrac{dy}{dx} = \dfrac{dy}{du} \times \dfrac{du}{dx} = 3u^2 \times 2 = 6u^2 = 6(2x + 5)^2$

Compare with,

$$y = (2x + 5)^3 = 8x^3 + 60x^2 + 150x + 125$$

$$\Rightarrow \quad \frac{dy}{dx} = 24x^2 + 120x + 150$$

$$= 6(4x^2 + 20x + 25) = 6(2x + 5)^2$$

Now that the rule has been established you can see why $\dfrac{d}{dx}(x + 1)^3 = 3(x + 1)^2$ appeared to obey the standard rule.

$$\text{If } y = (x + 1)^3, \text{ let } u = x + 1, \text{ giving } y = u^3$$

$$\therefore \frac{du}{dx} = 1 \quad \text{and} \quad \frac{dy}{du} = 3u^2$$

Using the chain rule, $\dfrac{dy}{dx} = \dfrac{dy}{du} \times \dfrac{du}{dx} = 3u^2 \times 1$

$$= 3(x + 1)^2$$

This would be true for any function in which $\dfrac{du}{dx} = 1$.

WORKED EXAMPLES

1 Find $\dfrac{dy}{dx}$ if $y = (x^2 - 5x + 7)^4$.

Let $u = x^2 - 5x + 7$, giving $y = u^4$ $\Rightarrow$ $\dfrac{du}{dx} = 2x - 5$ and $\dfrac{dy}{du} = 4u^3$

Using the chain rule, $\dfrac{dy}{dx} = \dfrac{dy}{du} \times \dfrac{du}{dx} = 4u^3 \times (2x - 5) = 4(2x - 5)(x^2 - 5x + 7)^3$

2 Find $\dfrac{dy}{dx}$ if $y = \left(\sqrt{x} - \dfrac{1}{\sqrt{x}} \right)^5$.

Let $u = \sqrt{x} - \dfrac{1}{\sqrt{x}} = x^{\frac{1}{2}} - x^{-\frac{1}{2}}$, giving $y = u^5$ $\Rightarrow$ $\dfrac{du}{dx} = \tfrac{1}{2}x^{-\frac{1}{2}} + \tfrac{1}{2}x^{-\frac{3}{2}}$ and $\dfrac{dy}{du} = 5u^4$

Using the chain rule, $\dfrac{dy}{dx} = \dfrac{dy}{du} \times \dfrac{du}{dx} = 5u^4 \times (\tfrac{1}{2}x^{-\frac{1}{2}} + \tfrac{1}{2}x^{-\frac{3}{2}})$

$$= \frac{5}{2}(x^{-\frac{1}{2}} + x^{-\frac{3}{2}})(x^{\frac{1}{2}} - x^{-\frac{1}{2}})^4$$

$$\text{or} \quad \frac{dy}{dx} = \frac{5}{2}\left[\frac{1}{\sqrt{x}} + \frac{1}{\sqrt{x^3}} \right]\left[\sqrt{x} - \frac{1}{\sqrt{x}} \right]^4$$

3 If $y = \dfrac{1}{(2-x^4)^3}$, find the gradient of the curve.

This must be written in the standard composite function form, as $y = (2 - x^4)^{-3}$.

Let $u = 2 - x^4$, giving $y = u^{-3}$

Hence, $\dfrac{du}{dx} = -4x^3$ and $\dfrac{dy}{du} = -3u^{-4}$

Using the chain rule, $\dfrac{dy}{dx} = \dfrac{dy}{du} \times \dfrac{du}{dx}$

$$= (-3u^{-4}) \times (-4x^3) = 12x^3 u^{-4} = 12x^3(2-x^4)^{-4} = \dfrac{12x^3}{(2-x^4)^4}$$

4 Differentiate $\dfrac{1}{1+\sqrt[3]{x}}$ with respect to x.

Let $y = \dfrac{1}{1+\sqrt[3]{x}} = (1 + \sqrt[3]{x})^{-1} = (1 + x^{\frac{1}{3}})^{-1}$

Let $u = 1 + x^{\frac{1}{3}}$, giving $y = u^{-1}$ $\Rightarrow$ $\dfrac{du}{dx} = \dfrac{1}{3}x^{-\frac{2}{3}}$ and $\dfrac{dy}{du} = -u^{-2}$

Using the chain rule, $\dfrac{dy}{dx} = \dfrac{dy}{du} \times \dfrac{du}{dx} = (-u^{-2}) \times \left(\dfrac{1}{3}x^{-\frac{2}{3}} \right) = -\dfrac{1}{3}x^{-\frac{2}{3}}(1 + x^{\frac{1}{3}})^{-2} = \dfrac{-1}{3\sqrt[3]{x^2}(1+\sqrt[3]{x})^2}$

Notice that in the chain rule, the result is given by the product of two differentials $\dfrac{du}{dx}$ and $\dfrac{dy}{du}$, i.e. the differential of y as though it were just a power of x and the differential of the substitution chosen for u.

This gives a rule of thumb approach which can be adopted when you have enough confidence to do so.

Thus, $\dfrac{d}{dx}(2-x^4)^{-3} = -3(2-x^4)^{-4} \times \dfrac{d}{dx}(2-x^4)$

$$= -3(2-x^4)^{-4} \times (-4x^3) = \dfrac{12x^3}{(2-x^4)^4}$$

$$\dfrac{d}{dx}(3x^2 - 8x)^7 = 7(3x^2 - 8x)^6 \times \dfrac{d}{dx}(3x^2 - 8x)$$

$$= 7(3x^2 - 8x)^6 \times (6x - 8) = 14(3x - 4)(3x^2 - 8x)^6$$

Exercise 9.1

1 If $f(x) = 1 + x^2$ and $g(x) = x^3$, find the composite function $gf(x)$ and differentiate the result with respect to x.

2 If $f(x) = x^{-1}$ and $g(x) = 1 + \sqrt{x}$, find the function $fg(x)$ and its differential with respect to x.

3 If $f(x) = 3x^2 - 6x + 1$ and $g(x) = \sqrt{x}$ find **(a)** $fg(x)$ and **(b)** $gf(x)$ and their differentials with respect to x.

4 Use the chain rule to differentiate the following functions with respect to x.

(a) $(x+2)^5$ (c) $(5-4x)^{-2}$ (e) $(3-4x)^{-\frac{1}{3}}$

(b) $(2x+3)^4$ (d) $(3x-1)^{\frac{1}{2}}$ (f) $(5x+7)^8$

5 Find $\dfrac{dy}{dx}$ if y equals,

(a) $\left(2x+\dfrac{1}{x}\right)^5$ (c) $(3x^2+4)^4$ (e) $\dfrac{1}{(x^3+1)^2}$

(b) $\left(x-\dfrac{1}{x}\right)^3$ (d) $\dfrac{1}{(1+x^2)}$ (f) $\dfrac{3}{\sqrt{4x-7}}$

6 Differentiate the following functions with respect to x.

(a) $(1+x^4)^3$ (c) $(3x^2-5x)^{\frac{2}{3}}$ (e) $(7x^2-8x+4)^{\frac{1}{2}}$

(b) $(4x^3-3x^2+2)^5$ (d) $(2x^2-4x)^{-3}$ (f) $\left(5x^3-\dfrac{1}{x^3}\right)^{\frac{3}{2}}$

7 Find $f'(x)$ if

(a) $f(x)=(4\sqrt{x}-4x)^3$ (c) $f(x)=\left(\sqrt[3]{x}+\dfrac{1}{x}\right)^{-1}$

(b) $f(x)=\left(\dfrac{1}{\sqrt{x}}+\sqrt{x}\right)^4$ (d) $f(x)=\sqrt{\left(1+\dfrac{1}{x}\right)}$

8 Differentiate with respect to x.

(a) $\dfrac{1}{(1-x^2)^3}$ (b) $\sqrt{(\sqrt{x}+3x^2)}$ (c) $\dfrac{1}{\sqrt{(x^2+1)^2+2}}$ (d) $\dfrac{2}{(1+x^{\frac{3}{2}})}$

9 If $f(x)=(3x+1)^4$, find $f'(x)$ and $f''(x)$.

10 Find the gradient of the tangent to the curve $y=\sqrt{\left(x+\dfrac{1}{x}\right)}$.

11 Find the gradient of the tangent to the curve $y=(x^2+1)^3$ at the point $(-1,8)$.

12 Find the equation of the tangent to the curve $y=\sqrt{5+x^2}$ at the point $(2,3)$.

13 Find the turning points of the curve $y=\dfrac{1}{4x-x^2}$. Hence sketch the curve.

14 Find the turning points of the curve $y=\dfrac{1}{(x^2-2x-3)}$ and hence sketch the curve.

Rates of change

If a child blows up a balloon both the balloon's volume, V, and its radius, r, will increase. The rate at which these variables change is defined by their differentials with respect to time, t.

i.e. rate of change of volume $= \dfrac{dV}{dt}$

rate of change of radius $= \dfrac{dr}{dt}$

These rates of change can be related using the chain rule.

e.g. $$\frac{dV}{dt} = \frac{dV}{dr} \times \frac{dr}{dt} \qquad\qquad \cdots\cdots(1)$$

Since the balloon can be assumed to be a sphere whose volume $V = \frac{4}{3}\pi r^3$, it is possible to differentiate to find $\dfrac{dV}{dr}$.

Hence, given $\dfrac{dr}{dt}$, we can find $\dfrac{dV}{dt}$ and vice versa.

Now $V = \frac{4}{3}\pi r^3 \quad\Rightarrow\quad \dfrac{dV}{dr} = 4\pi r^2$

Using the chain rule **(1)** above,

$$\frac{dV}{dt} = \frac{dV}{dr} \times \frac{dr}{dt} = 4\pi r^2 \times \frac{dr}{dt}$$

If the radius increases at $0.2\,\text{cm s}^{-1}$, then $\dfrac{dr}{dt} = 0.2$

Hence, $\dfrac{dV}{dt} = 4\pi r^2 \times 0.2 = 0.8\pi r^2$

The rate of change of the volume, $\dfrac{dV}{dt}$, is dependent upon the radius at a given time.

If $r = 5\,\text{cm}$, then $\dfrac{dV}{dt} = 0.8\pi(5)^2 = 20\pi = 62.84\,\text{cm}^3\,\text{s}^{-1}$.

Similarly, if a pen nib is left on a piece of blotting paper, an ink blot in the form of a circle spreads out so that its radius and area increase with time.

The rate of change of radius $= \dfrac{dr}{dt}$

The rate of change of area $= \dfrac{dA}{dt}$

These rates of change can be related by the chain rule, for example,

$$\frac{dA}{dt} = \frac{dA}{dr} \times \frac{dr}{dt} \cdots\cdots (1) \quad \text{or} \quad \frac{dr}{dt} = \frac{dr}{dA} \times \frac{dA}{dt} \cdots\cdots (2)$$

Now the area of a circle, $A = \pi r^2$ and hence $\dfrac{dA}{dr} = 2\pi r$.

Using the chain rule **(1)**

$$\frac{dA}{dt} = \frac{dA}{dr} \times \frac{dr}{dt} = 2\pi r \times \frac{dr}{dt}$$

If the radius is increasing at $0.1\,\text{cm s}^{-1}$ when $r = 3\,\text{cm}$, then the rate at which the area is increasing is

$$\frac{dA}{dt} = 2\pi r \times \frac{dr}{dt} = 2\pi(3)(0.1) = 1.89\,\text{cm}^2\,\text{s}^{-1}.$$

 WORKED EXAMPLES

1 A circular pool of water spreads so that its area increases at a rate of $2\pi\,\text{cm}^2\,\text{s}^{-1}$. Show that the rate of change of the radius of the pool is $\dfrac{1}{r}\,\text{cm s}^{-1}$, when r is the radius at time t seconds.

The question states that the rate of change of the area is $2\pi\,\text{cm}^2\,\text{s}^{-1}$, i.e. $\dfrac{dA}{dt} = 2\pi$. We need to find the rate of change of the radius i.e. $\dfrac{dr}{dt}$.

These can be related by using the chain rule in the form $\quad \dfrac{dr}{dt} = \dfrac{dr}{dA} \times \dfrac{dA}{dt}$

Since the area of a circle $A = \pi r^2$ and hence $\dfrac{dA}{dr} = 2\pi r$,

as $\dfrac{dA}{dr} = \dfrac{1}{\dfrac{dr}{dA}}$, then $\dfrac{dr}{dA} = \dfrac{1}{2\pi r}$ (see Implicit functions, later in this chapter)

$$\text{Hence,} \quad \frac{dr}{dt} = \frac{1}{2\pi r} \times 2\pi = \frac{1}{r}\,\text{cm s}^{-1}$$

The radius increases at a rate of $\dfrac{1}{r}\,\text{cm s}^{-1}$.

2 Sand is allowed to run out of the vertex of a funnel in the shape of an inverted cone of height 10 cm and base radius 2 cm. If the rate of change of the volume is $\pi/100\,\text{cm}^3\,\text{s}^{-1}$, find the rate at which the height of the sand is decreasing when it is **(a)** half way down and **(b)** three quarters of the way down the funnel.

The diagram shows the inverted cone with the sand running out at the vertex.

Let the depth of the sand be h cm at time t seconds. The surface of the sand is a circle of radius r cm.

The question states that the rate of change of volume is $-\dfrac{\pi}{100}\,\text{cm}^3\,\text{s}^{-1}$ and hence,

$$\frac{dV}{dt} = \frac{-\pi}{100}\cdots\cdots(1)$$

(The minus sign is required as the volume is decreasing.)

We require the rate of change of the depth of the sand, i.e. $\dfrac{dh}{dt}$.

These can be related by the chain rule

$$\frac{dh}{dt} = \frac{dh}{dV} \times \frac{dV}{dt}\cdots\cdots(2)$$

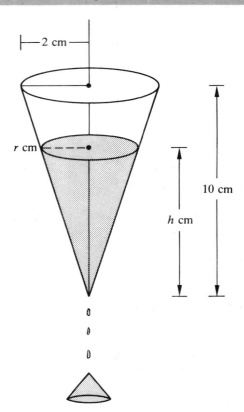

Now the sand is in the form of a cone whose volume V is given by $V = \frac{1}{3}\pi r^2 h$.

Since V is expressed in terms of two variables, r and h, it is necessary to eliminate one of them to give V as a function of h only before differentiating.

Considering the cross-section of the cone we have two similar triangles.
  Since the ratios of corresponding sides are equal,

$$\frac{r}{h} = \frac{2}{10} \quad\Rightarrow\quad r = \frac{1}{5}h$$

Hence $V = \frac{1}{3}\pi r^2 h = \frac{1}{3}\pi\left(\dfrac{h}{5}\right)^2 h \qquad\Rightarrow\qquad V = \dfrac{\pi h^3}{75}$

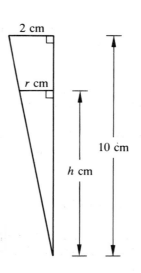

Differentiating with respect to h gives

$$\frac{dV}{dh} = \frac{3\pi h^2}{75} = \frac{\pi h^2}{25}\cdots\cdots(3)$$

Since $\dfrac{dh}{dV} = \dfrac{1}{\dfrac{dV}{dh}} = \dfrac{25}{\pi h^2}$ the chain rule **(2)** becomes

$$\frac{dh}{dt} = \frac{dh}{dV} \times \frac{dV}{dt} = \frac{25}{\pi h^2} \times \left(\frac{-\pi}{100}\right) = -\frac{1}{4h^2}$$

(The negative sign indicates that the height of the sand is decreasing.)

(a) When the sand is half way down the funnel, $h = 5\,\text{cm}$ and $\dfrac{dh}{dt} = \dfrac{-1}{(4 \times 25)} = -0.01\,\text{cm s}^{-1}$.

(b) When it is three quarters of the way down the funnel, $h = 2.5\,\text{cm}$ and $\dfrac{dh}{dt} = \dfrac{-1}{4(2.5)^2} = \dfrac{-1}{25} = -0.04\,\text{cm s}^{-1}$.

Exercise 9.2

---

1 The radius of a circle is increasing at a rate of $0.3\,\text{cm s}^{-1}$. Find the rate of increase of its area when the radius is $2\,\text{cm}$.

2 The area of a circle is increasing at a rate of $15\,\text{cm}^2\,\text{s}^{-1}$. Find the rate of increase of its radius when the radius is $5\,\text{cm}$.

3 A square of side x cm expands so that its edge increases at a rate of $3\,\text{cm s}^{-1}$. Find the rate at which its area increases when the side is $1\,\text{cm}$.

4 The side of a cube is increasing at a rate of $0.02\,\text{m s}^{-1}$. Find the rate of increase of the volume when the side is $4\,\text{m}$ in length.

5 The side of a cube decreases at a rate of $0.2\,\text{cm s}^{-1}$. Find the rate of change of its surface area when the side is $3\,\text{cm}$.

6 Find the rate of change of the volume of a sphere when the radius is $5\,\text{cm}$, if the radius is increasing at a rate of $0.5\,\text{cm s}^{-1}$.

7 If air is leaking from a balloon at a rate of $1\,\text{cm}^3\,\text{s}^{-1}$ find the rate of decrease of **(a)** the radius and **(b)** the surface area when the radius is $3\,\text{cm}$.

8 Sand falls on the ground at a rate of $200\,\text{cm}^3\,\text{s}^{-1}$ and forms a heap in the shape of a cone with its radius twice its height. Find the rate of change of the height when the height is $5\,\text{cm}$.

9 The volume of a cone is given by $V = \pi x^3$, where x cm is the radius of the base. If its volume is increasing at $9\,\text{cm}^3\,\text{s}^{-1}$, find an expression for the rate of change of the radius.

10 A ladder, $13\,\text{m}$ long, leans against a vertical wall with its foot on level ground x m from the wall. If h m is the distance of the top of the ladder from the ground, find the speed with which the top of the ladder moves if the foot of the ladder slips away from the wall at $0.2\,\text{m s}^{-1}$ when it is $5\,\text{m}$ from the base of the wall.

11 A solid cylinder is such that its height is six times its radius. Find the rate of change of its volume when the radius is $3\,\text{cm}$ if the rate of change of the radius is $1.5\,\text{cm s}^{-1}$.

12 A hemispherical bowl of radius $10\,\text{cm}$ is being filled with water such that the radius of the surface area of the water, r cm, is increasing at a rate of $0.5\,\text{cm s}^{-1}$. Find an expression for the depth of the water, h cm, in terms of r and hence find the rate at which the depth is changing when $h = 5$.

13 An inverted right-circular cone of height $10\,\text{cm}$ and radius $1\,\text{cm}$ is being filled with water at a rate of $2\,\text{cm}^3\,\text{s}^{-1}$. Find the rate at which the surface area of the water is increasing when the depth of the water is $5\,\text{cm}$.

Products and quotients

Using the chain rule to differentiate composite functions such as $(x + 3)^7$ obviated the need to multiply out brackets. A method can be developed to deal with the product of two terms; as examples, $(x^2 + 4)(x^3 - 2x^2 + 7x)$ or $(2x - 1)^3(x - 8)^7$.

Differentiation 2

In the first case you could expand the brackets and differentiate term by term, but you would hardly want to do that in the second example.

Product rule

Consider y to be the product of two terms each of which is a function of x. Thus, $y = f(x) \cdot g(x)$

$$\text{Let } f(x) = u \text{ and } g(x) = v \quad \Rightarrow \quad y = uv \dots \dots (1)$$

A small increment, δx, in x will cause increments of δu, δv and δy in u, v and y, respectively.

$$\text{Hence,} \quad y + \delta y = (u + \delta u)(v + \delta v) \dots \dots \dots \dots (2)$$

Subtracting **(1)** from **(2)**

$$\delta y = (u + \delta u)(v + \delta v) - uv$$

$$\Rightarrow \quad \delta y = uv + u\delta v + v\delta u + \delta u \, \delta v - uv$$

$$\Rightarrow \quad \delta y = u\delta v + v\delta u + \delta u \, \delta v$$

Dividing by δx,

$$\frac{\delta y}{\delta x} = u\frac{\delta v}{\delta x} + v\frac{\delta u}{\delta x} + \delta u\left(\frac{\delta v}{\delta x}\right)$$

In the limit as $\delta x \rightarrow 0$, δu, δv and δy all tend to zero.

$$\text{Hence,} \quad \lim_{\delta x \to 0}\left(\frac{\delta y}{\delta x}\right) = \lim_{\delta x \to 0}\left(u\frac{\delta v}{\delta x}\right) + \lim_{\delta x \to 0}\left(v\frac{\delta u}{\delta u}\right) + \lim_{\delta x \to 0}\left(\delta u \cdot \frac{\delta v}{\delta x}\right)$$

The last limit contains a separate δu and hence tends to zero.

$$\text{Hence,} \quad \boxed{\frac{dy}{dx} = u\frac{dv}{dx} + v\frac{du}{dx}}$$

This formula, called the **product rule**, gives a method for differentiating the product of two terms. Notice the pattern of the right-hand side:

The 1st term times the differential of the 2nd term added to the 2nd term times the differential of the 1st term.

We can now differentiate the examples at the beginning of this section.

WORKED EXAMPLES

1 Differentiate $y = (x^2 + 4)(x^3 - 2x^2 + 7x)$ with respect to x.

$$\text{Let} \quad u = x^2 + 4 \qquad \Rightarrow \quad \frac{du}{dx} = 2x$$

$$\text{and} \quad v = x^3 - 2x^2 + 7x \quad \Rightarrow \quad \frac{dv}{dx} = 3x^2 - 4x + 7$$

158

Using the product rule.

$$\text{If} \quad y = uv, \text{ then} \quad \frac{dy}{dx} = u\frac{dv}{dx} + v\frac{du}{dx}$$

$$\Rightarrow \quad \frac{dy}{dx} = (x^2 + 4)(3x^2 - 4x + 7) + (x^3 - 2x^2 + 7x)2x$$

$$\Rightarrow \quad \frac{dy}{dx} = 5x^4 - 8x^3 + 33x^2 - 16x + 28$$

In this case the product rule does not give a great advantage over expanding the original brackets.

$$y = (x^2 + 4)(x^3 - 2x^2 + 7x)$$

$$= x^5 - 2x^4 + 11x^3 - 8x^2 + 28x$$

$$\Rightarrow \quad \frac{dy}{dx} = 5x^4 - 8x^3 + 33x^2 - 16x + 28$$

However, the product rule saves pages of expansion in this next example because its use can be combined with the use of the chain rule.

2 Differentiate $y = (2x - 1)^3(x - 8)^7$ with respect to x.

$$\text{Let} \quad u = (2x - 1)^3 \quad \Rightarrow \quad \frac{du}{dx} = 3(2x - 1)^2(2) = 6(2x - 1)^2$$

$$\text{and} \quad v = (x - 8)^7 \quad \Rightarrow \quad \frac{dv}{dx} = 7(x - 8)^6(1) = 7(x - 8)^6$$

Note that $\dfrac{du}{dx}$ and $\dfrac{dv}{dx}$ are differentiated using the chain rule, if possible, using the rule of thumb discussed earlier in this chapter.

Using the product rule.

$$\text{If} \quad y = uv, \quad \text{then} \quad \frac{dy}{dx} = u\frac{dv}{dx} + v\frac{du}{dx}$$

$$\Rightarrow \quad \frac{dy}{dx} = (2x - 1)^3[7(x - 8)^6] + (x - 8)^7[6(2x - 1)^2]$$

The result should be simplified, if possible, by looking for common factors.

$$\text{Hence,} \quad \frac{dy}{dx} = 7(2x - 1)^3(x - 8)^6 + 6(x - 8)^7(2x - 1)^2$$

$$\Rightarrow \quad \frac{dy}{dx} = (2x - 1)^2(x - 8)^6[7(2x - 1) + 6(x - 8)]$$

$$= (2x - 1)^2(x - 8)^6[20x - 55]$$

$$\text{Hence,} \quad \frac{dy}{dx} = 5(2x - 1)^2(x - 8)^6(4x - 11)$$

3 Find $\dfrac{dy}{dx}$ if $y = \sqrt{(x^2 + 1)(1 - x)^3}$

Now, $y = (x^2 + 1)^{\frac{1}{2}}(1 - x)^{\frac{3}{2}}$

Let $u = (x^2 + 1)^{\frac{1}{2}}$ $\Rightarrow$ $\dfrac{du}{dx} = \dfrac{1}{2}(x^2 + 1)^{-\frac{1}{2}}(2x) = x(x^2 + 1)^{-\frac{1}{2}}$

and $v = (1 - x)^{\frac{3}{2}}$ $\Rightarrow$ $\dfrac{dv}{dx} = \frac{3}{2}(1 - x)^{\frac{1}{2}}(-1) = -\frac{3}{2}(1 - x)^{\frac{1}{2}}$

Using the product rule.

$$\text{If } \quad y = uv, \quad \text{then} \quad \frac{dy}{dx} = u\frac{dv}{dx} + v\frac{du}{dx}$$

$$\Rightarrow \quad \frac{dy}{dx} = (x^2 + 1)^{\frac{1}{2}}\left[-\frac{3}{2}(1 - x)^{\frac{1}{2}} \right] + (1 - x)^{\frac{3}{2}}[x(x^2 + 1)^{-\frac{1}{2}}]$$

$$= -\frac{3}{2}(x^2 + 1)^{\frac{1}{2}}(1 - x)^{\frac{1}{2}} + x(1 - x)^{\frac{3}{2}}(x^2 + 1)^{-\frac{1}{2}}$$

To simplify, remove the lowest power of $(x^2 + 1)$ and $(1 - x)$ i.e. $(x^2 + 1)^{-\frac{1}{2}}$ and $(1 - x)^{\frac{1}{2}}$.

$$\Rightarrow \quad \frac{dy}{dx} = (x^2 + 1)^{-\frac{1}{2}}(1 - x)^{\frac{1}{2}}\left[-\frac{3}{2}(x^2 + 1) + x(1 - x) \right]$$

$$= \frac{1}{2}(x^2 + 1)^{-\frac{1}{2}}(1 - x)^{\frac{1}{2}}[-3x^2 - 3 + 2x - 2x^2]$$

$$= \frac{1}{2}(x^2 + 1)^{-\frac{1}{2}}(1 - x)^{\frac{1}{2}}(-5x^2 + 2x - 3)$$

Hence, $\dfrac{dy}{dx} = \dfrac{-(5x^2 + 2x + 3)}{2}\sqrt{\dfrac{1 - x}{x^2 + 1}}$

4 Differentiate $y = \dfrac{3x}{1 - x^2}$ with respect to x.

The product rule can be used if the function is written as $y = 3x(1 - x^2)^{-1}$

Let $u = 3x$ $\Rightarrow$ $\dfrac{du}{dx} = 3$

and $v = (1 - x^2)^{-1}$ $\Rightarrow$ $\dfrac{dv}{dx} = -(1 - x^2)^{-2}(-2x) = 2x(1 - x^2)^{-2}$

Using the product rule.

$$\text{If } \quad y = uv, \quad \text{then} \quad \frac{dy}{dx} = u\frac{dv}{dx} + v\frac{du}{dx}$$

$$\Rightarrow \quad \frac{dy}{dx} = 3x[2x(1 - x^2)^{-2}] + (1 - x^2)^{-1}(3)$$

$$= 6x^2(1 - x^2)^{-2} + 3(1 - x^2)^{-1}$$

Hence, $\dfrac{dy}{dx} = 3(1 - x^2)^{-2}[2x^2 + (1 - x^2)] = \dfrac{3(1 + x^2)}{(1 - x^2)^2}$

Quotient rule

Investigation 2

In the last Worked Example, y was a quotient of two function of x.

So, if $y = \dfrac{f(x)}{g(x)}$ and if $u = f(x)$ and $v = g(x)$, then $y = \dfrac{u}{v}$.

By allowing x to have an increment of δx, use the method of the previous section to find an expression for δy in terms of $u, v, \delta u$ and δv. Hence form $\dfrac{\delta y}{\delta x}$ and find the limit as $\delta x \to 0$.

You should find that the result for $\dfrac{dy}{dx}$ is given by

$$\frac{dy}{dx} = \frac{v\dfrac{du}{dx} - u\dfrac{dv}{dx}}{v^2}$$

This is known as the **quotient formula**, and it is used for differentiating a quotient in the form $\dfrac{u}{v}$ where u and v are functions of x.

Notice that the order of the terms in the numerator is important because of the negative sign.

WORKED EXAMPLES

1 Repeat the last Worked Example, i.e. differentiate $y = \dfrac{3x}{1-x^2}$ with respect to x, using the quotient formula.

Let $u = 3x \quad \Rightarrow \quad \dfrac{du}{dx} = 3$

and $v = 1 - x^2 \quad \Rightarrow \quad \dfrac{dv}{dx} = -2x$

Using the quotient rule.

$$\text{If} \quad y = \frac{u}{v}, \quad \text{then} \quad \frac{dy}{dx} = \frac{v\dfrac{du}{dx} - u\dfrac{dv}{dx}}{v^2}$$

$$\text{Hence,} \quad \frac{dy}{dx} = \frac{(1-x^2)3 - 3x(-2x)}{(1-x^2)^2} = \frac{3 - 3x^2 + 6x^2}{(1-x^2)^2} = \frac{3(1+x^2)}{(1-x^2)^2}$$

2 Differentiate $y = \sqrt{\dfrac{x-2}{x^2+1}}$ with respect to x.

Let $u = \sqrt{x-2} = (x-2)^{\frac{1}{2}} \quad \Rightarrow \quad \dfrac{du}{dx} = \dfrac{1}{2}(x-2)^{-\frac{1}{2}}(1) = \dfrac{1}{2}(x-2)^{-\frac{1}{2}}$

and $v = \sqrt{x^2+1} = (x^2+1)^{\frac{1}{2}} \quad \Rightarrow \quad \dfrac{dv}{dx} = \dfrac{1}{2}(x^2+1)^{-\frac{1}{2}}(2x) = x(x^2+1)^{-\frac{1}{2}}$

Differentiation 2

Differentiation 2

Using the quotient rule.

$$\text{If} \quad y = \frac{u}{v}, \quad \text{then} \quad \frac{dy}{dx} = \frac{v\dfrac{du}{dx} - u\dfrac{dv}{dx}}{v^2}$$

$$\Rightarrow \quad \frac{dy}{dx} = \frac{(x^2+1)^{\frac{1}{2}}[\frac{1}{2}(x-2)^{-\frac{1}{2}}] - (x-2)^{\frac{1}{2}}[x(x^2+1)^{-\frac{1}{2}}]}{(x^2+1)}$$

$$= \frac{\frac{1}{2}(x^2+1)^{\frac{1}{2}}(x-2)^{-\frac{1}{2}} - x(x-2)^{\frac{1}{2}}(x^2+1)^{-\frac{1}{2}}}{(x^2+1)}$$

$$= \frac{(x^2+1)^{-\frac{1}{2}}(x-2)^{-\frac{1}{2}}[(x^2+1) - 2x(x-2)]}{2(x^2+1)}$$

$$= \frac{(x-2)^{-\frac{1}{2}}(1+4x-x^2)}{2(x^2+1)^{\frac{3}{2}}} = \frac{1+4x-x^2}{2\sqrt{(x-2)(x^2+1)^3}}$$

Exercise 9.3

---

1 Differentiate the following functions with respect to x by the product rule, simplifying your answers. Check by expanding the functions before differentiating.

    (a) $x^2(x-4)$ **(b)** $(x^3+4)(3x-5)$ **(c)** $(3x+1)^2(2x-1)$

Differentiate the functions, in questions **2** to **13**, with respect to x, simplifying your answers.

2 $(x+4)(x^3+6x^2)$	**6** $(x+1)^4(3x-5)^7$	**10** $(1-\sqrt{x})^2(1+\sqrt{x})^3$
3 $(x^2-7)(x-1)^2$	**7** $(1+x^2)(1-x^2)^2$	**11** $(2x+3)^3(4-x)^4$
4 $x^2(x+5)^2$	**8** $(x^2+1)\sqrt{(1-x^3)}$	**12** $\sqrt{(1+x^2)(1-x^2)}$
5 $(x+2)^3(x-1)^2$	**9** $x^2\sqrt{(1+4x)^3}$	**13** $\sqrt{(2x+1)^3(x-1)^5}$

14 Differentiate the following functions with respect to x by **(i)** using the quotient rule and **(ii)** using the product rule.

    (a) $\dfrac{2x}{3x+1}$ **(b)** $\dfrac{4}{x+5}$ **(c)** $\dfrac{x-1}{x+5}$ **(d)** $\dfrac{x^2}{x^3+1}$

Differentiate the functions in questions **15** to **26**, with respect to x, simplifying your answers.

15 $\dfrac{x}{x+1}$	**18** $\dfrac{x^2+1}{x^2-1}$	**21** $\dfrac{\sqrt{x}}{\sqrt{x+2}}$	**24** $\dfrac{(2x-1)^3}{2x+1}$
16 $\dfrac{x-2}{x+3}$	**19** $\dfrac{x^2+2x-1}{3-x}$	**22** $\dfrac{x^2-3x}{\sqrt{2x+1}}$	**25** $\sqrt{\dfrac{(1+x^2)^5}{1-x^2}}$
17 $\dfrac{1+x^2}{x-1}$	**20** $\dfrac{3x^2}{(x-1)^4}$	**23** $\sqrt{\dfrac{x+1}{x+5}}$	**26** $\dfrac{(3x^2+1)^3}{(4x^3-7)^2}$

27 Find the gradient of the curve $y = (x-1)(x+2)^2$ and hence find the co-ordinates of the turning points, distinguishing between them. Sketch the curve.

28 Find the co-ordinates of the points on the curve $y = (x+3)(2x-1)^2$ at which the gradient of the tangent is 5.

29 If $y = \dfrac{1}{1+x^2}$, find $\dfrac{dy}{dx}$ and $\dfrac{d^2y}{dx^2}$.

30 Show that the curve $y = \dfrac{x}{1+x}$ does not have any turning points.

31 Find the equation of the tangent to the curve $y = (3x+2)^2(x-1)^3$, at the point where $x = -1$.

32 Find $\dfrac{d}{dx}[(x+3)^2(x-1)]$ and use your result to differentiate $(x+3)^2(x-1)(2x+1)^3$ with respect to x.

Implicit functions

Functions such as $y = x^2$ and $y = (x^2+3)(x-1)^4$ define y solely in terms of x and are called **explicit functions**.

However, y may be expressed **implicitly** by a relation such as $xy = 2y - 1$, where neither variable is expressed only in terms of the other.

Sometimes the implicit function can be rearranged to give an explicit function of x or y, but not always.

$$xy = 2y - 1 \iff x = \frac{2y-1}{y} \quad \text{giving } x \text{ explicitly in terms of } y$$

or $\quad xy = 2y - 1 \iff xy - 2y = -1 \iff y(2-x) = 1$

$$\iff y = \frac{1}{2-x} \quad \text{giving } y \text{ explicitly in terms of } x.$$

However, $x^2 + xy + y^2 = 5$ cannot be rearranged to give x in terms of y or y in terms of x. Note that, strictly, $x^2 + xy + y^2 = 5$ is not a function since, by definition, a function assigns only one value to y for each value of x. However, this does not affect the differentiation of this equation.

If x **is an explicit function of** y, then $x = f(y)$ and a change of δy in y will cause a change of δx in x.

Since $\delta x, \delta y$ are only small finite quantities,

$$\frac{\delta x}{\delta y} = 1 \times \frac{\delta x}{\delta y} = 1 \div \frac{\delta y}{\delta x}$$

Now, as $\delta x \to 0$, $\delta y \to 0$ and so in the limit as $\delta x \to 0$

$$\lim_{\delta y \to 0} \left(\frac{\delta x}{\delta y} \right) = \lim_{\delta x \to 0} (1) \div \lim_{\delta x \to 0} \left(\frac{\delta y}{\delta x} \right)$$

Hence, $\qquad \boxed{\dfrac{dx}{dy} = 1 \Big/ \left(\dfrac{dy}{dx} \right) \quad \text{or} \quad \dfrac{dy}{dx} = 1 \Big/ \left(\dfrac{dx}{dy} \right)}$

> WORKED EXAMPLES

1 Find $\dfrac{dy}{dx}$ if $x = y^3$.

Method 1 Differentiate with respect to y.

$$x = y^3 \implies \frac{dx}{dy} = 3y^2 \quad \text{and} \quad \frac{dy}{dx} = 1 \Big/ \frac{dx}{dy} = \frac{1}{3y^2}$$

Method 2   Rearrange $x = y^3$ to give $y = x^{\frac{1}{3}}$ and differentiate with respect to x.

$$y = x^{\frac{1}{3}} \quad \Rightarrow \quad \frac{dy}{dx} = \frac{1}{3}x^{-\frac{2}{3}} = \frac{1}{3x^{\frac{2}{3}}} = \frac{1}{3(x^{\frac{1}{3}})^2} = \frac{1}{3y^2}$$

Method 3   Differentiate with respect to x.

$$x = y^3 \quad \Rightarrow \quad \frac{d}{dx}(x) = \frac{d}{dx}(y^3) \quad \Rightarrow \quad 1 = \frac{d}{dx}(y^3)\ldots\ldots\ldots(1)$$

To find $\dfrac{d}{dx}(y^3)$ use the chain rule.

Let $z = y^3 \quad \Rightarrow \quad \dfrac{dz}{dx} = \dfrac{dz}{dy} \times \dfrac{dy}{dx} = 3y^2\dfrac{dy}{dx}$ Hence, $\dfrac{d}{dx}(y^3) = 3y^2\dfrac{dy}{dx}$

Substituting in **(1)** $1 = 3y^2\dfrac{dy}{dx} \quad \Leftrightarrow \quad \dfrac{dy}{dx} = \dfrac{1}{3y^2}$

---

2   If $y^3 - 3x^2y = 6x$ find $\dfrac{dy}{dx}$ and show that $(y^2 - x^2)\dfrac{d^2y}{dx^2} + 2y\left(\dfrac{dy}{dx}\right)^2 - 4x\dfrac{dy}{dx} = 2y$

If neither x nor y can be expressed as an explicit function, we differentiate both sides of the equation with respect to x. This is called **differentiating implicitly**.

Differentiate implicitly with respect to x.

$$\frac{d}{dx}(y^3) - \frac{d}{dx}(3x^2y) = \frac{d}{dx}(6x)$$

$$\Rightarrow \quad 3y^2\frac{dy}{dx} - \left[3x^2\frac{dy}{dx} + 6xy\right] = 6$$

Note that $\dfrac{d}{dx}(3x^2y)$ is differentiated as a product of $3x^2$ and y.

$$\Rightarrow \quad (3y^2 - 3x^2)\frac{dy}{dx} = 6xy + 6$$

$$\Rightarrow \quad (y^2 - x^2)\frac{dy}{dx} = 2xy + 2\ldots\ldots\ldots(1)$$

If $y \neq x$, $\dfrac{dy}{dx} = \dfrac{2xy + 2}{y^2 - x^2}$

Differentiating **(1)** implicitly with respect to x.

$$\frac{d}{dx}\left[(y^2 - x^2)\frac{dy}{dx}\right] = \frac{d}{dx}(2xy) + \frac{d}{dx}(2)$$

$$\Rightarrow \quad (y^2 - x^2)\frac{d^2y}{dx^2} + \left[2y\frac{dy}{dx} - 2x\right]\frac{dy}{dx} = 2x\frac{dy}{dx} + 2y$$

Note that the first term is differentiated as a product of $(y^2 - x^2)$ and $\dfrac{dy}{dx}$.

$$\Rightarrow \quad (y^2 - x^2)\frac{d^2y}{dx^2} + 2y\left(\frac{dy}{dx}\right)^2 - 4x\frac{dy}{dx} = 2y$$

Investigation 3

In Chapter 4 we saw that $\frac{d}{dx}(ax^n) = nax^{n-1}$ for positive integral values of n. It is now possible to use the methods for differentiating composite functions and implicit functions to show that the result is also true if n is **(a)** negative and **(b)** fractional.

(a) Let $y = ax^n = ax^{-m}$ where m is positive.

Let $u = x^{-1}$, giving $y = au^m$. Form $\frac{du}{dx}$ and $\frac{dy}{du}$ and hence show that $\frac{dy}{dx} = -amx^{-m-1} = nax^{n-1}$.

(b) Let $y = ax^n = ax^{p/q}$ where p and q are integers. If $y = ax^{p/q}$, then $y^q = a^q x^p$. Differentiate implicitly with respect to x to show that $\frac{dy}{dx} = a^q(px^{p-1})/(qy^{q-1})$. Show that the right-hand side is equal to anx^{n-1}.

Investigation 4

The result in the section on implicit functions, gave $\frac{dy}{dx} = 1 / \left(\frac{dx}{dy}\right)$. This cannot be extended to second order differentials and $\frac{d^2y}{dx^2} \neq 1 / \left(\frac{d^2x}{dy^2}\right)$.

Consider $y = x^4$ and find $\frac{dy}{dx}$ and $\frac{d^2y}{dx^2}$ in terms of x.

Write $y = x^4$ as $x = y^{\frac{1}{4}}$ and show that $\frac{dx}{dy} = \frac{1}{4x^3}$. Find also $\frac{d^2x}{dy^2}$ and show that this is equal to $-\frac{3}{16x^7}$.

Deduce that, in this case, $\frac{d^2y}{dx^2} \neq 1 / \left(\frac{d^2x}{dy^2}\right)$ and hence **cannot** be true.

Exercise 9.4

1 Differentiate the following   **(a)** implicitly with respect to x **(b)** by expressing y explicitly in terms of x and   **(c)** by expressing x explicitly in terms of y. Show that your results for **(a)**, **(b)** and **(c)** are equivalent.

(i) $xy = 2$ **(ii)** $y^2 = 4x$ **(iii)** $x^2 + y^2 = 1$

2 Find $\frac{dy}{dx}$ in terms of x and y if,

(a) $3x^2 + 4y^2 = 8$ **(c)** $x^2y + y^2 = x(x+1)^2$
(b) $x^2 + y^2 - 3x - 4y + 5 = 0$ **(d)** $x^3 + y^3 = xy^2 + 3x$

3 Find the gradient of the following curves at the given points.

(a) $x^2 + y^2 = 4$ at $(1, \sqrt{3})$ **(b)** $2x^2 - 5y^2 = 3xy$ at $(-1, 1)$

4 At what points are the tangents to the curve $x^2 + y^2 - 3y - 4x = 0$ parallel to the x-axis?

5 Find the maximum and minimum values of y if $x^2 - 2y^2 + 6x - 3y + 18 = 0$.

6 Find the equation of the tangent to the curve $2x^2 + y^2 = 11$ at the point $(1, 3)$.

7 Find the equation of the tangent and normal to the curve $x^2 - 2y^2 = 8$ at $(4, 2)$.

8 Show that the gradient of the locus given by $x^2 + 2xy + y^2 = 5$ is constant. Explain the significance of this.

9 If $x^2 + 3xy + y^2 = 7$, find $\dfrac{dy}{dx}$ in terms of x and y and show that $(3x + 2y)\dfrac{d^2 y}{dx^2} + 2\left(\dfrac{dy}{dx}\right)^2 + 6\left(\dfrac{dy}{dx}\right) + 2 = 0$.

10 If $y\sqrt{1 + x^2} = x$, show that $\sqrt{1 + x^2}\,\dfrac{dy}{dx} + y^2 = 1$. Hence show that $\sqrt{1 + x^2}\,\dfrac{d^2 y}{dx^2} + 3y\dfrac{dy}{dx} = 0$

Parametric differentiation

Sometimes the Cartesian equation of a curve may be awkward and it is better to express x and y in terms of a third variable, say t or θ. The curve is then defined by two equations $x = f(t)$ and $y = g(t)$.

For example, $x = t^2$ and $y = 2t$.

To sketch the curve, select values of t to give particular points (x, y) on the curve.

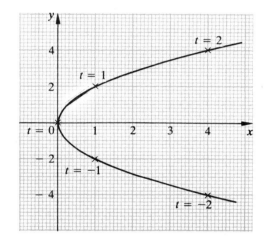

If $t = 0$ $\Rightarrow$ $(0, 0)$
If $t = 1$ $\Rightarrow$ $(1, 2)$
If $t = 2$ $\Rightarrow$ $(4, 4)$
If $t = -1$ $\Rightarrow$ $(1, -2)$

By eliminating the variable t we can find the Cartesian equation.

i.e. $t = \dfrac{y}{2}$ $\Rightarrow$ $x = \left(\dfrac{y}{2}\right)^2$ $\Leftrightarrow$ $y^2 = 4x$

which is the parabola shown in the figure.

The equation chosen here, $y^2 = 4x$, is hardly complicated, yet it is still sometimes useful to work in terms of an arbitrary third variable which is called a **parameter**.

To differentiate parametric equations, form $\dfrac{dx}{dt}$ and $\dfrac{dy}{dt}$ and use the chain rule.

Hence, $x = t^2$ $\Rightarrow$ $\dfrac{dx}{dt} = 2t$ $\Rightarrow$ $\dfrac{dt}{dx} = \dfrac{1}{2t}$

and $y = 2t$ $\Rightarrow$ $\dfrac{dy}{dt} = 2$

Using the chain rule, $\dfrac{dy}{dx} = \dfrac{dy}{dt} \times \dfrac{dt}{dx} = 2 \times \dfrac{1}{2t} = \dfrac{1}{t}$

Notice that the result will be in terms of the parameter, t. The gradient of the curve at $(4, 4)$ (i.e. where $t = 2$) will be $\frac{1}{2}$.

Care must be taken with the second differential in parametric form, as $\dfrac{dy}{dx}$ will be a function of t.

Hence,
$$\frac{d^2y}{dx^2}=\frac{d}{dx}\left(\frac{dy}{dx}\right)=\frac{d}{dt}\left(\frac{dy}{dx}\right)\times\frac{dt}{dx}$$

So if $\dfrac{dy}{dx}=\dfrac{1}{t}$ then $\dfrac{d^2y}{dx^2}=\dfrac{d}{dt}\left(\dfrac{1}{t}\right)\times\dfrac{dt}{dx}=\dfrac{-1}{t^2}\times\dfrac{1}{2t}=\dfrac{-1}{2t^3}$

WORKED EXAMPLE

Find $\dfrac{dy}{dx}$ and $\dfrac{d^2y}{dx^2}$ in terms of t, if $x=\dfrac{2t}{1+t^2}$ and $y=\dfrac{1-t^2}{1+t^2}$.

Differentiate x and y with respect to t by the quotient rule.

$$x=\frac{2t}{1+t^2}\quad\Rightarrow\quad\frac{dx}{dt}=\frac{(1+t^2)2-2t(2t)}{(1+t^2)^2}=\frac{2(1-t^2)}{(1+t^2)^2}$$

$$y=\frac{1-t^2}{1+t^2}\quad\Rightarrow\quad\frac{dy}{dt}=\frac{(1+t^2)(-2t)-(1-t^2)2t}{(1+t^2)^2}=\frac{-4t}{(1+t^2)^2}$$

Using the chain rule.

$$\frac{dy}{dx}=\frac{dy}{dt}\times\frac{dt}{dx}=\frac{-4t}{(1+t^2)^2}\times\frac{(1+t^2)^2}{2(1-t^2)}=\frac{-2t}{1-t^2}$$

Now, $\dfrac{d^2y}{dx^2}=\dfrac{d}{dx}\left(\dfrac{dy}{dx}\right)=\dfrac{d}{dt}\left(\dfrac{dy}{dx}\right)\times\dfrac{dt}{dx}=\dfrac{d}{dt}\left(\dfrac{-2t}{1-t^2}\right)\times\dfrac{dt}{dx}$

$$=\left[\frac{(1-t^2)(-2)-(-2t)(-2t)}{(1-t^2)^2}\right]\left[\frac{(1+t^2)^2}{2(1-t^2)}\right]$$

$$=\frac{(-2-2t^2)}{(1-t^2)^2}\times\frac{(1+t^2)^2}{2(1-t^2)}=-\frac{(1+t^2)^3}{(1-t^2)^3}.$$

Investigation 5

Using the equations in the above Worked Example, $x=\dfrac{2t}{1+t^2}$ and $y=\dfrac{1-t^2}{1+t^2}$, express t^2 in terms of y and substitute to find x in terms of y. Differentiate this Cartesian equation implicitly with respect to x to find its gradient. Verify that $\dfrac{dy}{dx}=-\dfrac{x}{y}$ and show that this is equivalent to $\dfrac{-2t}{1-t^2}$. Sketch the curve.

Exercise 9.5

1 Find $\dfrac{dy}{dx}$ if in terms of t if,

(a) $x=3t+1,\quad y=t^2-5$ (c) $x=1+t,\quad y=t^3$ (e) $x=(t+1)^2,\quad y=2t$

(b) $x=t+\dfrac{1}{t},\quad y=2t$ (d) $x=1-\dfrac{2}{t},\quad y=4t$ (f) $x=t(t+2),\quad y=t+2$

2 If $x=3t^2$, $y=6t$ are the parametric equations of a curve, find $\dfrac{dy}{dx}$ and $\dfrac{d^2y}{dx^2}$ in terms of t.

3 Find $\dfrac{dy}{dx}$ and $\dfrac{d^2y}{dx^2}$ in terms of t if,

    (a) $x = t^2$, $y = 1 - t^3$ **(b)** $x = 2 - t$, $y = t^2 - 1$

4 Using the equations in question **3**, eliminate t to find the Cartesian equations of the curve. Hence find the gradients and verify that the results are equivalent to those in question **3**.

5 Sketch the graphs of the following curves and find the gradient in terms of t.

    (a) $x = t + 1$, $y = \dfrac{1}{t}$ **(b)** $x = \dfrac{t}{1+t}$, $y = \dfrac{1}{1+t}$ **(c)** $x = t^2 - 2$, $y = t^4 - 1$

6 If $x = \dfrac{1}{\sqrt{1+t^2}}$ and $y = \dfrac{t}{\sqrt{1+t^2}}$, find $\dfrac{dy}{dx}$ in terms of t. Find the Cartesian equation and differentiate implicitly to obtain the gradient.

7 Find the gradient of the curve $x = 3t$, $y = 3/t$ in terms of t. Using this value of the gradient, find the equation of the tangent to the curve at the point $(3t, 3/t)$.

8 The parametric equations of a curve are $x = 1 - t^2$ and $y = t - t^3$. Find the gradient of the tangent in terms of t and hence find the co-ordinates of any turning points.

Small changes

The gradient of the tangent to a curve is defined as the limit of the gradient of the chord joining neighbouring points $P(x, y)$ and $Q(x + \delta x, y + \delta y)$ as $\delta x \to 0$.

$$\text{i.e. } \lim_{\delta x \to 0} \left(\frac{\delta y}{\delta x} \right) = \frac{dy}{dx}$$

So, if δx is small, if follows that $\dfrac{\delta y}{\delta x}$ is approximately equal to $\dfrac{dy}{dx}$, and we write $\dfrac{\delta y}{\delta x} \simeq \dfrac{dy}{dx}$.

$$\text{Hence,} \quad \boxed{\delta y \simeq \left(\frac{dy}{dx} \right) \delta x}$$

This result enables the change in one variable to be found, given the change in another variable, so long as it is possible to find $\dfrac{dy}{dx}$ from a relation between x and y.

WORKED EXAMPLE

Find the change in the area of a circle if the radius increases from 4 cm to 4.1 cm.

The area of a circle, $A = \pi r^2$ $\Rightarrow$ $\dfrac{dA}{dr} = 2\pi r$

$$\text{Hence,} \quad \delta A \simeq \left(\frac{dA}{dr} \right) \delta r = 2\pi r\, \delta r$$

When $r = 4$ and the change in the radius $\delta r = 0.1$ cm,

$$\delta A \simeq 2\pi(4)(0.1) = 2.51 \text{ cm}^2 \text{ (to two decimal places)}$$

Investigation 6

The result above giving δA as 2.51 cm² is approximate. If the radii before and after the increase are r cm and $(r + \delta r)$ cm, show that the true change in area is given by $\delta A = 2\pi r \, \delta r + \pi(\delta r)^2$.

Using $r = 4$ and $\delta r = 0.1$, find the true change in area and find the percentage error caused by the approximation.

WORKED EXAMPLES

1 An error of 2% is made in measuring the side of a cube. Find the percentage error in the volume.

Let the side of the cube be x cm and the volume V cm³.

Hence $V = x^3 \quad \Rightarrow \quad \dfrac{dV}{dx} = 3x^2$

Now, $\delta V \simeq \left(\dfrac{dV}{dx}\right)\delta x = 3x^2 \, \delta x$

Since a 2% error is made in measuring the side, the actual error, δx in x, is given by $\delta x = \dfrac{2}{100}x$.

$$\therefore \; \delta V \simeq 3x^2\left(\frac{2x}{100}\right) = \frac{6x^3}{100}$$

The percentage error in the volume $= \dfrac{\text{actual error}}{\text{true volume}} \times 100\%$.

$$= \frac{\delta V}{V} \times 100\% \simeq \frac{6x^3}{100} \times \frac{100}{V} = \frac{6x^3}{x^3} = 6\%$$

2 Find an approximation to $\sqrt[3]{8.02}$.

Since $\sqrt[3]{8} = 2$, the above method can be used since $\sqrt[3]{8.02}$ will not differ much from 2.

Let $y = \sqrt[3]{x} = x^{\frac{1}{3}} \quad \Rightarrow \quad \dfrac{dy}{dx} = \dfrac{1}{3}x^{-\frac{2}{3}} = \dfrac{1}{3(\sqrt[3]{x})^2}$

Hence, $\delta y \simeq \left(\dfrac{dy}{dx}\right)\delta x = \dfrac{1}{3(\sqrt[3]{x})^2}\,\delta x$

If $x = 8$ and $\delta x = 0.02$, $\quad \delta y \simeq \dfrac{1}{3(\sqrt[3]{8})^2} \times 0.02 = \dfrac{0.02}{12} = 0.0016$

$$\text{Hence,} \quad \sqrt[3]{8.02} \simeq 2.0017$$

Exercise 9.6

1 If $y = x^2 - 3x$, find the approximate change in y when x increases from 2 to 2.01.

2 If $y = x^4$, find the approximate decrease in y if x decreases from 2 to 1.96.

3 If $y = x^2(x - 1)$, find the approximate change in y if the value of x increases by 3%.

4 If $v = \dfrac{4}{t}$, find the percentage change in v if t increases by 2%.

5 If the surface area of a sphere is $4\pi r^2$ find **(a)** the approximate change in area and **(b)** the percentage change in area if the radius increases by 3%.

6 The volume of a sphere is $\frac{4}{3}\pi r^3$. Find the approximate change in volume if the radius increases from 3 cm to 3.01 cm. Calculate the actual volume at both radii and the actual change in volume. Find the percentage error in this change if the approximation is used rather than the true value.

7 Find approximate values for,

   (a) $\sqrt{4.02}$ **(b)** $\sqrt[3]{27.08}$ **(c)** $\sqrt[4]{16.3}$ **(d)** $\sqrt[3]{1004}$

8 The height of a solid cylinder is twice its radius. Find the approximate change in **(a)** the volume and **(b)** the surface area of the cylinder, if the radius is increased by 2%.

10 Trigonometry 2

Secant, cosecant, cotangent

Definitions

The secant (sec), cosecant (cosec) and cotangent (cot) of any angle θ are defined by

$$\sec\theta = \frac{1}{\cos\theta}, \quad \operatorname{cosec}\theta = \frac{1}{\sin\theta} \quad \text{and} \quad \cot\theta = \frac{1}{\tan\theta} = \frac{\cos\theta}{\sin\theta}$$

In $\triangle ABC$,

$$\cos\theta = \frac{c}{b} \qquad \sin\theta = \frac{a}{b} \qquad \tan\theta = \frac{a}{c}$$

$$\sec\theta = \frac{b}{c} \qquad \operatorname{cosec}\theta = \frac{b}{a} \qquad \cot\theta = \frac{c}{a}$$

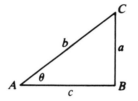

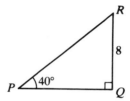

From $\triangle PQR$,

$$\frac{8}{PR} = \sin 40° \quad \Rightarrow \quad PR = \frac{8}{\sin 40°} = 8\operatorname{cosec} 40° = 12.44$$

$$\frac{8}{PQ} = \tan 40° \quad \Rightarrow \quad PQ = \frac{8}{\tan 40°} = 8\cot 40° = 9.53$$

$$\frac{PQ}{PR} = \cos 40° \quad \Rightarrow \quad PR = \frac{9.53}{\cos 40°} = 9.53\sec 40° = 12.44$$

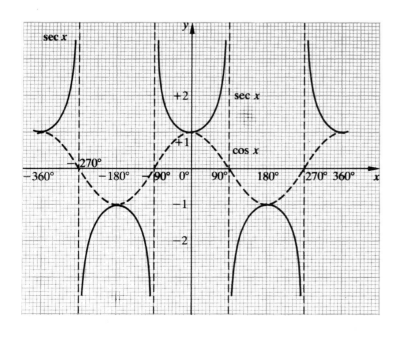

Graph of sec x

The figure, left, shows the graph of $\cos x$ (dotted).

1 $\quad |\cos x| \leqslant 1 \quad \Rightarrow \quad |\sec x| \geqslant 1$

2 $\quad \cos x$ **even** $\quad \Rightarrow \quad \sec x$ **even**

3 $\quad \cos x = 1 \quad \Rightarrow \quad \sec x = 1$
(when $x = 0°, \pm 360°$)

4 $\quad \cos x = -1 \quad \Rightarrow \quad \sec x = -1$
(when $x = \pm 180°$)

5 $\quad \cos x = 0 \quad \Rightarrow \quad \sec x = \pm \infty$

6 $\quad \cos x = -1 \quad \Rightarrow \quad \sec x = -1$

7 $\quad \cos x$ periodic $\quad \Rightarrow \quad \sec x$
periodic

(period 360°)

171

Graph of cosec x

The figure, right, shows the graph of sin x (dotted).

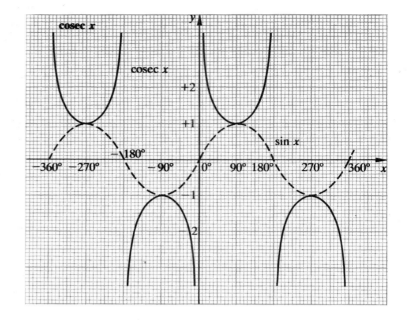

1 sin x **odd** $\Rightarrow$ cosec x **odd**

2 sin $x = 1$ $\Rightarrow$ cosec $x = 1$
(when $x = -270°, +90°$)

3 sin $x = 0$ $\Rightarrow$ cosec $x = \pm\infty$

4 sin $x = -1$ $\Rightarrow$ cosec $x = -1$
(when $x = -90°, +270°$)

5 $|\sin x| \leqslant 1$ $\Rightarrow$ $|\text{cosec } x| \geqslant 1$

6 sin x periodic $\Rightarrow$ cosec x periodic
(period 360°)

Graph of cot x

The figure, right, shows the graph of tan x (dotted).

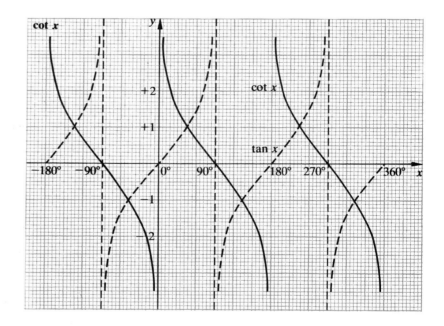

1 tan x **odd** $\Rightarrow$ cot x **odd**

2 tan $x = 0$ $\Rightarrow$ cot $x = \pm\infty$

3 tan $x = 1$ $\Rightarrow$ cot $x = 1$

4 tan $x = \pm\infty$ $\Rightarrow$ cot $x = 0$

5 tan $x = -1$ $\Rightarrow$ cot $x = -1$

6 tan x periodic $\Rightarrow$ cot x periodic
(period 180°)

Further trigonometrical formulae

$$\tan(A+B) = \frac{\sin(A+B)}{\cos(A+B)} = \frac{\sin A \cos B + \cos A \sin B}{\cos A \cos B - \sin A \sin B}$$

But to express $\tan(A+B)$ in terms of $\tan A$ and $\tan B$, divide each term by $\cos A \cos B$.

$$\tan(A+B) = \frac{\dfrac{\sin A \cos B}{\cos A \cos B} + \dfrac{\cos A \sin B}{\cos A \cos B}}{\dfrac{\cos A \cos B}{\cos A \cos B} - \dfrac{\sin A \sin B}{\cos A \cos B}} = \boxed{\frac{\tan A + \tan B}{1 - \tan A \tan B} = \tan(A+B)} \quad \dots\dots\dots\text{(1)}$$

Replacing B by $(-B)$ and using $\tan(-B) = -\tan B$, gives

$$\tan(A - B) = \frac{\tan A - \tan B}{1 + \tan A \tan B}$$

This can also be derived using the formulae for $\sin(A - B)$ and $\cos(A - B)$.

Putting $B = A$ in **(1)** gives $\tan(A + A) = \dfrac{\tan A + \tan A}{1 - \tan A \tan A}$, so

$$\tan 2A = \frac{2 \tan A}{1 - \tan^2 A}$$ (see t formulae, Chapter 6).

We can use $\sin^2 A + \cos^2 A = 1$ to derive two useful formulae involving $\sec x$, $\csc x$ and $\cot x$.

Divide $\sin^2 A + \cos^2 A = 1$ by $\cos^2 A$ to give $\dfrac{\sin^2 A}{\cos^2 A} + \dfrac{\cos^2 A}{\cos^2 A} = \dfrac{1}{\cos^2 A}$ $\Rightarrow$ $\boxed{\tan^2 A + 1 = \sec^2 A}$ .... **(2)**

Divide by $\sin^2 A$ to give $\dfrac{\sin^2 A}{\sin^2 A} + \dfrac{\cos^2 A}{\sin^2 A} = \dfrac{1}{\sin^2 A}$ $\Rightarrow$ $\boxed{1 + \cot^2 A = \csc^2 A}$

(2) is often used to change $\sec A$ into $\tan A$, and is worth remembering.

Factor formulae

Changing products into sums

$$\sin(A + B) = \sin A \cos B + \cos A \sin B$$
$$\sin(A - B) = \sin A \cos B - \cos A \sin B$$

Adding these equations gives $\quad \sin(A + B) + \sin(A - B) = 2 \sin A \cos B \quad \cdots\cdots$**(3)**

Subtracting gives $\quad\quad\quad\quad \sin(A + B) - \sin(A - B) = 2 \cos A \sin B \quad \cdots\cdots$**(4)**

Similarly, $\quad\quad\quad\quad\quad \cos(A + B) + \cos(A - B) = 2 \cos A \cos B \quad \cdots\cdots$**(5)**

$$\cos(A + B) - \cos(A - B) = -2 \sin A \sin B \quad \cdots\cdots\textbf{(6)}$$

For example $\quad \sin 3x \cos 2x = \tfrac{1}{2} \sin 5x + \tfrac{1}{2} \sin x \quad$ (using **(3)**)

$\quad\quad\quad\quad \cos 2x \sin 3x = \tfrac{1}{2} \sin 5x - \tfrac{1}{2} \sin(-x) = \tfrac{1}{2} \sin 5x + \tfrac{1}{2} \sin x \quad$ (since $\sin(-x) = -\sin x$)

$\quad\quad\quad\quad \cos 6x \cos 3x = \tfrac{1}{2} \cos 9x + \tfrac{1}{2} \cos 3x \quad$ (using **(5)**)

$\quad\quad\quad\quad \sin 5x \sin 3x = -\tfrac{1}{2} \cos 8x + \tfrac{1}{2} \cos 2x \quad$ (using **(6)**)

So products of sines or cosines can be changed into sums which may be dealt with more easily, especially when integrating.

Changing sums into products

Referring to equations **(3)** to **(6)**, write $\left.\begin{array}{l} P = A + B \\ Q = A - B \end{array}\right\}$ $\Rightarrow$ $P + Q = 2A, \quad P - Q = 2B$

$$\Rightarrow A = \frac{P + Q}{2}, \quad B = \frac{P - Q}{2}$$

(3) becomes $\sin P + \sin Q = 2\sin\dfrac{(P+Q)}{2}\cos\dfrac{(P-Q)}{2}$ $\cdots\cdots$ **(7)**

(4) becomes $\sin P - \sin Q = 2\cos\dfrac{(P+Q)}{2}\sin\dfrac{(P-Q)}{2}$ $\cdots\cdots$ **(8)**

These are known as the **factor formulae.**

(5) becomes $\cos P + \cos Q = 2\cos\dfrac{(P+Q)}{2}\cos\dfrac{(P-Q)}{2}$ $\cdots\cdots$ **(9)**

(6) becomes $\cos P - \cos Q = -2\sin\dfrac{(P+Q)}{2}\sin\dfrac{(P-Q)}{2}$ $\cdots\cdots$ **(10)**

WORKED EXAMPLE

Solve $\sin 3x + \sin 2x = 0$ for $0 \leqslant x \leqslant 360°$.

Using **(7)** gives $\sin 3x + \sin 2x = 2\sin\dfrac{(3x+2x)}{2}\cos\dfrac{(3x-2x)}{2} = 0$

$$\Rightarrow\quad 2\sin\frac{5x}{2}\cos\frac{x}{2} = 0$$

$$\Rightarrow\quad \sin\frac{5x}{2} = 0 \quad \text{or} \quad \cos\frac{x}{2} = 0$$

$$\Rightarrow\quad \frac{5x}{2} = 0°, 180°, 360°, 540°, 720°, 900° \quad \text{or} \quad \frac{x}{2} = 90°, 270° \cdots\cdots \textbf{(A)}$$

$$\Rightarrow\quad x = 0°, 72°, 144°, 216°, 288°, 360° \quad \text{or} \quad x = 180°$$

Complete solution $x = 0°, 72°, 144°, 180°, 216°, 288°, 360°$.

(Remember to consider all possible solutions at stage **(A)**, which will give a final solution.)

Products are useful when solving equations especially if they can be equated to zero.

Exercise 10.1

1 Solving $\cos 5x + \cos 3x = 0$ for $0 \leqslant x \leqslant 360°$.

2 Solve **(a)** $\cos 5x = \cos 4x$,   **(b)** $\sin 5x = \sin 4x$,   **(c)** $\sin 4x + \sin 2x = 0$,   for $0 \leqslant x \leqslant 360°$.

3 Use the values of sin, cos and tan of $30°, 45°, 60°$ and apply suitable formulae to find in surd form,

(a) $\sin 75° = \sin(45° + 30°)$,   **(b)** $\sin 15°$,   **(c)** $\cos 75°$,   **(d)** $\cos 15°$,   **(e)** $\tan 75°$,   **(f)** $\tan 15°$.

4 Check that $\sin^2 15° + \cos^2 15° = 1$ and $\sin^2 75° + \cos^2 75° = 1$, using the surd values found above.

5 Solve **(a)** $\sin x + \sin 2x + \sin 3x = 0$ and **(b)** $\cos x + \cos 2x + \cos 3x = 0$,   for $0 \leqslant x \leqslant 360°$.

6 Prove that $\cot(A+B) = \dfrac{\cot A \cot B - 1}{\cot A + \cot B}$.

7 Can you find formulae for **(a)** $\sec(A+B)$ in terms of $\sec A$ and $\sec B$ only?   **(b)** $\operatorname{cosec}(A+B)$ in terms of $\operatorname{cosec} A$ and $\operatorname{cosec} B$ only?

8 Solve **(a)** $\sec^2 x = 2\tan x$,   **(b)** $\sec^2 x = 2\tan^2 x$,   **(c)** $\sec x = \cot x$,   for $0 \leqslant x \leqslant 360°$.

9 Simplify **(a)** $\sqrt{1 - \sin^2 x}$, **(b)** $\sqrt{1 - \cos^2 \theta}$, **(c)** $\sqrt{1 + \tan^2 \alpha}$, **(d)** $\sqrt{\dfrac{1 - \sin^2 x}{1 - \cos^2 x}}$, **(e)** $\dfrac{1}{\sqrt{\csc^2 A - 1}}$.

10 If α is acute and $\tan \alpha = \dfrac{8}{15}$, find **(a)** $\sin \alpha$, **(b)** $\cos \alpha$, **(c)** $\cot \alpha$, **(d)** $\sec \alpha$, **(e)** $\csc \alpha$,

(f) $\tan 2\alpha$, **(g)** $\cos^4 \alpha - \sin^4 \alpha$, **(h)** $\cos 2\alpha$, **(i)** prove **(g)** = **(h)** in general.

11 Prove $\tan \theta + \cot \theta = \dfrac{1}{\sin \theta \cos \theta} = \frac{1}{2} \csc 2\theta$.

Radians

$$y = \sin x \quad \Rightarrow \quad \frac{dy}{dx} = ?$$

The gradient function, $g(x)$, for $y = \sin x$ can be plotted on the same graph.

$y = \sin x$ is periodic $\Rightarrow \dfrac{dy}{dx}$ is periodic.

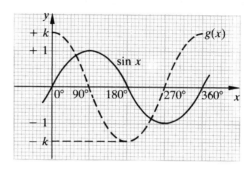

The maximum value of the gradient occurs when the graph is steepest i.e. at $x = 0°, 360°$.

If this maximum value of the gradient is k, then the value of $\dfrac{dy}{dx}$ at $x = 180°$ is $-k$.

The gradient of $\sin x$ changes more between $45°$ and $90°$ than between $0°$ and $45°$.

These facts suggest a gradient function $g(x)$ which looks like a cosine curve and if it is a cosine curve, $\dfrac{dy}{dx}$ will equal $k \cos x$, where k is the value of the gradient of $\sin x$ when $x = 0$.

Wouldn't it be convenient if we could arrange for k to be equal to 1? This is what we do by measuring angles in a different way, i.e. in **radians**.

We will define a **radian** and then prove that the gradient of $\sin x$ at the origin is equal to 1 when x is measured in radians.

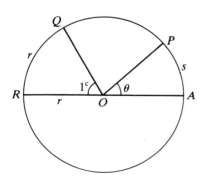

In the figure, left, AP is an arc (length s) of a circle (radius r) and AP subtends an angle θ at the centre, i.e. $P\hat{O}A = \theta$.

The ratio $\dfrac{s}{r} = \dfrac{\text{arc } AP}{\text{radius}}$ will be the same for every size of circle centre O for the same angle θ, and this ratio measures the angle in **radians**.

When the arc length is equal to the radius, i.e. $QR = r$, then $\dfrac{QR}{r} = \dfrac{r}{r} = 1^c$ (one radian).

From $\triangle OQR$, $1^c \simeq 60°$.

For the whole circle, the arc length is the circumference $2\pi r \quad \Rightarrow \quad 360°$ is equivalent to $\dfrac{2\pi r^c}{r} = 2\pi$ radians.

$$2\pi^c \equiv 360° \quad \Rightarrow \quad 1 \text{ radian} \equiv \frac{360}{2\pi} \simeq 57.3°, \text{ and } \pi^c = 180° \quad \Rightarrow \quad 1° = \frac{\pi^c}{180} \simeq 0.0175^c.$$

It is useful to remember that $\quad 90° \equiv \dfrac{\pi^c}{2}; \quad 60° \equiv \dfrac{\pi^c}{3}; \quad 45° \equiv \dfrac{\pi^c}{4} \quad$ and $\quad 30° \equiv \dfrac{\pi^c}{6}.$

Arc length of a sector of a circle

With θ measured in radians, $\dfrac{s}{r} = \theta^c \quad \Rightarrow \quad$ **arc length** $\quad \boxed{s = r\theta}$

Area of a sector of a circle

$2\pi^c$ produces the total area of the circle πr^2.

θ^c produces the **sector area**, $A = \dfrac{\theta}{2\pi} \times \pi r^2 \quad$ i.e. $\quad \boxed{A = \frac{1}{2}r^2\theta}$

WORKED EXAMPLE

A car travels around a bend of length 100 metres in the shape of the arc of a circle changing its direction of motion by 50°. Find the radius of the circle.

In travelling from P to Q, OP turns through 50° to OQ.

$PQ = 100$ m and $s = r\theta \quad \Rightarrow \quad r = \dfrac{s}{\theta} = \dfrac{100}{\theta^c}$

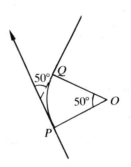

$50° = \dfrac{50\pi^c}{180} \times 0.872^c \quad \Rightarrow \quad r = \dfrac{100}{.872} = 115 \text{ m}$

The area of the sector POQ would be given by $A = \frac{1}{2}r^2\theta$.

$A = \frac{1}{2} \times 115^2 \times 0.872 = 6566 \text{ m}^2.$

Exercise 10.2

1 Complete the tables to change radians into degrees and vice versa.

(a)

Radians					$\pi/4$	1^c				2^c								
Degrees	1°	10°	15°	30°	45°		60°	75°	90°		120°	135°	150°	180°	270°	360°	720°

(b)

Radians	$\pi/10$	$\pi/8$	$\pi/6$	$\pi/5$	$\pi/4$	$\pi/3$	$\pi/2$	π	$3\pi/4$	$5\pi/6$	1.5^c				
Degrees												57.3°	70°	50° 30'	20° 26'	

2 Find the area of the minor segment of a circle of radius 10 cm formed by an arc of angle 100°.

3 Two circles of radius 6 cm and 8 cm cut orthogonally (at right angles). Find **(a)** the area of overlap, **(b)** the total area, **(c)** the perimeter of the two circles together.

4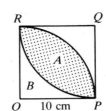

Radius 10 cm.

Find the shaded area.

5

Square side 10 cm.
Arcs *PR* have radii 10 cm.

Find **(a)** area *A* (shaded)
(b) area *B*.

6 A bicycle chain encloses two cog wheels of radii 10 cm and 4 cm whose centres are 40 cm apart. Find **(a)** the length of the chain and **(b)** the area enclosed within the chain.

7 Solve the following equations giving the general solutions in radians (set your calculator to radians).
(a) $\sin x = 0.3$ **(b)** $\cos x = 0.4$ **(c)** $\tan x = 0.5$ **(d)** $\cot x = 0.6$ **(e)** $\sec x = 2$

8 Sketch the graph of **(a)** $y = \sin x$ for $|x| \leqslant \pi/2$ and **(b)** $y = \sin^{-1} x = \arcsin x$ for $|x| < 1$, i.e. $-1 < x < 1$.

9 Sketch the graph of **(a)** $y = \cos x$ for $0 \leqslant x \leqslant \pi$ and **(b)** $y = \cos^{-1} x = \arccos x$ for $|x| < 1$.

10 Two circles of equal radii 10 cm overlap so that each centre lies on the circumference of the other circle. Find the area of overlap.

Small angles

Refer to the figure, right.

$OA = OP = 1$ unit; $T\hat{O}A = \theta^c$

In $\triangle ONP$, $PN = \sin \theta$.

Arc $AP = r\theta = \theta$ since $r = 1$.

In $\triangle OAT$, $AT = \tan \theta$.

$PN < \text{arc } AP < TA \quad \Rightarrow \quad \sin \theta < \theta^c < \tan \theta$

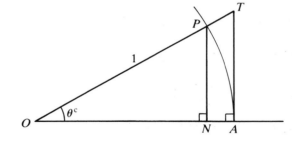

As $\theta \to 0$, the values of $\sin \theta$, θ and $\tan \theta$ approach the same value if θ is measured in radians, as you can see from the table of values, right,

When θ is small $\boxed{\sin \theta \simeq \theta}$

$\boxed{\tan \theta \simeq \theta}$

θ degrees	$\sin \theta$	θ^c	$\tan \theta$
10°	.1736	.1745	.1763
5°	.0872	.0873	.0875
3°	.052 33	.052 36	.052 41
1°	.017 45	.017 45	.017 46

To find an approximation for $\cos \theta$, use $\cos 2A = 1 - 2\sin^2 A \quad \Rightarrow \quad \cos \theta = 1 - 2\sin^2 \dfrac{\theta}{2}$.

$$\dfrac{\theta}{2} \text{ small} \quad \Rightarrow \quad \sin \dfrac{\theta}{2} \simeq \dfrac{\theta}{2} \quad \Rightarrow \quad \cos \theta = 1 - 2\sin^2 \dfrac{\theta}{2} \simeq 1 - 2\left(\dfrac{\theta}{2}\right)^2 = 1 - \dfrac{\theta^2}{2}$$

When θ is small $\boxed{\cos \theta \simeq 1 - \dfrac{\theta^2}{2}}$

WORKED EXAMPLE ➤

> Use $\sin(A + B) = \sin A \cos B + \cos A \sin B$ with $A = 30°$ and $B = 1°$ to find $\sin 31°$, also using $\sin B \simeq B$ and $\cos B \simeq 1$.

$\sin 31° = \sin(30° + 1°) = \sin 30° \cos 1° + \cos 30° \sin 1° = 0.5 + 0.866 \times 0.017\,45 = 0.5 + 0.015\,11 = 0.515\,11$

Exercise 10.3

1 Compare the values of $\cos\theta$ and $1 - \dfrac{\theta^2}{2}$ for **(a)** $\theta = 10°$, **(b)** $\theta = 5°$, **(c)** $\theta = 3°$, **(d)** $\theta = 1°$. Remember that for $1 - \dfrac{\theta^2}{2}$, θ is in radians.

2 Use the formula for $\cos(A + B)$ to find $\cos 31°$, with $\sin 30° = 0.5$, $\cos 30° = \dfrac{\sqrt{3}}{2}$, $\sin B \simeq B$ and $\cos B \simeq 1 - \dfrac{\theta^2}{2}$ (B in radians), as accurately as your calculator allows. Compare your result with the value your calculator gives for $\cos 31°$.

3 Use the formula for $\tan(A + B)$ to find $\tan 61°$, with $\tan A = \tan 60° = \sqrt{3}$ and $\tan B \simeq B$ ($B = 1°$).

4 Find an approximation for **(a)** $\dfrac{\sin 2\theta}{1 + \cos 2\theta}$ and **(b)** $\dfrac{\sin 2\theta}{1 - \cos 2\theta}$, when θ is small.

5 Find an approximation to the following expressions when θ is small, and their limit as $\theta \to 0$.

(a) $\dfrac{\sin 3\theta \tan\theta}{1 - \cos\theta}$ **(b)** $\dfrac{1 - \cos 2\theta}{1 + \cos 2\theta}$ **(c)** $\dfrac{\sin\theta}{\theta}$ **(d)** $\dfrac{\cos 4\theta - \cos\theta}{\cos 3\theta - \cos 2\theta}$ **(e)** $\dfrac{2\sin\theta - \tan^3\theta}{\cos\theta}$

(f) $\dfrac{\sin\theta}{1 - \cos\theta}$

Differentiation of trigonometrical functions

Derivative of sin *x and* cos *x from first principles*

The most convenient formula to use is $\dfrac{df}{dx} = f'(x) = \underset{h \to 0}{\text{limit}} \dfrac{f(x + h) - f(x - h)}{2h}$

For $f(x) = \sin x$,

$$f'(x) = \lim_{h \to 0} \frac{\sin(x + h) - \sin(x - h)}{2h} = \lim_{h \to 0} \frac{2\cos x \sin h}{2h} \quad \text{(using factor formula)}$$

$$= \lim_{h \to 0} \cos x \frac{\sin h}{h}$$

From our work an small angles, $\dfrac{\sin h}{h} \to 1$ as $h \to 0$, if h is in radians.

$$\Rightarrow \quad f'(x) = \cos x \qquad \text{(but remember that } x \text{ is measured in radians)}$$

$$g(x) = \cos x \quad \Rightarrow \quad \frac{dg}{dx} = g'(x) = \lim_{h \to 0} \frac{g(x+h) - g(x-h)}{2h} = \lim_{h \to 0} \frac{\cos(x+h) - \cos(x-h)}{2h}$$

$$= \lim_{h \to 0} \frac{-2\sin x \sin h}{2h} \quad \text{(using factor formula)}$$

$$= \lim_{h \to 0} - \sin x \frac{\sin h}{h} = - \sin x \quad \left(\text{since } \frac{\sin h}{h} \to 1 \text{ as } h \to 0\right)$$

We have now proved that $\boxed{\dfrac{d}{dx}(\sin x) = \cos x}$ and $\boxed{\dfrac{d}{dx}(\cos x) = -\sin x}$ for x in radians.

The rest of the trigonometrical ratios can be differentiated using these basic results together with the rules of differentiation of composite functions, the product rule and quotient rule.

WORKED EXAMPLE

Find the derivatives of **(a)** $\tan x$, **(b)** $\sec x$, **(c)** $\cos 2x$.

(a) $y = \tan x = \dfrac{\sin x}{\cos x}$ Using the quotient rule $\dfrac{dy}{dx} = \dfrac{\cos x \cos x - \sin x(-\sin x)}{\cos^2 x} = \dfrac{\cos^2 x + \sin^2 x}{\cos^2 x}$

Using $\cos^2 x + \sin^2 x = 1$ $\dfrac{dy}{dx} = \dfrac{\cos^2 x + \sin^2 x}{\cos^2 x} = \dfrac{1}{\cos^2 x} = \sec^2 x$

$$\therefore \quad \boxed{\frac{d}{dx}(\tan x) = \sec^2 x}$$

(b) $y = \sec x = \dfrac{1}{\cos x}$ Let $u = \cos x \quad \Rightarrow \quad \dfrac{du}{dx} = -\sin x$ and $y = \dfrac{1}{u}$ $\dfrac{dy}{dx} = \dfrac{dy}{du} \times \dfrac{du}{dx} = -\dfrac{1}{u^2} \times (-\sin x)$

$$= \frac{+\sin x}{\cos^2 x} = \sec x \tan x$$

$$\therefore \quad \boxed{\frac{d}{dx}(\sec x) = \sec x \tan x}$$

We usually express the derivatives of $\tan x$ and $\sec x$ in terms of themselves because they are well connected; remember $\sec^2 x = 1 + \tan^2 x$.

(c) $y = \cos 2x$ Put $u = 2x \quad \Rightarrow \quad \dfrac{du}{dx} = 2; \quad y = \cos u \quad \Rightarrow \quad \dfrac{dy}{du} = -\sin u$

$$\frac{dy}{dx} = \frac{dy}{du} \times \frac{du}{dx} = -\sin u \times 2 = -2 \sin 2x$$

$$\therefore \quad \boxed{\frac{d}{dx}(\cos 2x) = -2 \sin 2x}$$

Exercise 10.4

In questions **1**–**50**, differentiate the functions with respect to x.

1 $\cot x$	**11** $\sin^2 x$	**21** $\sin^8 x$	**31** $\sec^2 x$	**41** $1/\sin x$
2 $\operatorname{cosec} x$	**12** $\cos^2 x$	**22** $\cos^{13} x$	**32** $\operatorname{cosec}^2 x$	**42** $x\cos x$
3 $\sin 2x$	**13** $\sin^2 x + \cos^2 x$	**23** $\tan^4 x$	**33** $\cot^2 x$	**43** $x^2 \sin x$
4 $\sin 3x$	**14** $\sin^3 x$	**24** $\sec^5 x$	**34** $\sin x + \cos x$	**44** $(x^2 + 1)\tan x$
5 $\sin px$	**15** $\cos^3 x$	**25** $\cot^6 x$	**35** $\sin x \cos x$	**45** $\cot(3x + 2)$
6 $\sin(x + 3)$	**16** $\sin x^2$	**26** $\sin^2 3x$	**36** $\sin x \tan x$	**46** $x/\sin x$
7 $\cos(2x - 1)$	**17** $\tan x^3$	**27** $\sin^3 2x$	**37** $\cos x \tan x$	**47** $(\tan x)/x^2$
8 $\sin(x + \pi/2)$	**18** $\cos x^4$	**28** $\tan^3 2x$	**38** $\sin\sqrt{x}$	**48** $\operatorname{cosec} x/\cot x$
9 $\sin[(\pi/2) - x]$	**19** $\sec x^5$	**29** $\cos^4 4x$	**39** $\sqrt{\sin x}$	**49** $\sec x/\tan x$
10 $\tan(x + \pi/4)$	**20** $\operatorname{cosec} x^6$	**30** $\tan^2 x$	**40** $\sin\dfrac{1}{x}$	**50** $\tan x/\sec x$

51 Find the turning points on the graph of $y = \sin x + \cos x$ for $0 \leqslant x \leqslant 2\pi$. Sketch the graph of $y = \sin x + \cos x$ for $0 \leqslant x \leqslant 2\pi$.

52 Sketch the graph of $y = \sin x$ for $0 \leqslant x \leqslant 2\pi$. On the same graph sketch $y = \sin^2 x$. Find the maximum and minimum values of $y = \sin^2 x$ for $0 \leqslant x \leqslant 2\pi$.

53 Sketch the graph of $y = \sin x$ for $-\pi \leqslant x \leqslant \pi$. On the same graph draw $y = x$ and also $y = x + \sin x$. Find the maximum and minimum values of $x + \sin x$ for $-\pi \leqslant x \leqslant +\pi$.

54 Differentiate **(a)** $x \sin x - \cos x$ and **(b)** $\sin x - x \cos x$.

55 Using $x = 5\cos t$ and $y = 5\sin t$, find **(a)** $x^2 + y^2$, **(b)** $\dfrac{y}{x}$ in terms of t, **(c)** $\dfrac{dy}{dx}$ in terms of t, **(d)** $\dfrac{y}{x} \times \dfrac{dy}{dx}$ and interpret your answer; and **(e)** $\dfrac{d^2 y}{dx^2}$ in terms of t.

56 Show that $\dfrac{d}{dx}\left(\dfrac{x}{2} - \dfrac{1}{4}\sin 2x\right) = \sin^2 x$ and deduce that $\displaystyle\int_0^\pi \sin^2 x \, dx = \dfrac{\pi}{2}$. Confirm this result with reference to your graph in question **52**. What is the value of $\displaystyle\int_0^\pi \cos^2 x \, dx$?

57 Differentiate $\sin 3x$ and use the result to evaluate $\int \cos 3x \, dx$.

58 Differentiate $\sin ax$ (a constant) and use the result to evaluate $\int \cos ax \, dx$.

59 Differentiate $\cos ax$ (a constant) and use the result to evaluate $\int \sin ax \, dx$.

60 Differentiate $\tan ax$ (a constant) and use the result to evaluate **(a)** $\int \sec^2 ax \, dx$ and **(b)** $\int \tan^2 ax \, dx$.

61 Differentiate $\cot ax$ (a constant) and evaluate **(a)** $\int \operatorname{cosec}^2 ax \, dx$ and **(b)** $\int \cot^2 ax \, dx$.

Differentiation of inverse trigonometrical functions

The inverse function of $\sin x$ is $y = \sin^{-1} x$ where $-\pi/2 \leqslant y \leqslant +\pi/2$, otherwise $\sin^{-1} x$ is not a function because it becomes many-valued.

$$y = \sin^{-1} x \quad \Leftrightarrow \quad x = \sin y \quad \text{(see figure, right)}$$

Differentiating $x = \sin y$ with respect to y gives

$$\frac{dx}{dy} = \cos y \quad \Rightarrow \quad \frac{dy}{dx} = \frac{1}{\cos y}$$

Differentiating $x = \sin y$ with respect to x gives

$$1 = \cos y \frac{dy}{dx} \quad \Rightarrow \quad \frac{dy}{dx} = \frac{1}{\cos y} = \frac{1}{\sqrt{\cos^2 y}}$$

$\cos^2 y = 1 - \sin^2 y = 1 - x^2$, since $x = \sin y$

$$\therefore \quad \frac{dy}{dx} = \frac{1}{\sqrt{1 - x^2}} \quad \text{if} \quad y = \sin^{-1} x$$

$x = 0 \quad \Rightarrow \quad \dfrac{dy}{dx} = 1$ and $x = 1 \quad \Rightarrow \quad \dfrac{dy}{dx}$ is infinite, which the graph confirms.

This result has wider applications when you are integrating.

$$\frac{d}{dx}(\sin^{-1} x) = \frac{1}{\sqrt{1 - x^2}} \quad \Rightarrow \quad \boxed{\int \frac{1}{\sqrt{1 - x^2}} dx = \sin^{-1} x + k}$$

WORKED EXAMPLES

1 Differentiate $y = \tan^{-1} x$ and transform the result into the form of an integral.

$$y = \tan^{-1} x \quad \Rightarrow \quad x = \tan y \quad \Rightarrow \quad \frac{dx}{dy} = \sec^2 y \quad \Rightarrow \quad \frac{dy}{dx} = \frac{1}{\sec^2 y} = \frac{1}{1 + \tan^2 y} = \frac{1}{1 + x^2} \quad \text{since } x = \tan y$$

$$y = \tan^{-1} x \quad \Rightarrow \quad \frac{dy}{dx} = \frac{1}{1 + x^2} \quad \Rightarrow \quad \boxed{\int \frac{1}{1 + x^2} dx = \tan^{-1} x + k}$$

2 Differentiate **(a)** $\sin^{-1} ax$ and **(b)** $\sin^{-1} \dfrac{x}{b}$ and transform into integration results.

(a) $\quad y = \sin^{-1} ax \quad \Rightarrow \quad \dfrac{dy}{dx} = \dfrac{1}{\sqrt{1 - (ax)^2}} \times a = \dfrac{a}{\sqrt{1 - a^2 x^2}} \quad \Rightarrow \quad \displaystyle\int \frac{a}{\sqrt{1 - a^2 x^2}} dx = \sin^{-1} ax + k$

$$\text{or} \quad \int \frac{1}{\sqrt{1 - a^2 x^2}} dx = \frac{1}{a} \sin^{-1} ax + k$$

(b) $\quad y = \sin^{-1} \dfrac{x}{b} \quad \Rightarrow \quad \dfrac{dy}{dx} = \dfrac{1}{\sqrt{1 - (x/b)^2}} \times \dfrac{1}{b} = \dfrac{1}{b\sqrt{1 - x^2/b^2}} = \dfrac{1}{\sqrt{b^2 - x^2}} \quad \Rightarrow \quad \displaystyle\int \frac{1}{\sqrt{b^2 - x^2}} dx = \sin^{-1} \frac{x}{b} + k$

Exercise 10.5

---

1 Differentiate **(a)** $\cos^{-1} x$ **(b)** $\cos^{-1} 2x$ **(c)** $\cos^{-1} \dfrac{x}{3}$ **(d)** $\sec^{-1} x$ **(e)** $\cot^{-1} x$

2 Differentiate **(a)** $\tan^{-1} ax$ **(b)** $\tan^{-1} \dfrac{x}{b}$ and transform into integration results.

3 Evaluate **(a)** $\displaystyle\int_0^1 \dfrac{1}{\sqrt{1-x^2}}\,dx$ **(b)** $\displaystyle\int_0^1 \dfrac{1}{1+x^2}\,dx$

4 Differentiate $x\sin^{-1} x$ and use the result to evaluate $\displaystyle\int_0^1 \sin^{-1} x$. Check the result using the figure for $y = \sin^{-1} x$, on the previous page.

Parametric form using trigonometrical functions

> WORKED EXAMPLE

> Investigate the graph $x = \cos\theta$, $y = \sin\theta$ for $0 \leqslant \theta \leqslant 360°$.

The table of values noting that $|x| \leqslant 1$ and $|y| \leqslant 1$ for all θ is given below.

θ	0°	30°	60°	90°	120°	150°	180°	210°	240°	270°	300°	330°	360°
$x = \cos\theta$	1	0.87	0.5	0	-0.5	-0.87	-1	-0.87	-0.5	0	0.5	0.87	1
$y = \sin\theta$	0	0.5	0.87	1	0.87	0.5	0	-0.5	-0.87	-1	-0.87	-0.5	0

Eliminating $\theta \quad \Rightarrow \quad x^2 + y^2 = \cos^2\theta + \sin^2\theta = 1$

So $x^2 + y^2 = 1$ represents a circle, since if P is (x, y), $OP = 1$ for all points P i.e. for all values of θ.

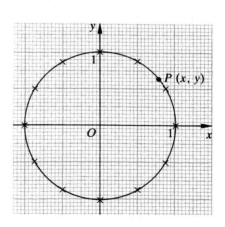

$$\frac{dy}{dx} = \frac{dy}{d\theta} \times \frac{d\theta}{dx} = \frac{dy/d\theta}{dx/d\theta} = \frac{\cos\theta}{-\sin\theta} = -\cot\theta = -\frac{1}{\tan\theta}$$

Gradient $OP = \dfrac{y}{x} = \dfrac{\sin\theta}{\cos\theta} = \tan\theta$, so the tangent $\left(\dfrac{dy}{dx}\right)$ is always perpendicular to the radius (OP) i.e. a **circle**.

Investigation 1

1 Plot $x = \cos^2\theta$, $y = \sin^2\theta$ for $0 \leqslant \theta \leqslant 90°$.

2 Find $\dfrac{dy}{dx}$ and interpret your result.

3 Find the x, y equation by elimination θ.

4 What happens for $90° \leqslant \theta \leqslant 180°$?

Investigation 2

1   Plot $x = 4\cos\theta$, $y = 3\sin\theta$ for $0 \leqslant x \leqslant 360°$.
2   Find $\dfrac{dy}{dx}$ in terms of θ and evaluate $\dfrac{dy}{dx}$ for $\theta = 0°, 90°, 180°, 270°, 360°$.
3   Note any symmetries of $\cos\theta$ and $\sin\theta$ (using the Worked Example, above) and the maximum values x and y can have.
4   Eliminate θ to form the x, y equation.

Investigation 3

1   Plot $x = \cos^3\theta$, $y = \sin^3\theta$, cubing the table of values in the Worked Example, above.
2   Find $\dfrac{dy}{dx}$ and compare with the Worked Example. Does this help? Evaluate $\dfrac{dy}{dx}$ when $\theta = 0°$, $45°$ and $90°$.
3   Is the graph circular between $\theta = 0°$ and $90°$?
4   What is the graph called?

Investigation 4

1   Plot $x = \theta - \sin\theta$, $y = 1 - \cos\theta$ for $0 \leqslant \theta \leqslant 360°$.   (Hints: $0 \leqslant x \leqslant 2\pi$; $0 \leqslant y \leqslant 2$)
2   Find $\dfrac{dy}{dx}$ and evaluate $\dfrac{dy}{dx}$ for $\theta = 0°, 90°, 180°, 270°, 360°$.
3   What happens for $360° \leqslant \theta \leqslant 720°$?
4   What is the curve called? (see Chapter 18).

Triangle centres

Each of the Investigations 5 to 8 highlights a particular centre of the triangle.

Investigation 5

1   Draw a triangle with sides $AB = 10$ cm, $AC = 9$ cm and $BC = 8$ cm.
2   Draw $CN \perp AB$ and $AM \perp BC$ to meet at H. Draw BH to meet AC at P.
    Is $BP \perp AC$? Can you prove it? H is called the **orthocentre** of triangle ABC (the meeting point of the altitudes).

   Hint for 2   Let the position vectors of ABC and H be $\mathbf{a}, \mathbf{b}, \mathbf{c}$ and $\mathbf{h}$. Using the scalar product express $\mathbf{CH}.\mathbf{AB} = 0$ in terms of $\mathbf{a}, \mathbf{b}, \mathbf{c}, \mathbf{h}$ and do the same for $\mathbf{AH}.\mathbf{BC}$.

Investigation 6

1   Draw another triangle with $AB = 10$ cm, $AC = 9$ cm and $BC = 8$ cm.
2   Mark in the mid-point of each side and join each vertex to the mid-point of the opposite side. These are called the medians of triangle ABC.
3   What centre do the medians define? Can you prove that the medians are concurrent?
4   The point of concurrence is called the centroid. The **centroid**, G, divides each median in a certain ratio. What is the ratio? Can you prove it?

Investigation 7

1   Draw △*ABC* with *AB* = 10 cm, *AC* = 9 cm and *BC* = 8 cm.
2   Mark in the mid-point *E* of *AB*, *D* of *BC* and *F* of *AC*.
3   Construct the mediators (perpendicular bisectors) of each side. (These should meet at *J*.)
     Can you prove that they meet at *J*? What centre is *J*?
4   How is *DE* related to *AC*? How is *DF* related to *AB*?
5   How is △*DEB* related to △*CAB*?
6   How is △*DEF* related to △*CAB*?
7   Where is the centroid of △*DEF*?
8   The four △s *DEF*, *AFE*, *BED* and *DCF* form the net for which solid?
9   Where is the centre of mass of this solid?

Investigation 8

1   Draw △*ABC* with *AB* = 10 cm, *AC* = 9 cm and *BC* = 8 cm.
2   Construct each angle bisector to meet at *I*. Can you prove that they meet at *I*? i.e. that the angle
     bisectors are concurrent.
3   Which centre is *I*? Can you prove it as well as demonstrating it on your diagram?

Investigation 9

1   Superimpose △s *ABC* for Investigations 5, 6 and 7.
2   What do you notice about *G*, *H* and *J*?
3   In what ratio does *G* divide *HJ*? Can you prove it?

Notes
(a)   It may help to confirm the results of Investigation 9 by drawing other triangles. Using graph paper with 1 cm
     squares divided into 2 mm squares, try △s *P*(0,0) *Q*(10,0) *R*(6,6) and *L*(1,0) *M*(3,0) *N*(0,3).
(b)   All the constructions can be performed by paper folding.

11 Kinematics

Displacement, velocity and acceleration

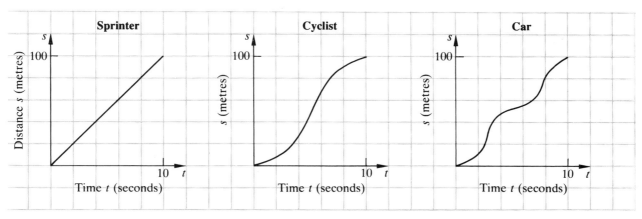

What is the same about the three situations above? What is different?

Consider the speeds in the different situations. When is the car travelling fastest? How can you tell when the speed is zero?

The gradient measures the change in distance per unit time i.e. the **speed**.

Below, are the graphs of speed against time.

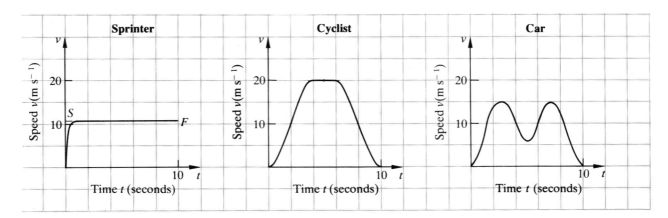

The sprinter, cyclist and car cover 100 metres in 10 seconds, so their **average speeds** are 10 metres per second (m s^{-1}).

The sprinter, once he gets into his stride, moves at a constant speed for the 100 metres. If he could run at 10 m s^{-1} for the whole 100 m, his speed-time graph would be a horizontal line, SF.

The cyclist gradually builds up speed to a maximum between $t = 4$ and $t = 6$ seconds and then slows down towards the end, stopping after 10 seconds.

The car starts from rest, builds up speed, then slows down (while changing gear) builds up speed again, then slows down coming to a stop at the end.

The gradient of the distance-time graphs represents the speed and $v = \dfrac{ds}{dt}$

The gradient functions are plotted in the second set of diagrams to represent how the speeds vary with time.

$v = \dfrac{ds}{dt} \quad \Rightarrow \quad s = \int v \, dt$ and the distance covered is the area under the velocity-time graph.

Acceleration is the rate of change of speed with time and appears as the gradient in these diagrams.

$$a = \frac{dv}{dt} = \frac{d^2s}{dt^2}$$

If the speed is decreasing (as the cyclist finishes) the cyclist is decelerating (by applying his brakes) and the acceleration is negative.

These three situations are **one-dimensional**. The acceleration and velocity (vectors) are in the direction of motion of the sprinter, cyclist or car, each of which does not change direction.

The figure, on the next page, shows the position of a boat whose displacement from O at various times t is given in the table.

Time t in seconds	0	1	2	3	4	t
Displacement $x\mathbf{i} + y\mathbf{j}$	$\mathbf{j}$	$\mathbf{i} + 2\mathbf{j}$	$2\mathbf{i} + 3\mathbf{j}$	$3\mathbf{i} + 4\mathbf{j}$	$4\mathbf{i} + 5\mathbf{j}$	$\mathbf{r}$

At time t, the displacement $\mathbf{r} = t\mathbf{i} + (t + 1)\mathbf{j} = \mathbf{j} + t(\mathbf{i} + \mathbf{j})$

During each second the boat's displacement is $\mathbf{i} + \mathbf{j}$ metres, so its velocity $\mathbf{v} = \mathbf{i} + \mathbf{j}$. The distance travelled each second is $|\mathbf{i} + \mathbf{j}| = \sqrt{2}\,\mathrm{m}$ and the boat's speed is $\sqrt{2}\,\mathrm{m\,s^{-1}}$.

The direction of the velocity vector $\mathbf{v} = \mathbf{i} + \mathbf{j}$ gives the direction of travel.

In this example, the boat travels in a straight line (one dimension) but the method of description allows us to develop methods for two-dimensional motion.

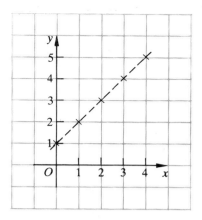

If the displacement of a particle from the origin is given by $\mathbf{r} = x\mathbf{i} + y\mathbf{j}$, where $x = f(t)$, $y = g(t)$ are functions of time.

$$\mathbf{r} = x\mathbf{i} + y\mathbf{j} = f(t)\mathbf{i} + g(t)\mathbf{j}$$

At time $t + \delta t$, the displacement $\mathbf{r} + \delta\mathbf{r} = f(t + \delta t)\mathbf{i} + g(t + \delta t)\mathbf{j}$ and the change in displacement $\delta\mathbf{r}$ is given by

$$\delta\mathbf{r} = f(t + \delta t)\mathbf{i} + g(t + \delta t)\mathbf{j} - \mathbf{r} = f(t + \delta t)\mathbf{i} + g(t + \delta t)\mathbf{j} - f(t)\mathbf{i} - g(t)\mathbf{j}$$

The change in displacement takes place in time δt so $\dfrac{\delta\mathbf{r}}{\delta t} = \left[\dfrac{f(t + \delta t) - f(t)}{\delta t}\right]\mathbf{i} + \left[\dfrac{g(t + \delta t) - g(t)}{\delta t}\right]\mathbf{j}$

The velocity is the rate of displacement per unit time so $\mathbf{v} = \dfrac{d\mathbf{r}}{dt} = \underset{\delta t \to 0}{\text{limit}}\,\dfrac{\delta\mathbf{r}}{\delta t} = f'(t)\mathbf{i} + g'(t)\mathbf{j}$, and to find the velocity vector we differentiate the components of $\mathbf{i}$ and $\mathbf{j}$.

In the example of the boat, $\mathbf{r} = t\mathbf{i} + (t + 1)\mathbf{j} \ \Rightarrow\ \mathbf{v} = \dfrac{d\mathbf{r}}{dt} = \mathbf{i} + \mathbf{j}$ so the velocity is in the direction $\mathbf{i} + \mathbf{j}$ with magnitude $|\mathbf{i} + \mathbf{j}| = \sqrt{2}\,\mathrm{m\,s^{-1}}$.

Similarly, the acceleration $\mathbf{a} = \dfrac{d\mathbf{v}}{dt} = \dfrac{d^2\mathbf{r}}{dt^2}$ and, for the boat, $\mathbf{a} = \mathbf{0}$ (zero vector) because the boat travels with constant velocity.

> **WORKED EXAMPLES**

1 Examine the motion of a particle whose position vector is $\mathbf{r} = (t^2 + 1)\mathbf{i} + t^2\mathbf{j}$

$t = 0 \ \Rightarrow\ \mathbf{r}_0 = \mathbf{i}$ $\qquad \mathbf{v} = \dfrac{d\mathbf{r}}{dt} = 2t\mathbf{i} + 2t\mathbf{j}$

$t = 1 \ \Rightarrow\ \mathbf{r}_1 = 2\mathbf{i} + \mathbf{j}$

$t = 2 \ \Rightarrow\ \mathbf{r}_2 = 5\mathbf{i} + 4\mathbf{j}$ $\qquad\qquad = 2t(\mathbf{i} + \mathbf{j})$

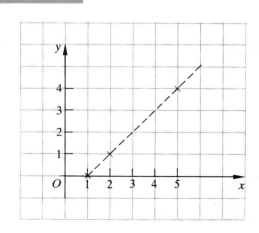

The particle travels in the direction of $\mathbf{i} + \mathbf{j}$ with increasing speed.

$\mathbf{a} = 2\mathbf{i} + 2\mathbf{j}$ and the particle moves with constant acceleration.

$|2\mathbf{i} + 2\mathbf{j}|$ gives the size of the acceleration as

$$\sqrt{8} = 2\sqrt{2} \simeq 2.83\,\mathrm{m\,s^{-2}}$$

2 Examine the motion represented by $\mathbf{r} = t\mathbf{i} + (2t - t^2)\mathbf{j}$

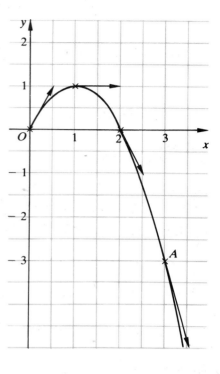

	$t =$	0	1	2	3
$\mathbf{r} = t\mathbf{i} + (2t - t^2)\mathbf{j}$	$\mathbf{r} =$	$\mathbf{0}$	$\mathbf{i} + \mathbf{j}$	$2\mathbf{i}$	$3\mathbf{i} - 3\mathbf{j}$
$\mathbf{v} = \mathbf{i} + (2 - 2t)\mathbf{j}$	$\mathbf{v} =$	$\mathbf{i} + 2\mathbf{j}$	$\mathbf{i}$	$\mathbf{i} - 2\mathbf{j}$	$\mathbf{i} - 4\mathbf{j}$
$\mathbf{a} = -2\mathbf{j}$	$\mathbf{a} =$	$-2\mathbf{j}$	$-2\mathbf{j}$	$-2\mathbf{j}$	$-2\mathbf{j}$

The values of $\mathbf{r}$ specify the position in the x-y plane.
The particle starts at O and at $t = 3$ is at A.

$\mathbf{v}$ gives the velocity vectors which, when drawn on the figure, right, give the direction of motion i.e. they lie along the tangent to the curve.

$\mathbf{r} = x\mathbf{i} + y\mathbf{j}$ where (x, y) are the co-ordinates of the particle.

$\mathbf{r} = t\mathbf{i} + (2t - t^2)\mathbf{j} \quad \Rightarrow \quad x = t$ and $y = 2t - t^2$

Eliminating t gives the equation of the curve as $y = 2x - x^2$ which represents an inverted parabola.

$$\frac{dy}{dx} = 2 - 2x \text{ and } \frac{dy}{dx} = \frac{dy/dt}{dx/dt} = \frac{2 - 2t}{1}$$

$$t = 0 \quad \Rightarrow \quad \frac{dy}{dx} = 2; \quad t = 1 \quad \Rightarrow \quad \frac{dy}{dx} = 0; \quad t = 2 \quad \Rightarrow \quad \frac{dy}{dx} = -2$$

which confirms that the velocity vectors lie along the tangent to the curve i.e. in the direction in which the particle moves.

The acceleration is always $-2\mathbf{j}$ and when you know Newton's Second Law (Chapter 15) $\mathbf{F} = m\mathbf{a}$, which relates force to acceleration. The force on the particle is in the direction $-\mathbf{j}$, so this motion is similar to a particle moving under the influence of gravity.

Dot notation

For a particle moving in three dimensions,

$$\mathbf{r} = x\mathbf{i} + y\mathbf{j} + z\mathbf{k} = f(t)\mathbf{i} + g(t)\mathbf{j} + h(t)\mathbf{k} \text{ and } \mathbf{v} = f'(t)\mathbf{i} + g'(t)\mathbf{j} + h'(t)\mathbf{k}$$

We shall be considering motion in one, two or three dimensions and will adopt the following notation for displacement, velocity and acceleration.

	One dimension	Two dimensions	Three dimensions
Displacement	s	$\mathbf{r} = x\mathbf{i} + y\mathbf{j}$	$\mathbf{r} = x\mathbf{i} + y\mathbf{j} + z\mathbf{k}$
Velocity	$v = \dfrac{ds}{dt}$	$\mathbf{v} = \dfrac{d\mathbf{r}}{dt} = \dfrac{dx}{dt}\mathbf{i} + \dfrac{dy}{dt}\mathbf{j} = \dot{x}\mathbf{i} + \dot{y}\mathbf{j}$	$\mathbf{v} = \dfrac{d\mathbf{r}}{dt} = \dot{x}\mathbf{i} + \dot{y}\mathbf{j} + \dot{z}\mathbf{k}$
Acceleration	$a = \dfrac{dv}{dt} = \dfrac{d^2s}{dt^2}$	$\mathbf{a} = \dfrac{d\mathbf{v}}{dt} = \dfrac{d^2x}{dt^2}\mathbf{i} + \dfrac{d^2y}{dt^2}\mathbf{j} = \ddot{x}\mathbf{i} + \ddot{y}\mathbf{j}$	$\mathbf{a} = \dfrac{d\mathbf{v}}{dt} = \dfrac{d^2\mathbf{r}}{dt^2} = \ddot{x}\mathbf{i} + \ddot{y}\mathbf{j} + \ddot{z}\mathbf{k}$

The **dot notation** is shorthand used in mechanics and kinematics for differentiation with respect to t, where t stands for time.

Thus $\dot{x} = \dfrac{dx}{dt}$, $\ddot{x} = \dfrac{d^2x}{dt^2}$; $\quad \dot{\mathbf{r}} = \dfrac{d\mathbf{r}}{dt} = \mathbf{v}$ and $\ddot{\mathbf{r}} = \dot{\mathbf{v}} = \mathbf{a}$.

Another useful fact is that $a = \dfrac{dv}{dt} = \dfrac{dv}{ds} \times \dfrac{ds}{dt} = \dfrac{dv}{ds} \times v = v\dfrac{dv}{ds}$, so $a = v\dfrac{dv}{ds} = \dfrac{dv}{dt} = \dfrac{d^2s}{dt^2}$

Motion in one dimension

> WORKED EXAMPLES

1 A particle moves in a straight line so that its displacement in metres from its starting position $O(t = 0)$ is given by $s = t^3 - 3t^2 + 2t$ (t in seconds). Find when it is subsequently at O and when and where it is momentarily at rest. Find the distance it covers during the first 2 seconds.

$s = t^3 - 3t^2 + 2t = t(t^2 - 3t + 2) = t(t-1)(t-2) \implies s = 0$ when $t = 0, 1,$ and 2 seconds.

When $t = 1$ and $t = 2$, the particle has returned to its starting point O.

$v = \dfrac{ds}{dt} = 3t^2 - 6t + 2 = 0$ when $t = \dfrac{6 \pm \sqrt{36-24}}{6} = \dfrac{3 \pm \sqrt{3}}{3} = 1 + \dfrac{1}{\sqrt{3}}$ or $1 - \dfrac{1}{\sqrt{3}} \simeq 1.577$ or 0.423 seconds.

Plotting displacement s against time t shows that the particle is momentarily at rest when the gradient (speed) is zero i.e. at $A\,(t = 0.42)$ and $B\,(t = 1.58)$.

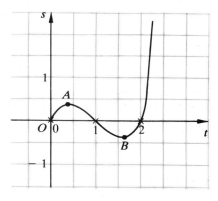

At A, $\quad s = (0.42)(-0.58)(-1.58) = 0.385\,\mathrm{m}$
At B, $\quad s = (1.58)(0.58)(-0.42) = -0.385\,\mathrm{m}$

During the first second the particle moves towards A and back to O. During the next second the particle moves backwards to B and back to O at time $t = 2s$.

During the first 2 seconds the total distance travelled is $4 \times 0.385 \simeq 1.54\,\mathrm{m}$, but s, the displacement, is zero at $t = 2$.

The acceleration $a = \dfrac{dv}{dt} = 6t - 6$ which is negative for $0 \leqslant t < 1$. So, during the first second, the particle slows down as it reaches A and then moves towards O increasing its speed. It passes through O ($t = 1$) with speed $1\,\mathrm{m\,s}^{-1}$ ($v = -1$) and then slows as it moves towards B as the acceleration is now positive (for $t > 1$). After the particle stops at B, it accelerates towards O and passes through O ($t = 2$) with speed $2\,\mathrm{m\,s}^{-1}$, increasing its speed thereafter.

2 A particle moves in a straight line with acceleration $a = 2t$. It starts from rest with $s = -3$. Find expressions for the velocity v and displacement s at time t.

$a = 2t \implies \dfrac{dv}{dt} = 2t \implies v = \int 2t\,dt = t^2 + c$ (c constant of integration)

$t = 0, v = 0$ (starts at rest) $\implies 0 = 0 + c \implies c = 0 \implies \boxed{v = t^2}$

$$v = t^2 \quad \Rightarrow \quad \frac{ds}{dt} = t^2 \quad \Rightarrow \quad s = \int t^2\, dt = -\frac{t^3}{3} + k \quad (k \text{ constant of integration})$$

$$t = 0,\ s = -3 \quad \Rightarrow \quad -3 = 0 + k \quad \Rightarrow \quad k = -3 \quad \Rightarrow \quad \boxed{s = \tfrac{1}{3}t^3 - 3}$$

Exercise 11.1

Each question refers to a particle moving in a straight line whose displacement is s, velocity v and acceleration a, at time t.

1 Find v and a, given that **(a)** $s = 6 + 6t + 6t^2$ **(b)** $s = (t-1)(t-2)(t-3)$ **(c)** $s = 3\cos t$.
Describe the motion in each case. It may help to sketch the graph of s against t.

2 Find a and s, given that $v = (1-t)(2-t)$ and initially $s = 1$. Find the distance travelled during the first 3 seconds.

3 Find v and s, when $a = 3 + 2t$ and at $t = 0$, $v = 2$ and $s = 3$. Find the distance moved during the first second and the speed when $t = 1$.

4 For $s = 20t - 5t^2$ find **(a)** v and a, when $t = 0$ **(b)** s and t, when $v = 0$ **(c)** v and t, when $s = 0$ **(d)** the distance travelled in the first 4 seconds.

5 For $a = -10$, find v and s at time t, given initially $s = 0$ and $v = 20$.

6 For $s = 5\sin t$, **(a)** find v and a at time t **(b)** draw graphs of s and v against t and **(c)** describe the motion.

7 For $a = 6$, find v and s, given that when $t = 0$, $v = 5$ and $s = 0$. Find t in terms of v and use this to express s in terms of v.

8 **(a)** Given that the acceleration a is constant, find v in terms of a and t if, when $t = 0$, $v = u$.
(b) Integrate your equation for v to find s, given that $s = 0$ when $t = 0$.
(c) Find a in terms of u, v and t.
(d) Find t in terms of v, u and a and substitute for t in your equation for s, to find an expression for v^2 in terms of u, a and s.

Uniform (or constant) acceleration

How deep is a well? How do you find out?
How high is a building? How tall is a tree?

If the well has a bucket on a rope, we can lower the bucket and measure the rope. If the building has several storeys we can measure one storey and multiply. We can draw a scale diagram of the tree (or use trigonometry) if we measure the angle of elevation of the tree top. Some people drop stones down wells; just for amusement?

A stone dropped down a well (or from the roof of a building) accelerates under gravity with constant (uniform) acceleration of $9.8\,\mathrm{m\,s^{-2}}$, i.e. its speed increases by $9.8\,\mathrm{m\,s^{-1}}$ each second.

Investigation 1

If a stone takes 5 seconds to fall down the well (or to reach the ground from the top of a building) how deep is the well? (or how high is the building?)

Theory   Acceleration is constant $\Rightarrow$ $a = \dfrac{dv}{dt} = 9.8$ $\Rightarrow$ $v = \int 9.8\,dt = 9.8t + c$

$t = 0,\ \ v = 0$ $\Rightarrow$ $c = 0$ $\Rightarrow$ $v = 9.8t$ $\Rightarrow$ $\dfrac{ds}{dt} = 9.8t$ $\Rightarrow$ $s = 4.9t^2 + k$

$s = 0,\ \ t = 0$ $\Rightarrow$ $k = 0$ $\Rightarrow$ $s = \frac{1}{2} \times 9.8t^2 = 4.9t^2$

$t = 5$ seconds $\Rightarrow$ $s = 4.9 \times 25 = 122.5\,\text{m}$ $\Rightarrow$ the well is 122.5 m deep or the building is 122.5 m high.

Practical   Drop a stone from the roof of a building and time its fall.

(a)   Using $s = 4.9t^2$, calculate the height of the building.
(b)   Measure the height of the building in some way and use $s = \frac{1}{2}gt^2$ to check g, the acceleration of gravity. Is it really $9.8\,\text{m s}^{-2}$?
(c)   What assumptions are you making in your experiment and calculation?
(d)   How accurate is your result?

This is just one example of motion with constant acceleration.

Constant acceleration formulae

If a is constant, $\dfrac{dv}{dt} = a$ $\Rightarrow$ $v = at + c$ $\cdots\cdots\cdots\cdots\cdots$ **(1)**

If the particle has a starting speed of u, then at $t = 0$, $v = u$ $\Rightarrow$ $u = 0 + c$

$$\Rightarrow \quad \boxed{v = u + at} \quad \cdots\cdots\cdots \quad \textbf{(A)}$$

$v = \dfrac{ds}{dt} = u + at$ $\Rightarrow$ $s = ut + \frac{1}{2}at^2 + k$ $\cdots\cdots\cdots\cdots$ **(2)**

If we measure displacement from $t = 0$, then $s = 0$ when $t = 0$ $\Rightarrow$ $0 = 0 + 0 + k$ in **(2)**

$$\Rightarrow \quad \boxed{s = ut + \tfrac{1}{2}at^2} \quad \cdots\cdots \quad \textbf{(B)}$$

$$
\begin{aligned}
v^2 = (u + at)^2 &= u^2 + 2uat + a^2t^2 \\
&= u^2 + 2a(ut + \tfrac{1}{2}at^2) \\
&= u^2 + 2as
\end{aligned}
\qquad \Rightarrow \quad \boxed{v^2 = u^2 + 2as} \quad \cdots\cdots \quad \textbf{(C)}
$$

From **(A)**, $at = v - u$, so **(B)** becomes $s = ut + \frac{1}{2}t \times at = ut + \frac{1}{2}t(v - u) = ut + \frac{1}{2}vt - \frac{1}{2}ut = \frac{1}{2}ut + \frac{1}{2}vt$

$$\Rightarrow \quad \boxed{s = \dfrac{(u + v)}{2}\,t} \quad \cdots\cdots\cdots \quad \textbf{(D)}$$

If we draw a graph of velocity against time, the acceleration is the gradient (rate of change of velocity with time) and since the acceleration is constant, the gradient will be constant and the graph will be a straight line.

If at time t the velocity is v, then $OT = t$, $TG = OF = v$.
$OS = u$ (starting velocity) and the gradient is a (acceleration).

$$a = \frac{GH}{HS} = \frac{v - u}{t} \quad \Rightarrow \quad v - u = at \quad \Rightarrow \quad v = u + at$$

Distance is given by the area under the graph,

$$\therefore \; s = \text{area of trapezium } OTGS = \frac{(u + v)}{2} \times t \text{ so } s = \tfrac{1}{2}(u + v)t$$

$$s = \text{area } OTHS + \text{area } \triangle SHG = ut + \tfrac{1}{2} \times t \times HG \quad \text{and} \quad \frac{HG}{HS} = a \quad \Rightarrow \quad HG = at$$

$$\Rightarrow \quad s = ut + \tfrac{1}{2} \times t \times at = ut + \tfrac{1}{2}at^2$$

By symmetry, area of $\triangle SGF = \tfrac{1}{2}at^2$ so $s = \text{area } OTGF - \text{area of } \triangle SGF$

$$= vt - \tfrac{1}{2}at^2$$

$$\Rightarrow \quad \boxed{s = vt - \tfrac{1}{2}at^2} \quad \ldots\ldots \quad \textbf{(E)}$$

The five formulae **(A)** to **(E)** are used for solving one-dimensional problems with constant acceleration. There are five unknowns: u, v, t, a and s and each formula omits one of these unknowns.

WORKED EXAMPLE

A ball is thrown **(a)** downwards and **(b)** upwards with a starting velocity of $5 \, \text{m s}^{-1}$ from a window $20 \, \text{m}$ above the ground. Find the difference in **(i)** the time taken to reach the ground and **(ii)** the velocity when the ball reaches the ground in each case (take $g = 10 \, \text{m s}^{-2}$).

(a) Downwards Take the downwards direction as positive.

$$a = 10 \, \text{m s}^{-2}; \quad u = 5 \, \text{m s}^{-1}; \quad s = 20 \, \text{m}; \quad t = ?; \quad v = ?$$

Using $s = ut + \tfrac{1}{2}at^2 \quad \Rightarrow \quad 20 = 5t + \tfrac{1}{2} \times 10t^2 = 5t + 5t^2 \quad \Rightarrow \quad t^2 + t - 4 = 0$

$$\Rightarrow \quad t = \frac{-1 \pm \sqrt{1 + 16}}{2} = \frac{-1 + \sqrt{17}}{2} = 1.56 \, \text{s (neglecting the negative value)}$$

Using $v^2 = u^2 + 2as \quad \Rightarrow \quad v^2 = 25 + 2 \times 10 \times 20 = 425 \quad \Rightarrow \quad v = 5\sqrt{17} \simeq 20.6 \, \text{m}$

(b) Upwards Remember that the upwards velocity is opposite to the downwards acceleration and displacement; take the downwards direction as positive.

$$a = +10 \, \text{m s}^{-2}; \quad u = -5 \, \text{m s}^{-1}; \quad s = +20 \, \text{m}; \quad t = ?; \quad v = ?$$

Using $s = ut + \tfrac{1}{2}at^2 \quad \Rightarrow \quad 20 = -5t + 5t^2 \quad \Rightarrow \quad t^2 - t - 4 = 0$

$$\Rightarrow \quad t = \frac{1 \pm \sqrt{1 + 16}}{2} = \frac{1 + \sqrt{17}}{2} = 2.56 \, \text{s}$$

Using $v^2 = u^2 + 2as$ $\Rightarrow$ $v^2 = (-5)^2 + 2 \times 10 \times 20 = 425$ $\Rightarrow$ $v = 20.6\,\text{m s}^{-1}$

(i) Difference in times $= 2.56 - 1.56 = 1$ second (ii) Velocities are the same.

Did you expect the velocities to be the same? Can you give a reason?
   How long would the ball in the second case take to go up and return to its original position? It may be instructive to plot the graph of v against t for both cases.

Exercise 11.2

In Questions **1** to **7**, u represents the starting velocity, v the finishing velocity, s displacement and a acceleration (constant).

1 Given $u = 6\,\text{m s}^{-1}$, $t = 10\,\text{s}$, $s = 20\,\text{m}$, find **(a)** a and **(b)** v.

2 Given $v = 6\,\text{m s}^{-1}$, $t = 5\,\text{s}$, $a = 4\,\text{m s}^{-2}$, find **(a)** u and **(b)** s.

3 Given $v = 6\,\text{m s}^{-1}$, $t = 4\,\text{s}$, $s = 30\,\text{m}$, find **(a)** u and **(b)** a.

4 Given $u = 3\,\text{m s}^{-1}$, $a = 4\,\text{m s}^{-2}$, $t = 5\,\text{s}$, find **(a)** s and **(b)** v.

5 Given $u = 3\,\text{m s}^{-1}$, $a = 4\,\text{m s}^{-1}$, $v = 5\,\text{m s}^{-1}$, find **(a)** t and **(b)** s.

6 Given $u = 6\,\text{m s}^{-1}$, $a = -2\,\text{m s}^{-2}$, $v = 4\,\text{m s}^{-1}$, find **(a)** t and **(b)** s.

7 Given $u = 10\,\text{m s}^{-1}$, $s = 20\,\text{m}$, $v = 5\,\text{m s}^{-1}$, find **(a)** a and **(b)** t.

8 A ball is thrown upwards with a velocity of $8\,\text{m s}^{-1}$ from a window. Taking $g = 10\,\text{m s}^{-2}$, find **(a)** the time taken to reach the ground and **(b)** the velocity with which the ball hits the ground, if the window is **(i)** 2 m **(ii)** 4 m **(iii)** 6 m above the ground.

9 A particle starting with a velocity of $6\,\text{m s}^{-1}$ accelerates at $7\,\text{m s}^{-2}$. Find the distances covered during the 4th and 5th seconds.

10 A particle covers distances of 1 m and 2 m in successive seconds travelling with constant acceleration. Find the acceleration and the velocities at the beginning and end of each second.

11 Find the time taken for a particle starting at $3\,\text{m s}^{-1}$, accelerating at $2\,\text{m s}^{-2}$, to cover 10 m.

12 An underground train accelerates uniformly from rest to reach a speed of $10\,\text{m s}^{-1}$, then travels at constant speed for 1 minute, then decelerates uniformly to stop after a total time of 2 minutes. Find the distance covered for each section and the time taken for the acceleration and deceleration if **(a)** the acceleration and deceleration are the same and **(b)** if the deceleration is double the acceleration.

Variable acceleration

Motion in two dimensions

In two dimensions we use the x-y plane to describe our motion using the co-ordinates x and y to represent displacements from the origin at time t and unit vectors $\mathbf{i}$ in the x direction and $\mathbf{j}$ in the y direction.

WORKED EXAMPLE

Examine the motion of a particle whose displacement from O is given by $\mathbf{r}=(t-2)\mathbf{i}+\dfrac{t}{3}(t-2)(t-4)\mathbf{j}$

Find the values of $\mathbf{r}$ for $t=0, 1, 2, 3, 4, 5$.

t	0	1	2	3	4	5
$\mathbf{r}$	$-2\mathbf{i}$	$-\mathbf{i}+\mathbf{j}$	0	$\mathbf{i}-\mathbf{j}$	$2\mathbf{i}$	$3\mathbf{i}+5\mathbf{j}$

Since $\mathbf{r}=x\mathbf{i}+y\mathbf{j}$, $x=t-2$ and $y=\dfrac{t}{3}(t-2)(t-4)$

Eliminating t gives the x, y equation of the locus

$$y=\frac{(x+2)}{3}(x)(x-2)=\frac{x}{3}(x^2-4)=\frac{x^3}{3}-\frac{4x}{3}$$

which is a cubic (x^3) curve ($t>0$ gives part of the cubic; allowing $t<0$ also, we get the whole curve).

$t=-1 \Rightarrow \mathbf{r}=-3\mathbf{i}-5\mathbf{j}$ and the curve is an odd function $[f(-x)=-f(x)]$.

The velocity $\mathbf{v}=\dfrac{d\mathbf{r}}{dt}=\mathbf{i}+\left(t^2-4t+\dfrac{8}{3}\right)\mathbf{j}$

The acceleration $\mathbf{a}=\dfrac{d\mathbf{v}}{dt}=(2t-4)\mathbf{j}$

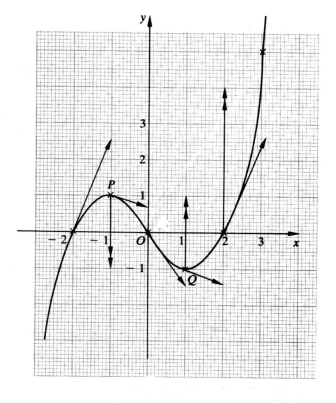

Find the velocity and acceleration at $t=0$ to $t=5$.

t	0	1	2	3	4	5
$\mathbf{v}$	$\mathbf{i}+\dfrac{8}{3}\mathbf{j}$	$\mathbf{i}-\dfrac{1}{3}\mathbf{j}$	$\mathbf{i}-\dfrac{4}{3}\mathbf{j}$	$\mathbf{i}-\dfrac{1}{3}\mathbf{j}$	$\mathbf{i}+\dfrac{8}{3}\mathbf{j}$	$\mathbf{i}+\dfrac{23}{3}\mathbf{j}$
$\mathbf{a}$	$-4\mathbf{j}$	$-2\mathbf{j}$	0	$2\mathbf{j}$	$4\mathbf{j}$	$6\mathbf{j}$

The velocities are plotted (single arrows) and appear to lie along the tangent to the curve. We can check this by calculating the gradient of the curve at each point given by $t=0, 1, 2\ldots$.

$$\frac{dy}{dx}=\frac{3x^2-4}{3}$$

t	0	1	2	3	4
x	-2	-1	0	1	2
$\dfrac{dy}{dx}$	$\dfrac{8}{3}$	$-\dfrac{1}{3}$	$-\dfrac{4}{3}$	$-\dfrac{1}{3}$	$\dfrac{8}{3}$

Since $\dfrac{dy}{dx} = \dfrac{dy/dt}{dx/dt} = \dfrac{\dot{y}}{\dot{x}}$, the components of the velocity lead to the gradient and the velocity is in the direction of the tangent to the curve.

The acceleration vectors (double arrows) lie on the 'inside' of the curves. Acceleration is in the same direction as the resultant force acting, so the force pulls the particle around the bend.

When is the particle moving most slowly? The x component of the velocity is always constant ($\dot{x} = 1$), so do we get minimum velocity when $\dot{y}$ is least?

$\dot{y} = t^2 - 4t + \dfrac{8}{3}$, which is least when $2t - 4 = 0$ i.e. at $t = 2$ $\Rightarrow$ $\mathbf{v} = \mathbf{i} - \dfrac{4}{3}\mathbf{j}$

However we are concerned with the magnitude of $\dfrac{dy}{dt}$, since a negative velocity implies a change in direction. So we look for the value of t for which $\dot{y} = 0$.

$\dot{y} = 0$ $\Rightarrow$ $t^2 - 4t + \dfrac{8}{3} = 0$ $\Rightarrow$ $3t^2 - 12t + 8 = 0$ $\Rightarrow$ $t = \dfrac{12 \pm \sqrt{144 - 96}}{6} = \dfrac{12 \pm \sqrt{48}}{6} = \dfrac{6 \pm 2\sqrt{3}}{3}$

$\Rightarrow$ $t = 2 \pm \dfrac{2}{\sqrt{3}}$ i.e. $\dfrac{2}{\sqrt{3}} = 1.15$ seconds either side of $t = 2$.

At these times (points P and Q on the graph) the velocity has magnitude 1 (from $\mathbf{v} = \mathbf{i}$) which is less than the speed at $t = 2$ when $\mathbf{v} = \mathbf{i} - \dfrac{4}{3}\mathbf{j}$ $\left(\text{magnitude } \dfrac{5}{3}\right)$.

Can you think of a practical situation for which this Example might be a model?

Exercise 11.3   (A computer graph plotting program might help!)

1 Plot the path followed by a particle whose position vector is $\mathbf{x} = (3 - 2t)\mathbf{i} + (t + 4)\mathbf{j}$, for values of t from 0 to 5. Find the x, y equation for the path the particle follows. Find $\mathbf{v}$ and $\mathbf{a}$ and a vector in the direction of motion.

   Give the position vector, $\mathbf{d}$, for a particle starting at the same point but travelling in the opposite direction.

2 Plot the path followed by a particle with position vector $\mathbf{r} = t\mathbf{i} + t^2\mathbf{j}$ for $-2 \leqslant t \leqslant 2$. Find the velocity and acceleration. When is the speed least? Find the x, y equation.

3 Repeat question 2 for $\mathbf{r} = t^2\mathbf{i} + t\mathbf{j}$.

4 Examine the motion for $\mathbf{r} = t^2\mathbf{i} + t^3\mathbf{j}$ for $-2 \leqslant t \leqslant +2$. What happens at the origin?

5 Examine the motion for $\mathbf{r} = t\mathbf{i} + t^3\mathbf{j}$. When is the particle moving most slowly?

6 Examine the motion for $\mathbf{r} = \cos t\,\mathbf{i} + \sin t\,\mathbf{j}$. Take values of $t = 0, \pi/6, \pi/4, \pi/3, \pi/2$ etc. to $t = 2\pi$. Find the x, y equation. Find $\mathbf{v}$ and compare it with $\mathbf{r}$. What can you say about $|\mathbf{v}|$, the speed? Find $\mathbf{a}$ and comment.

7 Repeat equation 6 for (a) $\mathbf{r} = 2\cos t\,\mathbf{i} + 2\sin t\,\mathbf{j}$ and (b) $\mathbf{r} = \cos 2t\,\mathbf{i} + \sin 2t\,\mathbf{j}$.

8 Examine the motion for $\mathbf{r} = \cos^2 t\,\mathbf{i} + \sin^2 t\,\mathbf{j}$. Find the x, y equation, velocity and acceleration. Can you produce the equation to make the particle describe a square?!

9 Examine the motion for $\mathbf{r} = \cos^3 t\,\mathbf{i} + \sin^3 t\,\mathbf{j}$. What happens to the velocity and acceleration when $t = \pi/2$?

10   Examine the motion for $\mathbf{r} = \cos t\mathbf{i} + \cos 2t\mathbf{j}$. Find the x, y equation and which t values produce the complete curve.

11   Examine the motion for $\mathbf{r} = \cos t\mathbf{i} + \cos 3t\mathbf{j}$.

12   Examine the motion for $\mathbf{r} = \cos t\mathbf{i} + \sin 2t\mathbf{j}$.

13   Examine the motion for $\mathbf{r} = 3\cos t\mathbf{i} + 4\sin t\mathbf{j}$. Find the greatest and least speeds.

14   Examine $\mathbf{r} = 5t\mathbf{i} + (20t - 5t^2)\mathbf{j}$. Find the x, y equation, velocity and acceleration.
   Find the speeds when $y = 0$. Find the maximum positive value for y, and the t value at which this occurs.

This last question leads into the next section on projectiles.

Projectiles

How many different types of water sprinklers have you seen?

Which type is used by a market gardener, a house gardener, a groundsman?

How is the whole area watered?

What path does the water follow in the air?

What patterns have you seen?

Which angle of water jet travels farthest?

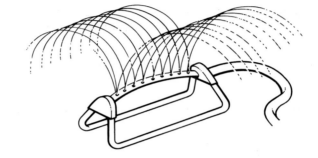

Each water droplet is a projectile, as are balls thrown in the air or the shot, discus and hammer used in athletics throwing events. Any particle moving under the action of gravity (its own weight) is a projectile, although we reserve the term to describe particles **projected** at some angle or speed. We also restrict ourselves to motion near the surface of the earth so that g (acceleration of gravity) can be regarded as constant.

At this stage we cannot consider **projectiles** where **air resistance** has a significant effect on the flight of the projectile. An athlete's javelin, a golf ball or a parachutist might be in this category. Can you say why air resistance affects their flight?

Investigation 2

If my garden hose delivers water at 10 m s^{-1}, can I rinse my car which stands 20 m from the end of my hose?

Assumption   Each drop of water has to travel 20 m horizontally, with the hose held at an angle of α above the horizontal.

Theory   Choose axes x horizontally and y vertically with hose at origin O.

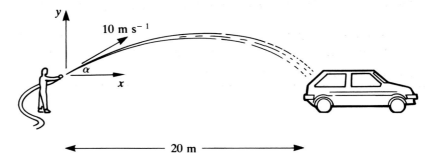

Let the displacement of each drop of water be given by $x = x\mathbf{i} + y\mathbf{j}$.

The starting velocity is 10 m s^{-1} at angle α above the horizontal

$t = 0 \Rightarrow \mathbf{v} = 10 \cos \alpha.\mathbf{i} + 10 \sin \alpha.\mathbf{j}$

The only force acting on the drop of water is its weight, so its acceleration $\mathbf{a} = 0\mathbf{i} - 9.8\mathbf{j}$.
   The acceleration of gravity g is 9.8 m s^{-2} vertically downwards.

$\mathbf{a} = \dfrac{d\mathbf{v}}{dt} = -9.8\mathbf{j} \Rightarrow \mathbf{v} = -9.8t\mathbf{j} + \mathbf{c}$ where $\mathbf{c}$ is the value of $\mathbf{v}$ when $t = 0$ i.e. starting velocity.

$\mathbf{v} = 10 \cos \alpha.\mathbf{i} + 10 \sin \alpha.\mathbf{j} - 9.8t\mathbf{j}$ and since $\mathbf{v} = \dfrac{d\mathbf{r}}{dt} \Rightarrow \mathbf{r} = \int \mathbf{v} \, dt.$

$\mathbf{r} = 10 \cos \alpha.t\mathbf{i} + 10 \sin \alpha.t\mathbf{j} - 9.8\dfrac{t^2}{2}\mathbf{j}.$

$\mathbf{r} = x\mathbf{i} + y\mathbf{j} \Rightarrow x = 10 \cos \alpha.t$ and $y = 10 \sin \alpha.t - 4.9t^2.$

To reach the car, $x = 20 \Rightarrow 20 = 10 \cos \alpha.t \Rightarrow t = \dfrac{2}{\cos \alpha}.$

$y = 0 \Rightarrow 0 = 10 \sin \alpha.t - 4.9t^2 \Rightarrow t = 0 \text{(start)}$ or $t = \dfrac{10 \sin \alpha}{4.9}.$

$\dfrac{2}{\cos \alpha} = \dfrac{10 \sin \alpha}{4.9} \Rightarrow 10 \sin \alpha \cos \alpha = 9.8 \Rightarrow \sin 2\alpha = \dfrac{9.8}{5} = 1.96,$ which has no solution for α.

So the water will not reach the car. But how far will it reach?

General theory of projectiles

Let a particle be projected with speed u at an angle α above the horizontal.

Horizontally there is no acceleration.

$$v_x = \dot{x} = u \cos \alpha \Rightarrow x = u \cos \alpha.t$$

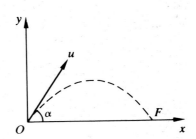

Vertically, the initial velocity is $u \sin \alpha$ and the acceleration $-g\,(-9.8\,\text{m s}^{-2})$ i.e. downwards, so

$$\frac{d^2y}{dt^2} = -g \quad \Rightarrow \quad \frac{dy}{dt} = -gt + c, \text{ where } c \text{ is the value of } \dot{y} \text{ when } t = 0 \quad \text{i.e. } c = u \sin \alpha.$$

$$\Rightarrow \quad \frac{dy}{dt} = \dot{y} = -gt + u \sin \alpha \quad \Rightarrow \quad y = u \sin \alpha . t - \tfrac{1}{2}gt^2 \text{ since } y = 0 \text{ when } t = 0.$$

In vector notation, $\quad r = x\mathbf{i} + y\mathbf{j} = u \cos \alpha . t \mathbf{i} + (u \sin \alpha . t - \tfrac{1}{2}gt^2)\mathbf{j}$

Time of flight (T)

The time the particle takes to reach F is called the time of flight.

At F, $\quad y = 0 \quad \Rightarrow \quad 0 = u \sin \alpha . t - \tfrac{1}{2}gt^2 \quad \Rightarrow \quad t = 0 \text{ (at } O\text{), or} \quad \boxed{T = \frac{2u \sin \alpha}{g}} \quad \text{at } F.$

Horizontal range (R) and greatest height (H)

The **horizontal range** (R) is the range of the particle horizontally i.e. how far it travels in the horizontal direction.

$$R \text{ is the value of } x \text{ when } T = \frac{2u \sin \alpha}{g} \quad \Rightarrow \quad R = u \cos \alpha . t = u \cos \alpha . \frac{2u \sin \alpha}{g} = \frac{u^2}{g} . 2 \sin \alpha \cos \alpha$$

$$\Rightarrow \quad \boxed{R = \frac{u^2 \sin 2\alpha}{g}}$$

The **maximum horizontal range** for a given starting velocity is achieved when $\sin 2\alpha$ is at its maximum value i.e. $\sin 2\alpha = 1 \quad \Rightarrow \quad 2\alpha = 90° \quad \Rightarrow \quad \alpha = 45°$ (angle of projection).

The maximum range of my water hose is $\dfrac{u^2}{g} = \dfrac{10^2}{9.8} = 10.2\,\text{m}$. To reach my car I need $\dfrac{u^2}{g} = 20 \quad \Rightarrow \quad u^2 = 20 \times 9.8 = 196 \quad \Rightarrow \quad u = 14\,\text{m s}^{-1}$.

The **greatest height** for a given angle of projection α, is achieved when $\dot{y} = 0$ i.e.

$$u \sin \alpha - gt = 0 \quad \Rightarrow \quad t = \frac{u \sin \alpha}{g} \quad \text{(this is also 'half-time' i.e. } t = \frac{u \sin \alpha}{g} = \tfrac{1}{2}T\text{)}$$

The **greatest height** (H) is the value of y when $t = \dfrac{u \sin \alpha}{g}$.

$$H = u \sin \alpha . t - \tfrac{1}{2}gt^2 = u \sin \alpha \frac{u \sin \alpha}{g} - \frac{g}{2} . \frac{u^2 \sin^2 \alpha}{g} = \frac{u^2 \sin^2 \alpha}{g} - \frac{u^2 \sin^2 \alpha}{2g} \quad \Rightarrow \quad \boxed{H = \frac{u^2 \sin^2 \alpha}{2g}}$$

By varying the angle, the greatest height is achieved when $\sin^2 \alpha = 1 \quad \Rightarrow \quad \alpha = 90°$ i.e. project upwards, rather obviously.

WORKED EXAMPLE

If my small son can throw his ball at $20\,\mathrm{m\,s^{-1}}$, at what angle must he throw the ball to just clear the garden wall, which is $20\,\mathrm{m}$ away and $2\,\mathrm{m}$ higher than his arm?

$$x = 20\,\mathrm{m}; \quad y = 2\,\mathrm{m}; \quad u = 20\,\mathrm{m\,s^{-1}}; \quad \alpha = ?; \quad t = ?.$$

$$x = u\cos\alpha.t \quad \Rightarrow \quad 20 = 20\cos\alpha.t \quad \Rightarrow \quad t = 1/\cos\alpha = \sec\alpha$$

$$y = u\sin\alpha.t - \tfrac{1}{2}gt^2 \quad \Rightarrow \quad 2 = 20\sin\alpha.t - \tfrac{1}{2}gt^2$$

Substituting $t = \sec\alpha \quad \Rightarrow \quad 2 = 20\sin\alpha\sec\alpha - \tfrac{1}{2}g\sec^2\alpha = 20\tan\alpha - \tfrac{1}{2}g(1 + \tan^2\alpha)$

Taking $g = 10\,\mathrm{m\,s^{-2}} \quad \Rightarrow \quad 2 = 20\tan\alpha - 5(1 + \tan^2\alpha) \quad \Rightarrow \quad 5\tan^2\alpha - 20\tan\alpha + 7 = 0$

$$\tan\alpha = \frac{20 \pm \sqrt{400 - 4\times 5\times 7}}{10} = \frac{20 \pm \sqrt{260}}{10} = \frac{20 \pm 16.12}{10} = 3.612 \text{ or } 0.388 \quad \Rightarrow \quad \alpha = 72.5° \text{ or } 21.2°.$$

There are two possible angles of projection to just clear the wall, $\alpha = 21.2°$ (low skimmer) or $72.5°$ (high diver).

Exercise 11.4   (Take $g = 10\,\mathrm{m\,s^{-2}}$)

1   A particle is projected with a speed of $20\,\mathrm{m\,s^{-1}}$ at an angle of $60°$ above the horizontal. Find its velocity and position **(a)** $1\,\mathrm{s}$,   **(b)** $2\,\mathrm{s}$,   **(c)** $3\,\mathrm{s}$ later. If it was projected from ground level, when does it hit the ground?

2   A particle is projected from ground level with speed $10\,\mathrm{m\,s^{-1}}$ at an angle of $30°$ above the horizontal. Find **(a)**   the greatest height,   **(b)**   the horizontal range and   **(c)**   the time of flight.

3   A ball is thrown with speed $20\,\mathrm{m\,s^{-1}}$ at an angle of $60°$ to clear a wall $2\,\mathrm{m}$ high. How far away is the wall?

4   A particle is projected from ground level with a velocity of $15\,\mathrm{m\,s^{-1}}$, and hits the ground 1 second later. Find the angle of projection, the greatest height and the range.

5   A particle is projected with a velocity of $10\,\mathrm{m\,s^{-1}}$ at an angle of $\alpha = \sin^{-1}\tfrac{4}{5}$ above the horizontal. It hits the ground at a point $4\,\mathrm{m}$ below its point of projection. Find the time the particle is in the air and the horizontal distance travelled.

6   A stone is thrown horizontally at $20\,\mathrm{m\,s^{-1}}$ from the top of a vertical cliff $50\,\mathrm{m}$ above sea level. How far out to sea does the stone hit the water?

7   A particle is projected from O with velocity $6\mathbf{i} + 8\mathbf{j}$. Find its velocity and displacement after **(a)** $1\,\mathrm{s}$,   **(b)** $2\,\mathrm{s}$, **(c)** $3\,\mathrm{s}$. Find the equation of the path it follows.

8   Find the angle of projection of a ball thrown at $30\,\mathrm{m\,s^{-1}}$ if it just clears a wall $3\,\mathrm{m}$ high when travelling horizontally.

9   A particle projected from O passes through the point with position vector $5\mathbf{i} + 2\mathbf{j}$ after 2 seconds. Find the initial velocity vector, the greatest height, the horizontal range and the equation of the path of the particle.

10   A cricketer can throw the ball at $30\,\mathrm{m\,s^{-1}}$. How long does the ball take to reach the wicket-keeper if the fielder throws in from the boundary ($70\,\mathrm{m}$ away)? Dare the batsmen take another run while he is throwing in?

11   An archer hits a target $90\,\mathrm{m}$ away when he fires his arrows at $50\,\mathrm{m\,s^{-1}}$. Find the time of flight.

12 Estimate the initial speed of **(a)** a short-putt of 20 m, **(b)** a discus thrown 60 m, **(c)** a hammer-throw of 75 m, **(d)** a rugby conversion from the half-way line, **(e)** a goalkeeper's soccer kick to the half-way line and **(f)** a golf drive of 300 m. Assume maximum range can be achieved by projecting at 45° above horizontal.

13 A soccer player 'chips' a free-kick 30 m out over the defenders 'wall' to hit the goal bar. What is the speed of his initial kick and how long elapses between the ball clearing the wall and hitting the bar.

14 If ski jumpers jump 90 m (measured down the slope of 30°) with a take-off speed of $30 \, \text{m s}^{-1}$, what is the angle of the take-off ramp?

15 With what speed does a quarter-back throw an American football for a touchdown pass of 50 m?

12 Relative motion

Relative velocity

Have you noticed that, if you stand on a station platform a passing express takes a very short time to go by, but if you are on a moving train an overtaking express takes considerably longer to pass?

How careful you have to be when you pull out to overtake another car when, in the distance, you can see a car approaching!

The gap seems to close so quickly.

These are both examples of relative motion, although in one case the trains are travelling in the same direction and the approaching cars are moving in opposite directions.

Investigation 1

An express train, travelling at $25\,\text{m s}^{-1}$, passes a signal. If the train is 200 m long, show that the time taken for the train to pass the signal is 8 seconds.

This train now catches and overtakes a freight train on a parallel track which is travelling at $15\,\text{m s}^{-1}$. If this train is also 200 m long, and the drivers of both trains pass a telegraph pole at the same moment, find how far the two drivers have moved in 20 seconds.

What can you say about the position of the back of the express and the front of the freight train after 20 seconds? How long does it take for the express to completely pass the freight train?

(a) How much faster than the freight train does the express travel?

(b) How much further does the express travel in passing the freight train?

Find the time taken to travel this distance at this speed. What do you notice about your answer?

Parallel motion in the same sense

Consider two trains, A and B, travelling on parallel tracks in the same direction, at $80\,\text{km h}^{-1}$ and $120\,\text{km h}^{-1}$ respectively.

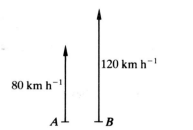

Effectively, train B is overtaking train A at an excess speed of $40\,\text{km h}^{-1}$.

This means that, to an observer on train A, train B appears to be moving at $40\,\text{km h}^{-1}$ in the same direction.

We say that the **velocity of B relative to A** is $40\,\text{km h}^{-1}$ in the direction of motion.

Remember that, to a passenger on train B, train A **appears** to be moving backwards, since train B is moving faster. This is equivalent to saying that train A appears to be moving with a velocity of $-40\,\text{km h}^{-1}$ and we say that **the velocity of A relative to B** is $40\,\text{km h}^{-1}$ in a direction opposite to the direction of travel.

Parallel motion in the opposite sense

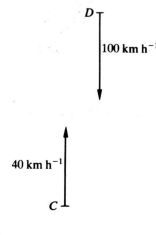

Consider two cars C and D travelling towards each other at speeds of $40\,\text{km h}^{-1}$ and $100\,\text{km h}^{-1}$ respectively.

Effectively, car D is approaching car C at a speed of $140\,\text{km h}^{-1}$.

The **velocity of D relative to C** is $140\,\text{km h}^{-1}$ in the direction of motion of D.

Similarly, the **velocity of C relative to D** is $140\,\text{km h}^{-1}$ in the direction of motion of C.

In general, if two particles, P and Q, have velocities $\mathbf{v}_1$ and $\mathbf{v}_2$ respectively in parallel directions, then

the velocity of P relative to Q is $\mathbf{v}_1 - \mathbf{v}_2$

and the velocity of Q relative to P is $\mathbf{v}_2 - \mathbf{v}_1$.

Remember that relative velocity is a **vector** quantity.

WORKED EXAMPLE

In a manoeuvre in a flying display, two pilots fly towards each other at $800\,\text{km h}^{-1}$ due east and $700\,\text{km h}^{-1}$ due west before veering off at the last moment. Find the relative speed of the aircraft in level flight.

Let $\mathbf{i}$ be the unit vector in the direction due east.

The velocity of the first aircraft can be written $\mathbf{v}_1 = 800\mathbf{i}$.

The velocity of the second aircraft is $\mathbf{v}_2 = -700\mathbf{i}$.

The velocity of the first aircraft to the second

aircraft $= \mathbf{v}_1 - \mathbf{v}_2 = 800\mathbf{i} - (-700\mathbf{i}) = 1500\mathbf{i}$

i.e. they close at a **relative speed** of $1500\,\text{km h}^{-1}$.

Non-parallel motion

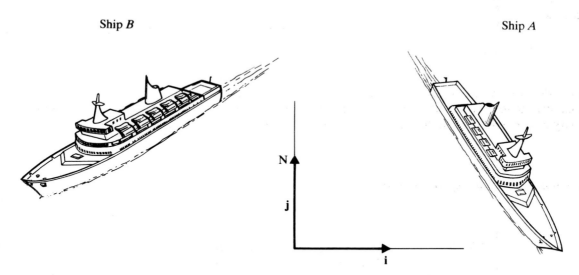

Ship B Ship A

Consider two ships, A and B, moving with velocities $\mathbf{v}_1$ and $\mathbf{v}_2$ respectively. If ship A has a velocity of 20 knots on a bearing of $150°$ and ship B has a velocity of 24 knots on a bearing of $240°$, then

$$\mathbf{v}_1 = 20\cos 60°\,\mathbf{i} - 20\sin 60°\,\mathbf{j} = 10\mathbf{i} - 10\sqrt{3}\mathbf{j}$$

$$\text{and } \mathbf{v}_2 = -24\cos 30°\,\mathbf{i} - 24\sin 30°\,\mathbf{j} = -12\sqrt{3}\mathbf{i} - 12\mathbf{j}$$

where $\mathbf{i}$ and $\mathbf{j}$ are unit vectors due east and due north, respectively.

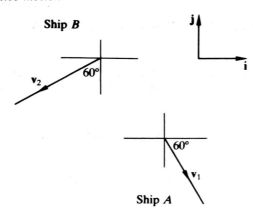

Ship *B*

Ship *A*

Since the ships are **not** moving in parallel directions as were the trains and cars in the first section, it is **not** possible to just add or subtract the velocities to find the relative velocities.

We use the fact that the relative velocity will be unchanged if any velocity is impressed on both ships.

We choose to impose a velocity of $-\mathbf{v}_2$ on both ships, which enables us to regard ship *B* as being at rest and ship *A* as moving with the resultant velocity of $\mathbf{v}_1$ and $-\mathbf{v}_2$, i.e. $\mathbf{v}_1 - \mathbf{v}_2$.

This is the velocity of *A* **relative to** $B = \mathbf{v}_1 - \mathbf{v}_2$.

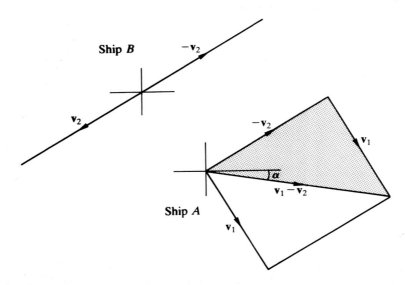

To find this resultant of $\mathbf{v}_1$ and $-\mathbf{v}_2$, draw a velocity vector diagram which shows the effect of imposing the velocity $-\mathbf{v}_2$ on both ships. The diagonal of the parallelogram formed with the vectors $\mathbf{v}_1$ and $-\mathbf{v}_2$ represents the velocity of *A* relative to *B*.

The magnitude and direction of this relative velocity can be found using vector methods or by solving the vector triangle using the sine and cosine rules, as the Worked Examples will show.

Notice that, at the moment, we have not mentioned the actual positions of the ships, and the diagrams do not indicate their displacements.

Hence, $\mathbf{v}_1 = 10\mathbf{i} - 10\sqrt{3}\mathbf{j}$ and $\mathbf{v}_2 = -12\sqrt{3}\mathbf{i} - 12\mathbf{j}$

$\Rightarrow$ the velocity of *A* relative to $B = \mathbf{v}_1 - \mathbf{v}_2 = (10 + 12\sqrt{3})\mathbf{i} + (12 - 10\sqrt{3})\mathbf{j} = 30.78\mathbf{i} - 5.32\mathbf{j}$

Hence the relative speed $= |\mathbf{v}_1 - \mathbf{v}_2| = \sqrt{(30.78)^2 + (-5.32)^2} = \sqrt{975.71} = 31.24 \, \text{knots}$

The direction is given by $\tan\alpha = \dfrac{5.32}{30.78} = 0.1728$

$$\Rightarrow \quad \alpha = 9.81° \text{ or } 9° \, 48'$$

Thus the **velocity of A relative to B** is 31.24 knots on a bearing of 099° 48'. This is the velocity of ship A as it **appears** to an observer on ship B.

The diagram below shows the effect of imposing a velocity $-\mathbf{v}_1$ on the system which reduces ship A to rest and gives ship B a velocity which is the resultant of $\mathbf{v}_2$ and $-\mathbf{v}_1$. The diagonal of the parallelogram represents the **velocity of B relative to $A = \mathbf{v}_2 - \mathbf{v}_1$**.

Since the dimensions and directions of this diagram are the same as those in the previous diagram, the calculation will be essentially the same.

The **velocity of B relative to A** is 31.24 knots on a bearing of 279° 48', i.e. $\mathbf{v}_2 - \mathbf{v}_1 = -30.78\mathbf{i} + 5.32\mathbf{j}$.

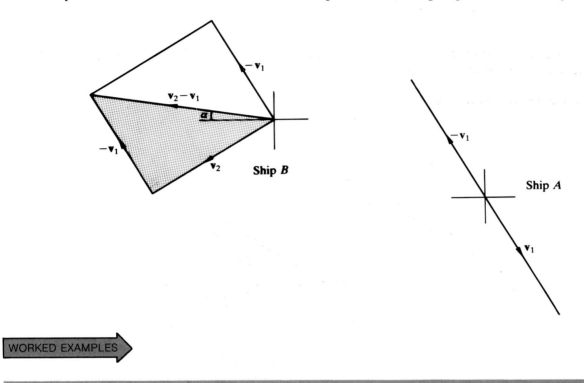

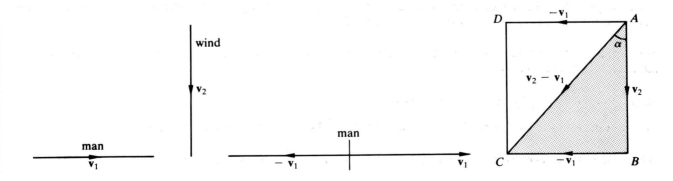

WORKED EXAMPLES

1 A man walks due east at 6 km h^{-1} with the wind blowing from the north at 8 km h^{-1}. Find the magnitude and direction of the velocity of the wind relative to the man.

From the diagram we see the actual velocities of the man and the wind represented by two vectors and the effect of imposing a velocity $-\mathbf{v}_1$ on the system. This effectively brings the man to rest and the vector represented by the diagonal AC of the rectangle $ABCD$ gives the velocity of the wind relative to the man.

Using Pythagoras' theorem in triangle ABC,

$$AC^2 = AB^2 + BC^2$$

But $|\mathbf{v}_2 - \mathbf{v}_1| = AC$, $\quad |\mathbf{v}_2| = AB = 8$ $\quad$ and $\quad |\mathbf{v}_1| = BC = 6$

Hence, $AC^2 = 6^2 + 8^2 = 100 \quad \Leftrightarrow \quad AC = 10$

Also in triangle ABC, $\quad \tan \alpha = \dfrac{BC}{AB} = \dfrac{6}{8} = 0.75 \quad \Leftrightarrow \quad \alpha = 36.87°$ or $36° 52'$

The velocity of the wind relative to the man is $10 \, \text{km h}^{-1}$ blowing from $036° 52'$.

Notice, the non-vector approach to the calculation. Alternatively, we could set up unit vectors $\mathbf{i}$ and $\mathbf{j}$, due east and due north, respectively. In which case,

The velocity of the man, $\mathbf{v}_1 = 6\mathbf{i}$ and the velocity of the wind, $\mathbf{v}_2 = -8\mathbf{j}$

$\therefore$ the velocity of the wind relative to the man $= \mathbf{v}_2 - \mathbf{v}_1 = -6\mathbf{i} - 8\mathbf{j}$

This gives $|\mathbf{v}_2 - \mathbf{v}_1| = \sqrt{(-6)^2 + (-8)^2} = 10$ and $\tan \alpha = \dfrac{3}{4} \quad \Leftrightarrow \quad \alpha = 36° 52'$.

2 To an observer on a destroyer moving with velocity $\mathbf{v}_1 = 2\sqrt{2}\mathbf{i} - 4\mathbf{j}$, a tanker appears to be moving with a velocity $-3\sqrt{2}\mathbf{i} - \mathbf{j}$, where $\mathbf{i}$ and $\mathbf{j}$ are the unit vectors due east and due north, respectively. Find the actual magnitude and direction of the velocity of the tanker, $\mathbf{v}_2$, if speeds are measured in knots.

In this example, the actual velocity of the tanker is not known and the relevant vector diagrams must be built up with care. If we impose a velocity of $-\mathbf{v}_1$ on both vessels, the destroyer will be at rest and the tanker will be subject to two velocities $\mathbf{v}_2$ and $-\mathbf{v}_1$, the resultant of which is $\mathbf{v}_2 - \mathbf{v}_1$, the relative velocity.

In triangle ABC, the sides $\overrightarrow{AB}$ and $\overrightarrow{AC}$ representing $-\mathbf{v}_1$ and $\mathbf{v}_2 - \mathbf{v}_1$, respectively can be drawn in the correct position and hence the third side, $\overrightarrow{BC}$, will represent $\mathbf{v}_2$, the velocity of the tanker.

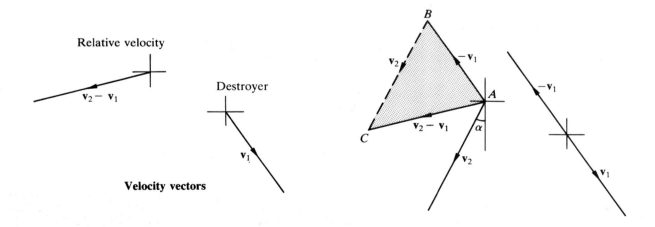

The velocity of the tanker relative to the destroyer $= \mathbf{v}_2 - \mathbf{v}_1$

$$\therefore -3\sqrt{2}\mathbf{i} - \mathbf{j} = \mathbf{v}_2 - (2\sqrt{2}\mathbf{i} - 4\mathbf{j})$$

$$\Leftrightarrow \quad \mathbf{v}_2 = -\sqrt{2}\mathbf{i} - 5\mathbf{j}$$

Hence, $\quad |\mathbf{v}_2| = \sqrt{(-\sqrt{2})^2 + (-5)^2} = \sqrt{27} = 5.2$

If α is the angle that $|\mathbf{v}_2|$ makes with due south, then $\tan\alpha = \dfrac{\sqrt{2}}{5}$.

$$\Leftrightarrow \quad \tan\alpha = 0.2828 \quad \Leftrightarrow \quad \alpha = 15.79° \quad \text{or} \quad 15°48'$$

The velocity of the tanker is 5.2 knots on a bearing of 195°48'.

---

3 A passenger on a bus travelling at $40\,\text{km}\,\text{h}^{-1}$ north-west sees a car which is actually moving at $80\,\text{km}\,\text{h}^{-1}$ in a direction N 60°E. Find the apparent velocity of the car to the passenger on the bus.

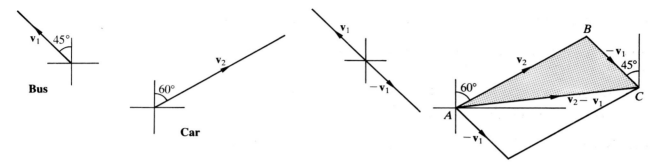

In this example, the actual velocities of the bus, $\mathbf{v}_1$, and the car, $\mathbf{v}_2$, are known, so we impose a velocity of $-\mathbf{v}_1$ on the system, which brings the bus to rest and has the car subject to two velocities $\mathbf{v}_2$ and $-\mathbf{v}_1$. The velocity of the car relative to the bus is represented by $\overrightarrow{AC}$ in the triangle ABC.

Since $A\hat{B}C = 60° + 45°$ (alternate angles) we have, using the cosine rule in triangle ABC,

$$AC^2 = AB^2 + BC^2 - (2\,.\,AB\,.\,BC \cos A\hat{B}C)$$
$$|\mathbf{v}_2 - \mathbf{v}_1|^2 = |\mathbf{v}_2|^2 + |\mathbf{v}_1|^2 - (2\,.\,|\mathbf{v}_2|\,.\,|\mathbf{v}_1| \cos 105°)$$
$$= 80^2 + 40^2 - (2 \times 40 \times 80 \cos 105°)$$
$$= 6400 + 1600 - (-1656.44) = 9656.44$$

$$\text{Hence,} \quad AC = |\mathbf{v}_2 - \mathbf{v}_1| = \sqrt{9656.44} = 98.27\,\text{km}\,\text{h}^{-1}$$

Using the sine rule in triangle ABC,

$$\frac{AC}{\sin A\hat{B}C} = \frac{BC}{\sin B\hat{A}C} \quad \Leftrightarrow \quad \sin B\hat{A}C = \frac{BC \sin A\hat{B}C}{AC}$$

$$\Rightarrow \quad \sin B\hat{A}C = \frac{40 \sin 105°}{98.27} = 0.3932$$
$$\Rightarrow \quad B\hat{A}C = 23.15° \quad \text{or} \quad 23°9'$$

Hence the velocity of the car relative to the bus is $98.27\,\text{km}\,\text{h}^{-1}$ in a direction N 83°09' E.

---

4 To a runner moving at $12\,\text{km}\,\text{h}^{-1}$ due north, the wind appears to blow from the east; but to a cyclist travelling at $20\,\text{km}\,\text{h}^{-1}$ south-west, the wind appears to blow from the south. Find the true velocity of the wind.

In this example the actual velocity of the wind is not known, so let it be represented by the vector $\mathbf{w} = a\mathbf{i} + b\mathbf{j}$ where $\mathbf{i}$ and $\mathbf{j}$ are the unit vectors due east and due north, respectively.

Two velocity vector diagrams must be drawn for each of the situations. Remember that the line representing the velocity of the wind will be the same in both.

In the vector diagram of the runner moving with velocity $\mathbf{v}_1$, impose a velocity $-\mathbf{v}_1$ on the system which will bring the runner to rest, and in the vector diagram of the cyclist moving with velocity $\mathbf{v}_2$, impose a velocity $-\mathbf{v}_2$ on the system.

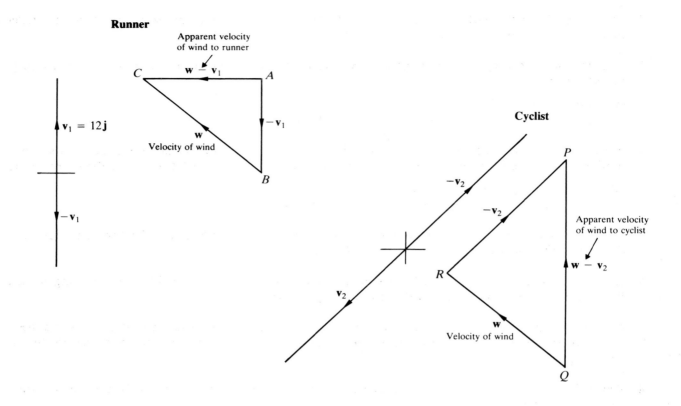

The velocity of the wind is represented by $\overrightarrow{BC}$ in triangle ABC and by $\overrightarrow{QR}$ in triangle PQR.

Since $\mathbf{v}_1 = 12\mathbf{j}$ and $\mathbf{w} = a\mathbf{i} + b\mathbf{j}$, the velocity of the wind relative to the runner $= \mathbf{w} - \mathbf{v}_1$

$$= (a\mathbf{i} + b\mathbf{j}) - 12\mathbf{j} = a\mathbf{i} + (b - 12)\mathbf{j}$$

Since the direction of the relative velocity is due west (i.e. a multiple of $-\mathbf{i}$), $b - 12 = 0 \quad \Rightarrow \quad b = 12$.

Since $\mathbf{v}_2 = -20\cos 45°\mathbf{i} - 20\sin 45°\mathbf{j} = -10\sqrt{2}\mathbf{i} - 10\sqrt{2}\mathbf{j}$,

the velocity of the wind relative to the cyclist $= \mathbf{w} - \mathbf{v}_2$

$$= (a\mathbf{i} + b\mathbf{j}) - (-10\sqrt{2}\mathbf{i} - 10\sqrt{2}\mathbf{j})$$

$$= (a + 10\sqrt{2})\mathbf{i} + (b + 10\sqrt{2})\mathbf{j}$$

Since the direction of the relative velocity is due north (i.e. a multiple of $\mathbf{j}$),

$$a + 10\sqrt{2} = 0 \quad \Rightarrow \quad a = -10\sqrt{2}$$

The true velocity of the wind $\mathbf{w} = a\mathbf{i} + b\mathbf{j} = -10\sqrt{2}\mathbf{i} + 12\mathbf{j}$,

or $\sqrt{344} = 18.55 \,\mathrm{km\,h^{-1}}$ at an angle $\tan^{-1}\left(\dfrac{10\sqrt{2}}{12}\right) = 49.7°$ to north. i.e. $18.55\,\mathrm{km\,h^{-1}}$ from a direction $130.3°$.

1 A police car travelling at $200\,\mathrm{km\,h^{-1}}$ is pursuing a car moving at $160\,\mathrm{km\,h^{-1}}$ on a motorway. What is their relative speed?

2 Two trains, A and B, travelling at $180\,\mathrm{km\,h^{-1}}$ and $100\,\mathrm{km\,h^{-1}}$ respectively, approach each other on parallel tracks. Find the velocity of train A relative to train B.

3 Two particles, P and Q, move with velocities $\mathbf{v}_1$ and $\mathbf{v}_2$, respectively. Find the velocity of P relative to Q in vector form, if

(a) $\mathbf{v}_1 = 2\mathbf{i} + 3\mathbf{j}$ (b) $\mathbf{v}_1 = \mathbf{i} - 2\mathbf{j}$ (c) $\mathbf{v}_1 = -5\mathbf{i} + 2\mathbf{j} + 3\mathbf{k}$
     $\mathbf{v}_2 = -\mathbf{i} + 4\mathbf{j}$ $\mathbf{v}_2 = 4\mathbf{i} + 3\mathbf{j}$ $\mathbf{v}_2 = \mathbf{i} - 4\mathbf{j} - \mathbf{k}$

4 A walker moves with velocity $\mathbf{v}_1$ and a cyclist with velocity $\mathbf{v}_2$. If $\mathbf{i}$ and $\mathbf{j}$ are unit vectors due east and due north respectively, find the magnitude and direction of the velocity of the cyclist relative to the walker, when,

(a) $\mathbf{v}_1 = 3\mathbf{i} + 4\mathbf{j}$ (b) $\mathbf{v}_1 = \mathbf{i} - \mathbf{j}$ (c) $\mathbf{v}_1 = -\mathbf{i} + 7\mathbf{j}$
     $\mathbf{v}_2 = -4\mathbf{i} + 3\mathbf{j}$ $\mathbf{v}_2 = -\mathbf{i} - \mathbf{j}$ $\mathbf{v}_2 = 3\mathbf{i} + 4\mathbf{j}$

5 A man walks at $4\,\mathrm{km\,h^{-1}}$ due north in a wind blowing from the east at $5\,\mathrm{km\,h^{-1}}$. Find the magnitude and direction of the velocity of the wind relative to the man.

6 A ship, P, is sailing at 20 knots on a bearing of $315°$ and a ship, Q, is sailing at 15 knots on a bearing of $225°$. Find the magnitude and direction of the velocity of Q relative to P.

7 A pleasure steamer moves with a speed of 10 knots on a bearing of $170°$ and a sailing boat moves with a speed of 8 knots on a bearing of $230°$. Find the magnitude and direction of the velocity of the steamer relative to the sailing boat.

8 Rain falls vertically at $4\,\mathrm{m\,s^{-1}}$. Find the direction in which the raindrops run down the window of a train moving at $54\,\mathrm{km\,h^{-1}}$.

9 If rain, which would fall vertically in still air at $3\,\mathrm{m\,s^{-1}}$, actually falls when a wind blows at $4\,\mathrm{m\,s^{-1}}$ from the south, find the magnitude and direction to the vertical of the velocity of the rain. If a cyclist moves at $18\,\mathrm{km\,h^{-1}}$ on level ground due south, find the velocity of the rain relative to the cyclist.

10 A passenger, on an open-top bus travelling at $36\,\mathrm{km\,h^{-1}}$ feels the wind which appears to blow at right angles to the path of the bus at $20\,\mathrm{km\,h^{-1}}$. Find the actual velocity of the wind.

11 Two particles, P and Q, have velocities given by $2\mathbf{i} + \mathbf{j}$ and $4\mathbf{i} - 3\mathbf{j}$, respectively. If the velocity of a third particle, R, relative to P, is $\mathbf{i} - 3\mathbf{j}$, find the velocity of R and the velocity of Q relative to R.

12 Two particles, A and B have velocities $3\mathbf{i} + \mathbf{j} - \mathbf{k}$ and $2\mathbf{i} - \mathbf{j} + 2\mathbf{k}$, respectively. If the velocity of a particle, C, relative to B, is $\mathbf{i} + 4\mathbf{j} - \mathbf{k}$, find the velocity of C and the velocity of C relative to A.

13 To an observer on a ship A, sailing at 20 knots due east, a ship B appears to be sailing at 20 knots in a direction N $30°$ W. Find the actual speed and bearing of ship B.

14 To a particle P, travelling with velocity $\mathbf{i} + 2\mathbf{j}$, a particle Q appears to be moving in a direction $2\mathbf{i} + 3\mathbf{j}$. To a particle R, travelling with velocity $-2\mathbf{i} + 4\frac{1}{2}\mathbf{j}$, Q appears to be moving in a direction $\mathbf{i} - 2\mathbf{j}$. Find the velocity of Q in vector form.

15 An aircraft, A, is flying due east at $900\,\mathrm{km\,h^{-1}}$ and a second aircraft, B, is flying at $600\,\mathrm{km\,h^{-1}}$ on a bearing $210°$. A third aircraft, C, appears to the pilot of A to be flying due south and to the pilot of B appears to be flying due east. Find the actual speed and bearing of the aircraft C.

Relative motion

Relative displacement and closest approach

In many problems the actual positions of the moving objects are included.

What course and speed should the two pilots of the Red Arrows display team set to rejoin the main group?

How long will it take the destroyer on the right to rendezvous with the ship on the left?

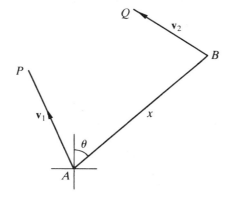

If the two ships cannot intercept, given their present velocities, what is the shortest distance between them and when does this occur?

We can find the answers to questions like these either by **(a)** considering their relative motion or **(b)** finding their actual displacement in vector form.

Using relative velocities and displacements

Consider two particles, P and Q, moving with constant velocities $\mathbf{v}_1$ and $\mathbf{v}_2$ respectively. If they are initially x m apart such that the bearing of Q from P is N $\theta°$ E, then we can draw a **space diagram**.

The space diagram shows the actual positions and paths of the particles.

For example, particle P starts from A and moves in the direction of its velocity vector $\mathbf{v}_1$.

If we impose a velocity of $-\mathbf{v}_2$ on both particles, then Q will be at rest and P will move with the resultant of $\mathbf{v}_1$ and $-\mathbf{v}_2$, i.e. $\mathbf{v}_1 - \mathbf{v}_2$.

If this resultant velocity is such that its direction is along AB, then P and Q will collide.

Thus, when solving this situation, the direction of the relative velocity $\mathbf{v}_1 - \mathbf{v}_2$ is N $\theta°$ E.

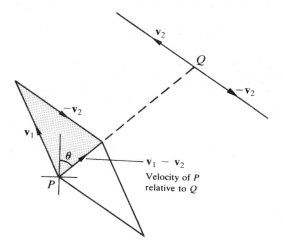

Velocity of P relative to Q

On the other hand, for some velocities $\mathbf{v}_1$ and $\mathbf{v}_2$, it may not be possible to complete the triangle of velocities with the relative velocity $\mathbf{v}_1 - \mathbf{v}_2$ along AB.

In this case, P and Q cannot collide. The dotted line in the figure, left, shows the path of P **relative** to Q, and BN represents the minimum displacement between P and Q.

The time taken to reach this minimum distance is given by

$$t = \frac{\text{relative distance}}{\text{relative speed}} = \frac{AN}{|\mathbf{v}_1 - \mathbf{v}_2|}$$

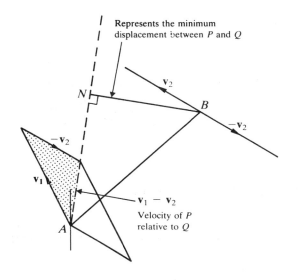

Represents the minimum displacement between P and Q

Velocity of P relative to Q

Using the actual displacements

Alternatively, it is possible to calculate the position vectors of P and Q after t seconds.

i.e. if $\mathbf{v}_1$ and $\mathbf{v}_2$ are constant velocities,

$$\mathbf{r}_p = \mathbf{r}_A + t\mathbf{v}_1$$

and $\mathbf{r}_q = \mathbf{r}_B + t\mathbf{v}_2$

If the particles collide, $\mathbf{r}_p = \mathbf{r}_q$ and a value of t can be found; or if they do not collide, their displacement is $\mathbf{r}_p - \mathbf{r}_q$ and the minimum can be found by differentiation.

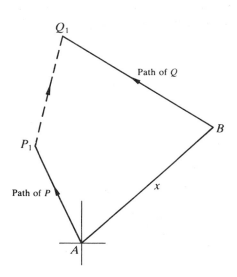

Path of Q

Path of P

WORKED EXAMPLES

1 A Tanker Ship, *T*, must rendezvous with a Cruiser which is steaming N 30° E at 20 km h^{-1}. If the Tanker is 20 km, S 50° E of the Cruiser and has a maximum speed of 30 km h^{-1}, find the course it must set and the time it takes to reach the rendezvous.

Draw a space diagram and a **separate** velocity vector diagram showing the velocity of the Tanker relative to the Cruiser. If the ships are to rendezvous, the relative velocity vector must lie along the direction of the initial displacement $\overrightarrow{TC}$.

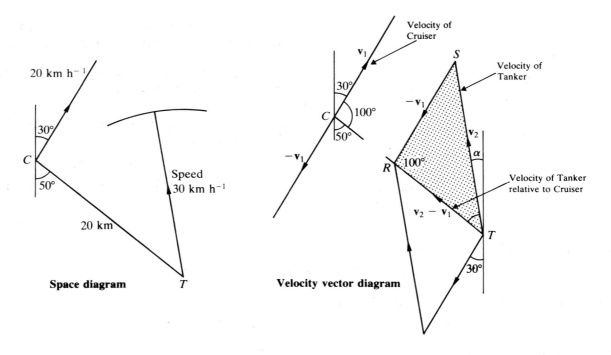

Space diagram   **Velocity vector diagram**

Let the course set by the Tanker be N α° W, as shown.

Using the sine rule in triangle *RST*

$$\frac{ST}{\sin 100°} = \frac{SR}{\sin R\hat{T}S} \quad \Rightarrow \quad \frac{|\mathbf{v}_2|}{\sin 100°} = \frac{|\mathbf{v}_1|}{\sin(50° - \alpha)}$$

$$\Rightarrow \quad \sin(50° - \alpha) = \frac{|\mathbf{v}_1|\sin 100°}{|\mathbf{v}_2|} = \frac{20\sin 100°}{30} = 0.6565$$

$$\Rightarrow \quad (50° - \alpha) = 41.04° \quad \Leftrightarrow \quad \alpha = 8.96° \text{ or } 8°58'$$

Hence the Tanker sets a course of N 8°58' W (351°2').

Now the relative speed is represented by $|\overrightarrow{TR}|$. Using the sine rule in triangle *RST*,

$$\frac{TR}{\sin(30° + \alpha)} = \frac{ST}{\sin 100°} \quad \Rightarrow \quad \frac{|\mathbf{v}_2 - \mathbf{v}_1|}{\sin 38°58'} = \frac{|\mathbf{v}_2|}{\sin 100°}$$

$$\Rightarrow \quad |\mathbf{v}_2 - \mathbf{v}_1| = \frac{|\mathbf{v}_2|\sin 38°58'}{\sin 100°} = \frac{30\sin 38°58'}{\sin 100°} = 19.16 \text{ km h}^{-1}$$

Thus, the Tanker has a speed of $19.16 \, \text{km h}^{-1}$ relative to the Cruiser.

$$\text{The time taken} = \frac{\text{relative distance}}{\text{relative speed}} = \frac{20}{19.16} = 1.04 \, \text{hours}$$

$$= 1 \, \text{hour} \, 3 \, \text{minutes, to the nearest minute.}$$

2 A boy rolls a marble with a velocity of $3\mathbf{i} + \mathbf{j}$ from a point A, whose position vector is $-5\mathbf{i} + 5\mathbf{j}$ relative to the origin O. At the same moment a friend rolls a second marble from the origin with a speed of $3 \, \text{m s}^{-1}$, so that the marbles collide. If $\mathbf{i}$ and $\mathbf{j}$ are the unit vectors along the x and y axes, respectively, find the direction in which the second marble must be rolled and the time taken before they collide.

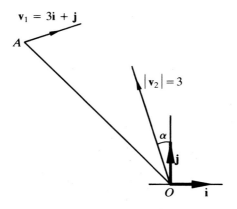

Although this question is given in vector form it is essentially the same problem as in the previous Example.

We shall solve it by finding the actual position vectors of both marbles after t seconds.

Since the speed of the second marble is $3 \, \text{m s}^{-1}$, its velocity can be written

$$\mathbf{v}_2 = -3 \sin \alpha \, \mathbf{i} + 3 \cos \alpha \, \mathbf{j}$$

where α is the angle which $\mathbf{v}_2$ makes with the $\mathbf{j}$ vector, as shown.

The position vector of the first marble after t s is

$$\mathbf{r}_1 = (-5\mathbf{i} + 5\mathbf{j}) + t(3\mathbf{i} + \mathbf{j}) = (3t - 5)\mathbf{i} + (t + 5)\mathbf{j}$$

The position vector of the second marble after t s is

$$\mathbf{r}_2 = t(-3 \sin \alpha \, \mathbf{i} + 3 \cos \alpha \, \mathbf{j}) = -3t \sin \alpha \, \mathbf{i} + 3t \cos \alpha \, \mathbf{j}$$

If the marbles collide, $\mathbf{r}_1 = \mathbf{r}_2$

$$\Rightarrow \quad 3t - 5 = -3t \sin \alpha \quad \Rightarrow \quad t(3 + 3 \sin \alpha) = 5 \quad \cdots\cdots\cdots \quad \textbf{(1)}$$

$$\text{and} \quad t + 5 = 3t \cos \alpha \quad \Rightarrow \quad t(1 - 3 \cos \alpha) = -5 \quad \cdots\cdots \quad \textbf{(2)}$$

Dividing **(1)** by **(2)** $\quad \dfrac{t(3 + 3 \sin \alpha)}{t(1 - 3 \cos \alpha)} = -1 \quad \Leftrightarrow \quad \dfrac{3 + 3 \sin \alpha}{1 - 3 \cos \alpha} = -1$

Hence, $\quad 3 + 3 \sin \alpha = -1 + 3 \cos \alpha \quad \Rightarrow \quad 3 \cos \alpha - 3 \sin \alpha = 4$

Divide both sides by $\sqrt{3^2 + 3^2} = 3\sqrt{2}$

$$\frac{1}{\sqrt{2}} \cos \alpha - \frac{1}{\sqrt{2}} \sin \alpha = \frac{4}{3\sqrt{2}}$$

Now, since $\cos 45° = \sin 45° = \dfrac{1}{\sqrt{2}}$, this equation can be written $\quad \cos 45° \cos \alpha - \sin 45° \sin \alpha = \dfrac{4}{3\sqrt{2}}$

$$\text{or,} \quad \cos(45° + \alpha) = 0.9428$$

$$\Rightarrow \quad 45° + \alpha = 19.47° \quad \Rightarrow \quad \alpha = -25.53° \text{ or } -25°32'$$

The negative sign implies that, in the diagram, the vector representing v_2 should be on the other side of the unit vector j. The second marble should be rolled at an angle of $25°32'$ to the y-axis.

Using equation (1) $t = \dfrac{5}{3 + 3\sin\alpha} = \dfrac{5}{3 + 3\sin(-25.53)°} = 2.93\,\text{s}$

The marbles will collide after 2.93 seconds.

---

3 A man driving a car at $90\,\text{km}\,\text{h}^{-1}$ due south, sees a train due east of him travelling at $150\,\text{km}\,\text{h}^{-1}\,\text{S}\,60°\,\text{W}$ on a track which meets the road at a level crossing at an angle of $60°$. If the car and the train are initially $1\,\text{km}$ and $2\,\text{km}$ from the crossing, respectively, find the distance between them when they are closest together and the time that elapses before that point is reached.

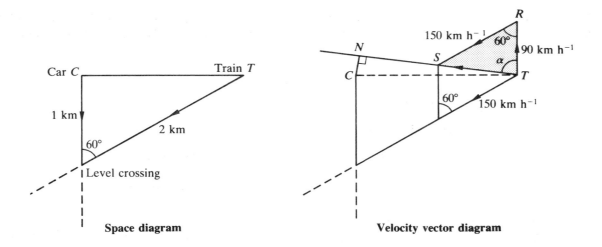

Space diagram   **Velocity vector diagram**

Impose a velocity of $90\,\text{km}\,\text{h}^{-1}$ due north on the car and the train, thus effectively bringing the car to rest. The velocity of the train relative to the car can be found, its direction being along TN at $\alpha°$ west of north.

Using the vector velocity diagram and the cosine rule in triangle RST,

$$TS^2 = SR^2 + RT^2 - (2 \times SR \times RT \times \cos R)$$
$$= 150^2 + 90^2 - (2 \times 150 \times 90 \times \cos 60°)$$
$$= 22\,500 + 8100 - 13\,500$$
$$\therefore TS^2 = 17\,100$$

The speed of the train relative to the car is $|\overrightarrow{TS}| = \sqrt{17\,100} = 130.77\,\text{km}\,\text{h}^{-1}$.

Using the sine rule in triangle SRT,

$$\frac{150}{\sin\alpha} = \frac{|\overrightarrow{TS}|}{\sin 60°} \quad \Rightarrow \quad \sin\alpha = \frac{150\sin 60°}{130.77}$$

$$\Rightarrow \quad \sin\alpha = 0.9934 \quad \Rightarrow \quad \alpha = 83.4° \quad \text{or} \quad 83°24'$$

The velocity of the train relative to the car is $130.77\,\text{km}\,\text{h}^{-1}$, N $83.4°$ W.

Since the car is at rest and the train is moving relatively along the line TN, the minimum distance apart is represented by CN.

214

Now, $CT^2 = 2^2 - 1^2 = 3 \quad \Rightarrow \quad CT = \sqrt{3}$ (by Pythagoras' theorem).

Thus, in $\triangle NCT$, $\quad \sin N\hat{T}C = \dfrac{CN}{CT} = \dfrac{CN}{\sqrt{3}}$

$$\Rightarrow \quad CN = \sqrt{3}\sin N\hat{T}C = \sqrt{3}\sin(90° - 83°24') = 0.2\,\text{km}$$

The minimum separation is $0.2\,\text{km}$, or $200\,\text{m}$.

The time taken is given by, $\quad \dfrac{NT}{\text{relative speed}} = \dfrac{\sqrt{3}\cos 6°36'}{130.77} = 0.013\,\text{h} = 47\,\text{seconds}$

Alternatively, this can be solved using vector methods.

Let the level crossing be the origin and $\mathbf{i}$ and $\mathbf{j}$ be unit vectors due east and due north, respectively. If X and Y, (see diagram), with position vectors $\mathbf{r}_1$ and $\mathbf{r}_2$, respectively, are the points to which the car and the train move in t hours, then

$$\mathbf{r}_1 = \overrightarrow{LC} + \overrightarrow{CX} = \mathbf{j} - 90t\,\mathbf{j} = (1 - 90t)\mathbf{j}$$
$$\text{and} \quad \mathbf{r}_2 = \overrightarrow{LT} + \overrightarrow{TY}$$
$$\Rightarrow \quad \mathbf{r}_2 = (2\sin 60°\,\mathbf{i} + 2\cos 60°\,\mathbf{j}) - (150\sin 60°\,\mathbf{i} + 150\cos 60°\,\mathbf{j})t$$
$$= (\sqrt{3}\,\mathbf{i} + \mathbf{j}) - (75\sqrt{3}\,\mathbf{i} + 75\mathbf{j})t = (\sqrt{3} - 75\sqrt{3}t)\mathbf{i} + (1 - 75t)\mathbf{j}$$

Hence, the displacement is $\overrightarrow{XY}$, where
$$\overrightarrow{XY} = \mathbf{r}_2 - \mathbf{r}_1$$
$$= (\sqrt{3} - 75\sqrt{3}t)\mathbf{i} + [(1 - 75t) - (1 - 90t)]\mathbf{j}$$
$$= (\sqrt{3} - 75\sqrt{3}t)\mathbf{i} + 15t\,\mathbf{j}$$

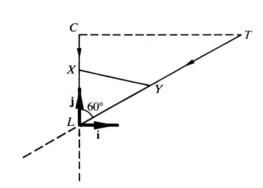

If the $|\overrightarrow{XY}| = s$, then
$$s^2 = (\sqrt{3} - 75\sqrt{3}t)^2 + (15t)^2$$
$$\Rightarrow \quad s^2 = 3 - 450t + 16\,875t^2 + 225t^2$$
$$\Rightarrow \quad s^2 = 3 - 450t + 17\,100t^2 \dots\dots\dots\,(1)$$

Since s is a minimum when s^2 is a minimum, we find $\dfrac{d}{dt}(s^2) = 0$.

$$\dfrac{d}{dt}(s^2) = -450 + 34\,200t = 0$$

$$\Rightarrow \quad t = \dfrac{450}{34\,200} = 0.013\,\text{h} = 47\,\text{seconds}$$

Since $\dfrac{d^2}{dt^2}(s^2) = 34\,200 > 0$, this value of t gives the time to reach the minimum displacement.

Substituting into equation **(1)**, above $\quad s^2 = 3 - 450(0.013) + 17\,100(0.013)^2 = 0.04$

The minimum distance $= \sqrt{0.04} = 0.2\,\text{km}$

1 Two particles, A and B, moving with constant velocities $\mathbf{v}_1$ and $\mathbf{v}_2$, respectively, start from the points P and Q with position vectors $\mathbf{r}_1$ and $\mathbf{r}_2$. Find the position vectors of A and B after t seconds and hence find the time before they collide.

(a) $\mathbf{v}_1 = 2\mathbf{i} + \mathbf{j}; \quad \mathbf{r}_1 = 0\mathbf{i} + 0\mathbf{j}$ (b) $\mathbf{v}_1 = -\mathbf{i} + 3\mathbf{j}; \quad \mathbf{r}_1 = \mathbf{i} + \mathbf{j}$
      $\mathbf{v}_2 = \mathbf{i} + 2\mathbf{j}; \quad \mathbf{r}_2 = 4\mathbf{i} - 4\mathbf{j}$ $\mathbf{v}_2 = \mathbf{i} + 4\mathbf{j}; \quad \mathbf{r}_2 = -\mathbf{i}$

2 A particle P starts from $2\mathbf{i} - 9\mathbf{j}$ and moves with constant velocity $-2\mathbf{i} + 5\mathbf{j}$. A second particle Q starts from $-3\mathbf{i} - \mathbf{j}$ one second later and moves with constant velocity $\mathbf{i} + 2\mathbf{j}$. Show that the particles collide and find the time that elapses before the moment of impact.

3 A stone starts from the point with position vector $4\mathbf{i} + 3\mathbf{j} + a\mathbf{k}$ and moves with velocity $\mathbf{i} + 2\mathbf{j} - \mathbf{k}$. A pebble starts 1 second earlier from a point with position vector $-\mathbf{i} + \mathbf{j} + 3\mathbf{k}$ and moves with velocity $2\mathbf{i} + 2\mathbf{j} + \mathbf{k}$. If the particles collide, find the value of a.

4 Two roads cross at right angles at A. When a man walking along one road towards A at $6\,\mathrm{km\,h^{-1}}$ is $\frac{1}{2}\,\mathrm{km}$ from A, he sees a boy at A moving along the other road away from A at $8\,\mathrm{km\,h^{-1}}$. Find **(a)** The velocity of the man relative to the boy and hence deduce when they are closest together and **(b)** using vector methods, the actual distance at time t and the minimum value of this distance.

5 A submarine travelling due east at $20\,\mathrm{km\,h^{-1}}$, sees a frigate $10\,\mathrm{km}$ ahead which is sailing due north at $25\,\mathrm{km\,h^{-1}}$. What is the least distance apart of the vessels in the subsequent motion? Find the time taken to reach this point.

6 A pilot must rendezvous with another aircraft $40\,\mathrm{km}$ away on a bearing of $120°$. The second aircraft is flying at $200\,\mathrm{km\,h^{-1}}$ on a course of $210°$. If the first aircraft can fly at $270\,\mathrm{km\,h^{-1}}$, find the course the pilot must set and the time taken to rendezvous.

7 Two straight paths intersect at P at an angle of $60°$. Two girls are running, one along each path, towards P at $16\,\mathrm{km\,h^{-1}}$ and $12\,\mathrm{km\,h^{-1}}$ respectively. If they both start $2\,\mathrm{km}$ from P, find the time when they are closest together.

8 A dog sees a cat $10\,\mathrm{m}$ away due south, running on a bearing of $135°$ with a speed of $4\,\mathrm{m\,s^{-1}}$. If the dog can run at a maximum speed of $5\,\mathrm{m\,s^{-1}}$, find the direction in which the dog must run to catch the cat, and the time taken, if both animals move only in straight lines.

9 A man can swim at $3\,\mathrm{km\,h^{-1}}$ in still water. Find the time taken to swim across a river to a point directly opposite, if the river is $200\,\mathrm{m}$ wide and there is a current flowing at $2\,\mathrm{km\,h^{-1}}$.

10 A girl can swim at $3\,\mathrm{km\,h^{-1}}$ in still water. She swims across a river $300\,\mathrm{m}$ wide, that has a current flowing at $5\,\mathrm{km\,h^{-1}}$. What is the angle that her actual course makes with the normal to the bank if she aims to reach the opposite bank as little downstream as possible?

11 In a game of cricket, a fielder stands $22\,\mathrm{m}$ from the batsman, who hits the ball so that it runs along the ground at $7\,\mathrm{m\,s^{-1}}$ at an angle of $70°$ to the line joining the fielder to the batsman. If the fielder immediately sets off to chase the ball by running in a straight line at $6\frac{2}{3}\,\mathrm{m\,s^{-1}}$, find the time taken for him to field the ball. Show that, if the boundary is $60\,\mathrm{m}$ from the batsman, the fielder will stop the ball before it reaches the boundary.

13 Permutations and combinations

Introduction

Everyday we place objects in a specific order or select items from a group of objects.

In how many ways can six playing cards be placed in a row?

If you have three flags to fly on three flag poles, in how many different ways can this be done?

Four items of crockery are placed on a table. In how many ways can they be arranged in a line, if the jug must go on the right-hand end?

Have you played games with numbers and letters of a car number plate? How many formations can you make **(a)** using three of the four letters **(b)** using three letters and two digits, if the digits must be next to each other?

How many teams of five players can be formed from eight names on the team notice board?

These, and similar questions about selections, can be solved using mathematical reasoning. The questions fall into two main categories.

(a) when the order of selection is important – called **permutations**, and
(b) when the order is unimportant and only the composition of the group matters – called **combinations**.

In the examples shown, the arrangements of the cards, the flags, the crockery and the digits and letters of the number plate are all permutations since the order matters. In the team selection, only the group of five players chosen is of concern and this is, thus, a combination.

Permutations

A **permutation** is an arrangement of a number of chosen items in a particular order.

Consider the **three flags** mentioned in the Introduction. If we call these *A*, *B* and *C*, it is not too much trouble to write out all the possible arrangements,

$$ABC \qquad BAC \qquad CAB$$
$$ACB \qquad BCA \qquad CBA$$

Thus there are six permutations of the three flags. Now, whilst it is possible to list all the arrangements with a small number of items, it would be impracticable with a large number and so we approach the problem from a different angle.

If three flags are to be placed in a row, we can choose any one of the three to be on the left-hand flagpole, i.e. in three ways. Having chosen the first flag, two remain and thus, either of these two could be put on the centre flagpole, i.e. two ways. This leaves one flag to place on the last flagpole i.e. no choice – only one way.

Now, for each of the three options for the first flag, there are two ways of choosing the second flag. This gives six ways (3×2 ways) of filling the first two positions. For each of the six possibilities of filling these two places, there is only one way of filling the last place.

The total number of arrangements $= 3 \times 2 \times 1 = 6$ ways.

We can write this out simply in the following format,

The 1st flag can be chosen in 3 ways

The 2nd flag can be chosen in 2 ways

The 3rd flag can be chosen in 1 way

$$\text{Total} = 3 \times 2 \times 1 = 6 \text{ ways}$$

Four different books are to be placed on a bookshelf. In how many different ways can this be done?

The solutions can be written out as above.

The 1st book can be chosen in 4 ways

The 2nd book can be chosen in 3 ways

The 3rd book can be chosen in 2 ways

The 4th book can be chosen in 1 way

$$\text{Total} = 4 \times 3 \times 2 \times 1 = 24 \text{ ways}$$

Factorials

Since in many problems of this type products of the form $4 \times 3 \times 2 \times 1$ occur frequently, we use a **special notation** to denote the product of all integers between a given value and 1.

We write, $4 \times 3 \times 2 \times 1 = 4!$ which is called a **factorial**

'Five factorial' or 'factorial five' is denoted by 5! and is a shorthand notation for $5 \times 4 \times 3 \times 2 \times 1 = 120$. Sometimes we use a **.** instead of $\times$ for the multiplication sign.

Hence, $3! = 3 \times 2 \times 1 = 3.2.1 = 6$

In general, $n! = n(n-1)(n-2)(n-3) \cdots\cdots 4.3.2.1$

Investigation 1

Use the method of the Worked Example to find the number of arrangements of the following.

(a)   5 people in a taxi queue,
(b)   6 playing cards placed in a row,
(c)   7 different coins placed in a line,
(d)   10 different shoes on a rack during a sale.

Express your answers in factorial notation and evaluate the value of these factorials. Deduce a general result for the number of arrangements of n different items.

> The number of permutations of n unlike objects is $n!$

If we select only **some items from a group of different objects**, the same approach can be used.

WORKED EXAMPLES

1 Find the number of different arrangements that can be formed by choosing three letters from the word SPECIAL.

There are 7 different letters in the word, none of which is repeated.

The 1st letter can be chosen in 7 ways

The 2nd letter can be chosen in 6 ways

The 3rd letter can be chosen in 5 ways

$$\text{Total} = 7 \times 6 \times 5 = 210 \text{ ways}$$

We can express this in factorial notation,

$$7 \times 6 \times 5 = \frac{7 \times 6 \times 5 \times 4 \times 3 \times 2 \times 1}{4 \times 3 \times 2 \times 1} = \frac{7!}{4!} \quad \text{or} \quad \frac{7!}{(7-3)!}$$

In general,

> The number of arrangements of r items selected from n different items is $\dfrac{n!}{(n-r)!}$ $(r \leqslant n)$.
>
> We use the notation, $^nP_r = \dfrac{n!}{(n-r)!}$

Notes

(i) ^nP_r is sometimes written $_nP_r$.

(ii) If all the items are chosen (i.e. $r = n$), then $^nP_n = \dfrac{n!}{(n-n)!} = \dfrac{n!}{0!}$

Since n items can be arranged in $n!$ ways, it follows that we must define $0! = 1$.

---

2 A code on a bicycle lock consists of three digits. How many different codes are possible if only the digits between 1 and 9 inclusive are used?

---

3 digits can be chosen from 9 in $^9P_3$ ways

$$= \frac{9!}{6!} = 9.8.7 = 504 \text{ ways}$$

Hence, 504 different codes are possible.

Exercise 13.1

1 Find the total number of arrangements of the following, giving your answer in factorials or numerically, as appropriate.

    (a) 7 books on a shelf   **(b)** 4 cars on a forecourt   **(c)** 5 garden tools

    (d) the 13 cards in a hand at bridge

2 Find the total number of arrangements of the following. Give your answers in factorial notation, evaluating this where practicable.

    (a) 4 letters chosen from 7 letters   **(b)** 10 trees chosen from 15 trees   **(c)** 2 cats chosen from 5 cats

    (d) 5 books chosen from 25 books   **(e)** 2 pictures chosen from 120 pictures

3 How many different 4-digit numbers can be made from the digits $5, 6, 7$ and 8 if no digit may be repeated?

4 How many different arrangements can be made from the letters of the word FATHER?

5 There are twelve contestants in a competition. In how many ways can the first three places be filled?

6 A cricket team of 11 players contains 5 batsmen who fill the first five places in the batting order. In how many ways can these five places be filled?

7 How many 3-digit numbers can be made from the digits $2, 3, 5, 7, 8, 9$ if no digit may be repeated?

8 How many different permutations can be made by choosing 4 letters from the word PECULIAR?

The factorial notation

It is useful to be able to manipulate factorials and simplify them where necessary.

WORKED EXAMPLES

1 Evaluate $\dfrac{12!}{9!3!}$

Expanding each of the factorials,

$$\frac{12!}{9!3!} = \frac{12.11.10.9.8.7.6.5.4.3.2.1}{(9.8.7.6.5.4.3.2.1)(3.2.1)} = \frac{12.11.10}{3.2.1} = 220$$

In practice, one would not necessarily write in the product for 9! as, clearly, it is going to cancel.

2 Express $52 \times 51 \times 50 \times 49 \times 48$ in factorial notation.

We arrange to make the given product into 52! and divide by the extra terms included.

$$52.51.50.49.48 = 52.51.50.49.48.\frac{(47.46.45\cdots\cdot 3.2.1)}{(47.46.45\cdots\cdot 3.2.1)} = \frac{52!}{47!}$$

3 Simplify **(a)** $8! + 9!$ **(b)** $\dfrac{10!}{7!3!} + \dfrac{11!}{8!2!}$ leaving answers in factorials.

(a)   Since $9! = 9 \times 8!$, we have $8! + 9! = 8! + 9.8! = 8!(1 + 9) = 10(8!)$

(b)   Rearrange $\dfrac{10!}{7!3!}$ by multiplying the numerator and denominator by 8 to give $\dfrac{8(10!)}{8.7!3!} = \dfrac{8 \times 10!}{8!3!}$

Similarly, $\dfrac{11!}{8!2!} = \dfrac{3 \times 11!}{8!3!}$

Hence, $\dfrac{10!}{7!3!} + \dfrac{11!}{8!2!} = \dfrac{8 \times 10!}{8!3!} + \dfrac{3 \times 11!}{8!3!} = \dfrac{8 \times 10!}{8!3!} + \dfrac{3 \times 11 \times 10!}{8!3!}$

$$= \frac{10!}{8!3!}(8 + 33) = \frac{41 \times 10!}{8!3!}$$

Exercise 13.2

1 Evaluate,

(a) $\dfrac{5!}{2!}$ **(b)** $\dfrac{11!}{8!}$ **(c)** $\dfrac{20!}{15!6!}$ **(d)** $\dfrac{4! \times 5!}{3! \times 2!}$

(e) $(4!)^2$ **(f)** $\dfrac{(5!)^2}{4!3!}$ **(g)** $\dfrac{4! \times 2!}{5 \times 3!}$ **(h)** $\dfrac{10!}{6!4!2!}$

2 Express in factorial notation,

(a) $8 \times 7 \times 6$ **(b)** $15.14.13.12$ **(c)** $\dfrac{8 \times 7}{3 \times 2}$

(d) $\dfrac{7.6.5.4}{3.2.1}$ **(e)** $\dfrac{14 \times 13 \times 12 \times 11}{4 \times 3 \times 2 \times 1}$ **(f)** $n(n-1)(n-2)$

3 Simplify the following leaving your answers in factorials.

(a) $4! + 3!$

(b) $12! - 10!$

(c) $(2 \times 5!) - 4!$

(d) $\dfrac{8!}{2!} + \dfrac{8!}{3!}$

(e) $\dfrac{10!}{2!} - \dfrac{10!}{3!}$

(f) $\dfrac{14!}{3!11!} + \dfrac{14!}{12!2!}$

(g) $n! - (n-1)!$

(h) $(n+2)! + 2(n+1)!$

(i) $\dfrac{7!}{4!3!} + \dfrac{7!}{5!}$

Permutations under conditions

In quite a number of selection problems, there are conditions imposed which limit the freedom of choice. For example, when finding how many **even** numbers can be made from the digits 2, 3, 5, 7, the last digit must be the 2.

If we arrange the letters of the word SEE, there will be fewer permutations than with the word SEA, since the E is repeated and some arrangements will be repeats of others.

There are several standard techniques which can be used.

Objects placed in a circle

If 5 people are to sit at a round table for a meal, then their actual position relative to the table is unimportant, since they could all move round one place, without changing the formation of the people.

Thus, if the first person sits in chair 1, the other four people can be arranged in the remaining chairs in 4! ways.

The number of arrangements of 5 people at a round table is $4! = 24$ ways.

In general, n objects can be placed in a circle in $(n-1)!$ different ways.

Some restriction of choice

To find how many even numbers greater than 4000 can be made from the digits 2, 3, 6, 8, we use the method of the section on Permutations, although the order in which we consider the choice of digits, is different.

The first digit must be 6 or 8, as the number is greater than 4000. The last digit must be the 2 or whichever of the 6 or 8 was not used for the first digit.

The 1st digit can be chosen in 2 ways

The last digit can be chosen in 2 ways

The 2nd digit can be chosen in 2 ways

The 3rd digit can be chosen in 1 way

Total $= 2 \times 2 \times 2 \times 1 = 8$ ways

These are,

6382	6832	8632	8362
6328	6238	8236	8326

Subtraction of the opposite condition from unrestricted choice

Consider 10 items of grocery to be passed across the price scanner at the checkout. In how many different orders can this be achieved, if the two packets of biscuits must not be next to each other?

We solve this problem by finding the number of arrangements when the packets of biscuits are together, and subtracting the result from the number of arrangements of 10 items with unrestricted choice.

With no restriction, there are 10! arrangements. If we regard the two packets of biscuits as one item (since they must be together) there are effectively 9 items to place in order and this can be done in 9! ways.

However, the two packets of biscuits could be reversed in order and still be together. This doubles the number of permutations.

$\therefore$ Total number of arrangements with these items together $= 2 \times 9!$

$\therefore$ Number of arrangements when these items are not together is

$$10! - (2 \times 9!) = 9!(10 - 2) = 8 \times 9!$$

Repeated items

If there are identical objects in a group from which a selection is to be made, it will reduce the number of arrangements.

Consider the number of arrangements of all the letters of the word DEFENCE.

Since there are three Es, think of them as E_1, E_2 and E_3, in which case the 7 different letters can be arranged in 7! ways.

But in some of these choices, only the Es will be interchanged leaving the other letters in the same positions.

e.g. $FE_1NE_2E_3CD$ and $FE_2NE_1E_3CD$ etc.

Since there are three Es, they can be arranged amongst themselves in 3! ways.

$\therefore$ The number of **different** arrangements $= \dfrac{7!}{3!}$

Splitting a selection into parts

Sometimes, when selecting a particular item from a group, it affects our choice for later items.

Consider how many odd numbers greater than 50 000 can be formed from the digits 3, 4, 5, 6 and 7, if each digit is used only once in each number.

The difficulty here is that the choice of the last digit depends upon whether an odd or an even number has been selected for the first digit. To overcome this we divide the solution into two parts,

   (a)   when an odd number is chosen for the first digit

and   **(b)**   when an even number is chosen for the first digit

The results from these two parts will be added together to give the final result, since these parts are independent i.e. they have no effect on each other.

(a) If the first digit is odd (the 5, or 7),

> The 1st digit can be chosen in 2 ways
>
> The last digit can be chosen in 2 ways
>
> The 2nd digit can be chosen in 3 ways
>
> The 3rd digit can be chosen in 2 ways
>
> The 4th digit can be chosen in 1 way
>
> Total $= 2 \times 2 \times 3 \times 2 \times 1 = 24$ ways

(b) If the first digit is even (only the 6),

> The 1st digit can be chosen in 1 way
>
> The last digit can be chosen in 3 ways
>
> The 2nd digit can be chosen in 3 ways
>
> The 3rd digit can be chosen in 2 ways
>
> The 4th digit can be chosen in 1 way
>
> Total $= 1 \times 3 \times 3 \times 2 \times 1 = 18$ ways

Since these parts are independent, the total of odd numbers greater than 50 000 using these five digits is $24 + 18 = 42$.

Exercise 13.3

1 A girl has five counters of different colours. How many arrangements can she make, if the red one must be at the centre of the row?

2 In how many ways can the letters of the word INSTALL be arranged?

3 Twelve volumes of a book, two of which form an index, are placed on a shelf. How many arrangements can be made if the two index volumes must be **(a)** together and **(b)** one at each end?

4 A man chooses eight differently coloured rose bushes which he will plant in a row. How many arrangements can he make, if the yellow and white roses must not be next to each other?

5 How many arrangements can be made of four letters chosen from the word CUPBOARD, if the P must be included?

6 How many arrangements of five letters chosen from the word MERCHANT can be made if **(a)** the last letter must be a vowel and **(b)** the first letter must be a consonant?

7 How many numbers divisible by 5 can be made from the digits 2, 4, 5, 6, 8, if each digit can be used once only in any number? (Remember you can use some, or all, of the digits in any selection.)

8 How many even numbers greater than 4000 can be made from the digits 2, 4, 5, 6, 9 by using some or all of the digits, each only once in each number?

9 A railway carriage has two bench seats, each able to seat four passengers. In how many ways can a group of six men and two women be seated if the women must sit in an end position?

10 How many different permutations can be formed when **(a)** all the letters, and **(b)** 4 of the letters, are chosen from the following words,

(i) PETER **(ii)** ELEMENT **(iii)** ONION

11 Four boys and two girls sit on a bench. In how many ways can they sit, if the two girls do not sit next to each other?

12 Four boys and three girls stand in a line so that no two girls stand next to each other. Find the number of arrangements.

13 A group of eight carol singers stand in a circle. In how many different ways can they be arranged?

14 A group of five ladies and two men sit down at a round table. How many different arrangements are there if the two men must not sit next to each other?

15 How many 5 digit odd numbers less than 40 000 can be formed from the digits 1, 2, 3, 6, 7, 8, if no digit is repeated in any number?

16 How many odd 5 digit or 6 digit numbers greater than 60 000 can be made from the digits 4, 5, 6, 7, 8, 0, if **(a)** repetitions are allowed, and **(b)** each digit can be used only once?

17 In how many ways can the letters of the word EXCELLENCE be arranged?

18 A group of six men and two ladies sit at a round table. In how many ways can they be seated if the two ladies must not sit directly opposite each other?

19 In how many ways can five footballs, four rugby balls and five basketballs be arranged in a row? In how many of these arrangements will the different types of balls each be kept together? Assume that the balls of each type are distinguishable.

20 Six different coffee cups are to be placed on six matching saucers. In how many ways can this be done so that every cup is not matched with the correct saucer?

21 These are twenty seats in a row. In how many ways can ten married couples be seated, if each couple must sit together?

Combinations

A **combination** is a selection of a group of items from a given set irrespective of the order of the items.

Consider the team selection given in the Introduction. The main point of interest is which five players will be chosen from the eight available and not the order in which they are listed.

Permutations and combinations

If we choose 5 players from 8, taking account of the order of selection, we could obtain

$$^8P_5 = \frac{8!}{3!} \quad \text{permutations}$$

However, for each group of 5 players, they can be rearranged amongst themselves in 5! ways. All these 5! arrangements would represent the same team, and hence the number of different teams that can be formed will be obtained by dividing the number of permutations by 5!

$$\therefore \text{ The number of combinations is } \quad \frac{8!}{5!3!} = \frac{8.7.6}{3.2.1} = 56$$

In general, the number of combinations of r objects chosen from n different objects, will be obtained by dividing the number of permutations $^nP_r = \frac{n!}{(n-r)!}$ by the number of ways of arranging the r objects amongst themselves (i.e. $r!$).

> The number of groups of r items chosen from n different items, is $\dfrac{n!}{(n-r)!r!}$ $\quad (r \leqslant n)$
>
> We use the notation $^nC_r = \dfrac{n!}{(n-r)!r!}$

Notes

(i) ^nC_r is sometimes written $_nC_r$ or $\begin{pmatrix} n \\ r \end{pmatrix}$.

(ii) If $r = n$, then $^nC_n = \dfrac{n!}{0!n!} = 1$, since $0! = 1$.

Thus, only 1 group can be made when n objects are chosen from a set of n objects.

Investigation 2

(a) Write ^nC_r and $^nC_{n-r}$ in factorial notation and hence show that, for $r \leqslant n$, $^nC_r = {}^nC_{n-r}$.

(b) Write ^nC_r and $^nC_{r-1}$ in factorial notation and use the methods of the section on Factorial Notation to find $^nC_r + {}^nC_{r-1}$ in factorials.

(c) Write $^{n+1}C_r$ in factorials and show that it is equal to the result obtained in **(b)**.

(d) Test your result for **(i)** $n = 8$ and $r = 5$ **(ii)** $n = 10$ and $r = 4$.

You should have proved two useful results involving the ^nC_r notation, i.e.

$$^nC_r = {}^nC_{n-r} \quad \text{and} \quad {}^nC_r + {}^nC_{r-1} = {}^{n+1}C_r$$

WORKED EXAMPLES

1 If a lift can take 6 people, find the number of ways the first load can be formed from 9 waiting passengers.

The order within the chosen group of 6 is unimportant. Thus, this is a combination.

$$\therefore \ 6 \text{ people can be chosen from 9 in } {}^9C_6 = \frac{9!}{6!3!} \text{ ways}$$

$$= \frac{9.8.7}{3.2.1} = 84 \text{ ways}$$

2 A party of 4 boys and 4 girls is to be chosen from a group of 6 boys and 7 girls. In how many ways can the party be formed?

The order is unimportant and hence, this is a combination. The selection of **(a)** the boys, and **(b)** the girls, can be evaluated separately, since the choice of either does not affect the choice of the other group.

$$\therefore \ 4 \text{ boys can be chosen from 6 boys in } {}^6C_4 = \frac{6!}{2!4!} \text{ ways}$$

$$\text{and 4 girls can be chosen from 7 girls in } {}^7C_4 = \frac{7!}{3!4!} \text{ ways}$$

Since for each selection of the boys, we could have each of the groups of girls, the total number of ways of forming the party is obtained by **multiplying** these results.

$$\text{Total number of groups} = {}^6C_4 \times {}^7C_4$$

$$= \frac{6!}{2!4!} \times \frac{7!}{3!4!}$$

$$= \frac{6.5}{2.1} \times \frac{7.6.5}{3.2.1} = 525 \text{ ways}$$

Exercise 13.4

1 Evaluate **(a)** ${}^{10}C_3$ **(b)** ${}^5C_2$ **(c)** ${}^8C_4$ **(d)** ${}^5C_3$.

2 Show that ${}^9C_6 = {}^9C_3$ by evaluating both sides.

3 Find the value of ${}^9C_4 + {}^9C_3$, giving your answer in factorials. Express this result in the ${}^nC_r$ notation.

4 Find the number of ways the following groups of items can be chosen.

  (a) 2 hats from 5 hats on a shelf.

  (b) 5 birthday cards from 8 birthday cards.

  (c) A 6-a-side football team from 11 players.

  (d) 3 snooker balls from the 6 balls of different colours.

5 In how many ways can,

   (a) a group of 4 people be chosen from 9 people to travel in a taxi,

   (b) 4 leaves be chosen from a collection of 7 different leaves,

   (c) 10 stamps be chosen from a packet containing 15 different stamps,

   (d) 8 pens be chosen from a box of 12 unlike pens?

6 Ten people turn up to a games evening. If the first game needs four players, in how many ways can this group of four be chosen?

7 At a six-a-side cricket match there are twelve players present. If the teams are formed randomly, find the number of different teams that can be made.

8 If ten people are to travel by two taxis, each with five seats, find the number of ways the groups can be arranged.

9 If ten people travel by taxis, each able to take a maximum of four passengers, find the number of ways the group can be divided if the first two taxis each take four people.

10 Find the number of different groups of four letters that can be made from the letters of the word REPEAT. How many of these contain the letter T?

11 A committee consists of 3 men and 4 women. If 8 men and 6 women are eligible to serve, find the number of different committees that can be formed.

12 Six players are required for a badminton team at a club. If the captain and secretary must play, find the number of ways the team can be chosen if the club has 11 players (inclusive of the captain and secretary) eligible.

13 If ten coins are spun, find the numbers of ways that exactly four heads can be obtained.

14 A panel of 6 is formed from 12 boys and 7 girls. In how many ways can the panel be selected if **(a)** it must contain 3 boys and 3 girls, and **(b)** it must contain at least 1 girl.

15 Two lifts can each hold 8 people. In how many ways can a waiting group of 9 men and 7 women be carried if no lift may take 1 man or 1 woman on their own.

16 A cricket team of 11 players is to be chosen from 6 batsmen, 6 bowlers and 2 wicket-keepers. In how many ways can the team be chosen if it must contain **(a)** at least 1 wicket-keeper and **(b)** at least 4 batsmen.

14 Sequences and series

Introduction

A man sitting in a railway carriage noticed the following advertisement:

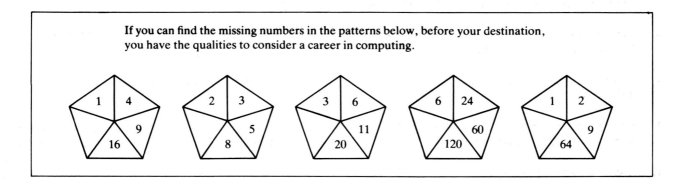

If you can find the missing numbers in the patterns below, before your destination, you have the qualities to consider a career in computing.

In effect, the reader was being urged to find the next term in a given set of numbers, where each element is evaluated by a definite rule.

> A set of numbers written in a definite order with a constant relationship between the terms, is called a **sequence**.

So, $1, 3, 5, 7, 9, \ldots\ldots$ is a sequence of odd integers.

$1^3, 2^3, 3^3, 4^3, 5^3, \ldots\ldots$ is a sequence of the cubes of natural numbers.

Did you obtain the answers in the advertisement? The sequences should be,

$1, 4, 9, 16, 25$ —   the sequence of squares $1^2, 2^2, 3^2, 4^2, 5^2, \ldots\ldots$

$2, 3, 5, 8, 13$ —   each term being the sum of the previous two

$3, 6, 11, 20, 37$ —   each term of the form $2^n + n$ for $n = 1, 2, 3, \ldots\ldots$

$6, 24, 60, 120, 210$ —   the sequence $(1 \times 2 \times 3), (2 \times 3 \times 4), (3 \times 4 \times 5), \ldots\ldots$

$1, 2, 9, 64, 625$ —   the sequence $1^0, 2^1, 3^2, 4^3, 5^4, \ldots\ldots$

If you did not manage to find all the solutions do not be too dismayed, for it is sometimes difficult to spot a pattern from a few terms.

Defining a sequence

A sequence can be defined mathematically in a variety of ways.

By writing out the terms.

(a) If there are only a few terms, then you can simply list each term.

$$\text{e.g.} \quad 1, 5, 9, 13, 17, 21, 25, 29$$

$$\text{or} \quad -3, 6, -12, 24, -48$$

A sequence which ends after a given number of terms, is called a **finite** sequence.

(b) For sequences with a large or infinite number of terms, it is usual to write sufficient terms to indicate the pattern and, maybe, the last term or number of terms.

$$\text{e.g.} \quad 4, 6, 8, 10, \ldots, 100$$

$$1, \tfrac{1}{3}, \tfrac{1}{9}, \tfrac{1}{27}, \ldots$$

$$2, 4, 8, 16, \ldots \text{to 20 terms}$$

In the case of $1, \tfrac{1}{3}, \tfrac{1}{9}, \tfrac{1}{27}, \ldots$, it is assumed that the sequence does not end and that another term can always be found. Such a sequence is called **infinite**.

Using an algebraic formula

In the sequence $1^2, 2^2, 3^2, 4^2, 5^2, \ldots$, it is clear that each term is a perfect square. The third term is 3^2, the fourth term 4^2, and so on. Thus, the eighteenth term will be 18^2 and, in general the rth term will be r^2 and the kth term will be k^2.

Using the notation $u_1, u_2, u_3, u_4, \ldots u_k, \ldots, u_n$ to represent a finite sequence, the first term is u_1 and the last term is u_n.

Thus, the general term u_k could be written as k^2.

The sequence could be defined by giving an algebraic formula for the general term u_k.

So the sequence $1^2, 2^2, 3^2, 4^2, \ldots, 19^2$ can be defined as

$$u_k = k^2 \quad \text{for } k = 1, 2, 3, \ldots, 19$$

Substituting successive values of k, produces the terms of the sequence.

Using an inductive definition

Consider the sequence $1, 5, 9, 13, 17, 21, 25, 29$. The difference between any pair of successive terms is 4, and so,

$$u_2 = 5, \quad u_1 = 1 \quad \Rightarrow \quad u_2 = u_1 + 4$$

$$u_3 = 9, \quad u_2 = 5 \quad \Rightarrow \quad u_3 = u_2 + 4 \quad \text{etc.}$$

In fact, this pattern is constant and each term is greater than the previous one by 4. If successive terms are denoted by u_r and u_{r+1}, this relation could be written,

$$u_{r+1} = u_r + 4 \cdots\cdots (1)$$

This enables any term to be evaluated from the previous term but it does not give a starting point. The sequence will be defined completely by stating relation **(1)** together with a value for u_1 and the highest value of r.

$$\text{i.e.} \quad u_1 = 1$$
$$u_{r+1} = u_r + 4 \qquad \text{for } r = 1, 2, 3, \ldots \ldots 7$$

Many sequences can be defined by more than one of the above means but for some it may be difficult or impossible to find, say, an algebraic definition, although an inductive definition is straightforward.

WORKED EXAMPLES

1 Write down the first five terms of a sequence defined inductively by $u_1 = 2$, $u_{r+1} = 5u_r + 1$ for all $k \in \mathbb{N}$.

Using the relation $u_{r+1} = 5u_r + 1$, it is possible to calculate each term from the previous one.

$$\text{Since } u_1 = 2, \qquad u_2 = 5u_1 + 1 = 5(2) + 1 \qquad \Rightarrow \quad u_2 = 11$$
$$\text{Since } u_2 = 11, \qquad u_3 = 5u_2 + 1 = 5(11) + 1 \qquad \Rightarrow \quad u_3 = 56$$
$$\text{Since } u_3 = 56, \qquad u_4 = 5u_3 + 1 = 5(56) + 1 \qquad \Rightarrow \quad u_4 = 281$$
$$\text{Since } u_4 = 281, \qquad u_5 = 5u_4 + 1 = 5(281) + 1 \qquad \Rightarrow \quad u_5 = 1406$$

The first five terms are $2, 11, 56, 281, 1406$.

2 A sequence is given by $u_n = 2n + 1$ for $n = 1, 2, 3, \ldots \ldots$. Write down the first three terms and the 20th term.

The expression for any term $u_n = 2n + 1$ will produce the first term when $n = 1$ is substituted, the second term for $n = 2$, and so on.

Thus,
$$u_1 = 2(1) + 1 = 3 \qquad u_2 = 2(2) + 1 = 5$$
$$u_3 = 2(3) + 1 = 7 \qquad u_{20} = 2(20) + 1 = 41$$

The first three terms are $3, 5, 7$ and the 20th term is 41.

3 If the first few terms of a sequence are $5, 17, 53, 161, \ldots \ldots$, find **(a)** an inductive definition and **(b)** an algebraic definition.

There is no hard and fast rule for finding formulae when only a few terms are given. It requires a good knowledge of how sequences behave and usually needs careful thought.

(a) The key is to notice that each term is **nearly** three times the previous one.

i.e. $3 \times 5 = 15$, $\quad 3 \times 17 = 51$ etc.

In fact, if each term is multiplied by 3, and 2 added to the result, the next term is obtained.

i.e. $(3 \times 5) + 2 = 17$, $\quad (3 \times 17) + 2 = 53$ etc.

Thus, the inductive definition would be

$$u_1 = 5, \quad u_{k+1} = 3u_k + 2 \quad \text{for } k \in \mathbb{N}$$

Now, as n becomes larger, the term $\dfrac{2}{2n+1}$ becomes smaller, and when n is large it can be neglected. So the value of u_n approaches 2 and you say that the sequence tends to a limit as n becomes large.

In mathematical notation you write,

$$\lim_{n \to \infty} u_n = \lim_{n \to \infty} \left[\frac{4n}{2n+1} \right] = \lim_{n \to \infty} \left[2 - \frac{2}{2n+1} \right] = 2$$

Not all sequences tend to a limit and the other sequence, $1, 4, 9, 16, \ldots$, is such a case. Sequences may also oscillate between finite or infinite values.

WORKED EXAMPLES

1 Discuss the behaviour of the sequence given by $u_n = 2 + (-1)^n$, stating what happens as n tends to infinity.

You can clearly see the pattern of this sequence by considering a few terms (figure, right).

$u_1 = 2 + (-1)^1 = 1$

$u_2 = 2 + (-1)^2 = 3$

$u_3 = 2 + (-1)^3 = 1$

$u_4 = 2 + (-1)^4 = 3$

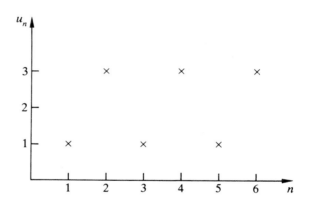

i.e. the terms will successively take the values of 1 and 3. The sequence oscillates finitely. As $n \to \infty$ $(-1)^n$ still takes only values of $+1$ or -1 and hence u_n can only be 1 or 3.

2 Find the limit of the sequence given by $u_n = \dfrac{n+4}{n^2+1}$, as n becomes large.

Rewrite u_n by dividing the numerator and denominator by n^2.

$$\therefore u_n = \frac{n+4}{n^2+1} = \left(\frac{n}{n^2} + \frac{4}{n^2} \right) \Big/ \left(\frac{n^2}{n^2} + \frac{1}{n^2} \right) = \left(\frac{1}{n} + \frac{4}{n^2} \right) \Big/ \left(1 + \frac{1}{n^2} \right)$$

As $n \to \infty$, $\dfrac{1}{n} \to 0$ and $\dfrac{1}{n^2} \to 0$

Hence,

$$\lim_{n \to \infty} \left[\frac{n+4}{n^2+1} \right] = \lim_{n \to \infty} \left[\left(\frac{1}{n} + \frac{4}{n^2} \right) \Big/ \left(1 + \frac{1}{n^2} \right) \right] = 0$$

Exercise 14.2

1 Find the behaviour of the following sequences, as n becomes large.

(a) $u_n = \dfrac{1}{n}$

(d) $u_n = \dfrac{n}{n+2}$

(g) $u_n = \dfrac{n^2}{n+2}$

(i) $u_n = \dfrac{(n+2)(n+3)}{(n+1)}$

(b) $u_n = 2n - 3$

(e) $u_n = \dfrac{n^2 - 1}{n^2 + 1}$

(h) $u_n = 3 + \dfrac{2}{n^2}$

(j) $u_n = \dfrac{n-5}{2n+1}$

(c) $u_n = 4 - \dfrac{1}{n}$

(f) $u_n = 1 + (-1)^n$

2 Find the limits (if any) of the sequences defined inductively, as n tends to infinity.

(a) $u_{r+1} = \tfrac{1}{2}u_r$; $u_1 = 2$

(b) $u_{r+1} = u_r + 1$; $u_1 = 1$

(c) $u_{r+1} = -u_r$; $u_1 = 2$

3 State whether the following sequences tend to a limit or oscillate as n becomes large, and give the value of the limit.

(a) $u_n = \left(-\dfrac{1}{2}\right)^n$

(d) $u_n = 1 + \dfrac{1}{n}$

(g) $u_n = n + (-1)^n n^2$

(b) $u_n = \sqrt{n}$

(e) $u_n = \cos n\pi$

(h) $u_n = \dfrac{3n^2}{(n-1)(n-2)}$

(c) $u_n = \dfrac{3n+2}{n-1}$

(f) $u_n = (-1)^n$

Series

If the terms of a sequence are added to form a sum, the result is called a **series**.

e.g. the sequence 1, 3, 5, 7, 9 would form the finite series $1 + 3 + 5 + 7 + 9$,

and the sequence 2, -4, 8, -16, 32,... would form the series $2 - 4 + 8 - 16 + 32 - \cdots$.

The sum of n terms of a series is denoted by S_n and any term by u_r, where r takes all values between 1 and n.

$$\therefore S_n = u_1 + u_2 + u_3 + u_4 + \cdots + u_r + \cdots + u_n$$

As we have seen in an earlier chapter, mathematicians use the Greek letter Σ (sigma) to denote a summation, and the above series is written as,

$$S_n = \sum_{r=1}^{r=n} u_r \quad \text{or} \quad \sum_{r=1}^{n} u_r$$

WORKED EXAMPLES

1 Find the terms of the series $\displaystyle\sum_{r=1}^{5} \dfrac{2}{3r+1}$

Substituting successive values of r between 1 and 5, will produce each term of the series.

$$\text{Since} \quad u_r = \frac{2}{3r+1}, \quad u_1 = \frac{2}{4}, \quad u_2 = \frac{2}{7} \quad \text{etc.}$$

$$\therefore \quad S_n = \frac{1}{2} + \frac{2}{7} + \frac{1}{5} + \frac{2}{13} + \frac{1}{8}$$

2 Write the following series using the Σ notation, $1 + 5 + 9 + 13 + \cdots + 41$.

It is necessary to find a general expression for the general term u_r.

Since the value of each term increases by 4, you need an expression for u_r which produces multiples of 4.

Now, $u_r = 4r$ gives 4, 8, 12, 16,... as $r = 1, 2, 3, \ldots$.

You must also arrange for the first term to be correct.

If $u_r = 4r - 3$, then the terms $1, 5, 9, 13$ will be produced.

The last term is 41, which occur when $r = 11$ (since $4r - 3 = 41 \Rightarrow r = 11$).

$$\therefore S_n = \sum_{r=1}^{11} 4r - 3$$

The use of the Σ notation is not unique and it would also be possible to express this series as

$$\therefore S_n = \sum_{r=0}^{10} (4r + 1)$$

The arithmetical series (arithmetical progression)

This series is formed by summing the terms of a sequence in which each term is produced from the previous one by adding or subtracting a constant value. It is called an **arithmetical progression** (A.P.).

Investigation 1

Write down a series in which the first term is 2 and each successive term is formed by adding 3 to the previous term.

State the value of the 5th term, the 7th term and the 10th term.

Can you see a method by which you could state the value of the 10th term without writing down all the terms before it?

Try this for other examples!

If the first term is a and the value of the constant to be added is d, then the general form can be written

$$S_n = a + (a + d) + (a + 2d) + (a + 3d) + \cdots\cdots$$

d is called the **common difference.**

The **value of any term** can be found from observation. The value of the 2nd term is $a + d$, the value of the 3rd term is $a + 2d$, and so on.

$$\boxed{\text{The value of the } n\text{th term} = a + (n - 1)d}$$

The **sum of the series** can be found as follows. If the first n terms are written, you have

$$S_n = a + (a + d) + (a + 2d) + (a + 3d) + \cdots [a + (n-1)d]$$

In the reverse order, $\qquad S_n = [a + (n-1)d] + [a + (n-2)d] + [a + (n-3)d] + \cdots + a$

Adding, you obtain $\qquad 2S_n = [2a + (n-1)d] + [2a + (n-1)d] + \cdots [2a + (n-1)d]$

On the right-hand side there are n terms $\quad \Rightarrow \quad 2S_n = n[2a + (n-1)d]$

The sum of an A.P. to n terms, $S_n = \dfrac{n}{2}[2a + (n-1)d]$

If the last term of the series is denoted by l, then $l = a + (n-1)d$.

$$\text{Thus,} \quad S_n = \frac{n}{2}[a + \{a + (n-1)d\}] = \frac{n}{2}(a + l)$$

The sum of an A.P. to n terms, $S_n = \dfrac{n}{2}(a + l)$

These three results should be learnt, so that they can be applied quickly and easily.

WORKED EXAMPLES

1 Find the value of the 21st term of the series $49 + 45 + 41 + 37 + \cdots$.

This is an A.P. whose first term $a = 49$ and whose common difference $d = -4$ (note the minus sign).

$$\text{The value of the } n\text{th term} = a + (n-1)d$$
$$\Rightarrow \quad \text{The value of the 21st term} = 49 + (21 - 1)(-4)$$
$$= 49 + (20)(-4) = -31$$

2 Find the number of terms in the series $231 + 225 + 219 + \cdots + 135$.

Using the result for the value of the nth term with $a = 231$ and $d = -6$, you obtain

$$\text{The value of the } n\text{th term} = a + (n-1)d$$
$$135 = 231 + (n-1)(-6)$$
$$\text{Rearranging sides,} \quad \Rightarrow \quad 6(n-1) = 231 - 135$$
$$\Rightarrow \quad 6(n-1) = 96 \quad \Rightarrow \quad n - 1 = 16 \quad \Rightarrow \quad n = 17$$

There are 17 terms in this series.

3 Find the sum of the first **(a)** 20 terms and **(b)** n terms of the series $3 - 4 - 11 - 18 - \cdots$.

The sum of an A.P. to n terms, $S_n = \dfrac{n}{2}[2a + (n-1)d]$

(a) In this case, $a = 3$, $d = -7$ and $n = 20$.

$$\therefore S_{20} = \frac{20}{2}[6 + (20 - 1)(-7)]$$

$$= 10(6 - 133) = -1270$$

(b) In this case, $a = 3$, $d = -7$ and the number of terms is n.

$$\therefore S_n = \frac{n}{2}[6 + (n - 1)(-7)]$$

$$= \frac{n}{2}(6 - 7n + 7) = \frac{n}{2}(13 - 7n)$$

4 Find the sum of the series $3 + 7 + 11 + 15 + \cdots + 79$.

To use the formula for the sum, S_n, you need values for the first term a, the common difference d, and the number of terms, n. In this example, n must be evaluated first.

Using, the value of the nth term $= a + (n - 1)d$

$$\Rightarrow \quad 79 = 3 + (n - 1)4$$

$$\Rightarrow \quad 76 = 4(n - 1)$$

$$\Rightarrow \quad 19 = n - 1 \qquad \text{giving } n = 20$$

Using the sum to n terms,

$$S_n = \frac{n}{2}(a + l)$$

$$\Rightarrow \quad S_{20} = \frac{20}{2}(3 + 79) = 10 \times 82 = 820$$

The sum of the series is 820.

5 The sum of the 3rd and 4th terms of an A.P. is 1 and the sum of the first 7 terms is -7. Find the first term and the common difference.

The 3rd term and 4th term can be written as $a + 2d$ and $a + 3d$, respectively, where a is the first term and d is the common difference. The sum of these terms is 1.

Hence, $(a + 2d) + (a + 3d) = 1 \quad \Rightarrow \quad 2a + 5d = 1 \ldots \ldots \ldots \ldots$ **(1)**

The sum of the first seven terms is -7

Using $S_n = \dfrac{n}{2}[2a + (n - 1)d]$

$$-7 = \frac{7}{2}[2a + 6d] \quad \Rightarrow \quad 2a + 6d = -2 \ldots \ldots \ldots \ldots \ldots$$ **(2)**

Solving equations **(1)** and **(2)** by subtracting, gives $d = -3$.

Substituting $d = -3$ into equation **(1)** gives $\quad 2a + 5(-3) = 1 \quad \Rightarrow \quad a = 8$

The first term is 8 and the common difference is -3.

Exercise 14.3

1 Write out the terms of the following series in full.

    (a) $\displaystyle\sum_{r=1}^{5} (r^2 + 1)$
    (b) $\displaystyle\sum_{r=3}^{7} \frac{(r+1)^2}{(r-1)^2}$
    (c) $\displaystyle\sum_{r=0}^{4} (2r - 1)$
    (d) $\displaystyle\sum_{r=1}^{6} \frac{1}{r+1}$

    (e) $\displaystyle\sum_{r=0}^{5} (-1)^{r+1}(2^r + 1)$
    (f) $\displaystyle\sum_{r=n}^{n+3} r(r+1)$

2 Write in the Σ notation.

    (a) $1 + 2 + 3 + \cdots + 99$
        (e) $(2 \times 3) + (3 \times 4) + (4 \times 5) + \cdots + (16 \times 17)$

    (b) $1^3 + 2^3 + 3^3 + \cdots + 20^3$
        (f) $-1 + 2 - 3 + 4 - 5 + \cdots + 20$

    (c) $2 + 7 + 12 + \cdots + 97$
        (g) $16 + 14 + 12 + \cdots$ to n terms

    (d) $-2 + 4 + 10 + \cdots + 64$
        (h) $5^2 + 9^2 + 13^2 + \cdots$ to $(2n+1)$ terms

3 Which of the following series are arithmetical progressions?

    (a) $6 + 4 + 2 + 0 - 2 - \cdots$
        (d) $\frac{1}{2} + \frac{4}{3} + \frac{13}{6} + 3 + \cdots$

    (b) $1 + \frac{1}{2} + \frac{1}{4} + \frac{1}{8} + \cdots$
        (e) $-7 - 10 - 13 - 16 - \cdots$

    (c) $1 - 1 + 1 - 1 + \cdots$
        (f) $2.2 + 2.22 + 2.222 + \cdots$

4 Find the value of the terms indicated in the following series.

    (a) $2 + 5 + 8 + \cdots$ 15th, 30th
        (e) $13 + 17 + 21 + \cdots$ 10th, nth

    (b) $-32 - 30 - 28 - \cdots$ 10th, 17th
    (f) $51 + 49 + 47 + \cdots$ 25th, nth

    (c) $100 + 97 + 94 + \cdots$ 13th, 40th
    (g) $4 + 6\frac{1}{2} + 9 + \cdots$ nth, $(n+1)$th

    (d) $1 + \frac{7}{4} + \frac{5}{2} + \cdots$ 9th, 15th
    (h) $20 + 30 + 40 + \cdots$ $(2n)$th

5 Find the number of terms in the series.

    (a) $1 + 6 + 11 + \cdots + 81$
        (e) $a + 2a + 3a + \cdots + 21a$

    (b) $-18 - 15 - 12 - \cdots + 39$
    (f) $3 + 3.01 + 3.02 + \cdots + 3.17$

    (c) $2.5 + 2.8 + 3.1 + \cdots + 6.7$
    (g) $\frac{1}{2} + \frac{5}{8} + \frac{3}{4} + \cdots + 1\frac{3}{4}$

    (d) $-25 - 35 - 45 - \cdots - 105$

6 Find the sum of each of the following series.

    (a) $4 + 3 + 2 + \cdots - 16$
        (e) $1 + 2 + 3 + \cdots$ to 100 terms

    (b) $13 + 15 + 17 + \cdots$ to 10 terms
    (f) $73 + 69 + 65 + \cdots - 23$

    (c) $-10 - 7 - 4 - \cdots + 47$
        (g) $1 + 3 + 5 + 7 + \cdots$ to n terms

    (d) $2 + 4 + 6 + \cdots$ to 50 terms

7 Find the sum of the even numbers between 2 and 200 inclusive.

8 If the first and last terms of an A.P. are 20 and 113 and the common difference is 3, find the number of terms.

9 Find the sum of 10 terms of an A.P. in which the first term is 52 and the last term is -60.

10 The first term of an A.P. is 8 and the sum to 10 terms is 170. Find the common difference and the sum to n terms.

11 If the 4th term of an A.P. is 15 and the sum of the first 6 terms is 66, find the common difference, the first term and the sum to 20 terms.

12 In an A.P. the 2nd term is five times the 6th term. If the 8th term is -4, find the first term and the sum of the first 12 terms.

13 If the sum of the first 9 terms of an A.P. is 63 and the sum of the first four terms is 48, find the sum to 16 terms.

14 The 3rd term of an A.P. is -28 and the 7th term is 42. Find the sum of the first 10 terms.

The geometrical series (geometrical progression)

This series is formed by summing the terms of a sequence in which each term is produced by multiplying the previous term by a constant factor. Such a series is called a **geometrical progression** (G.P.).

> *Investigation 2*
>
> Write down the terms of a series in which the first term is 2 and each term is produced from the previous one by multiplying by 4.
>
> State the value of the 3rd term, the 5th term and the 6th term.
>
> Can you see a way by which you could give the value of the 8th term without writing out every term of the series?
>
> Try this for other examples!

If the first term is denoted by a and the constant factor (called the **common ratio**) by r, then the general form of the series can be written as

$$S_n = a + ar + ar^2 + ar^3 + \cdots.$$

Since the 3rd term is ar^2, the 4th term is ar^3 etc. it is clear that the nth term can be written as ar^{n-1}.

> The value of the nth term $= ar^{n-1}$

The sum of the series can be found as follows. Taking n terms, the sum S_n is

$$S_n = a + ar + ar^2 + ar^3 + \cdots + ar^{n-1} \ldots\ldots\ldots (1)$$

Multiply by r,
$$rS_n = \quad ar + ar^2 + ar^3 + \cdots + ar^{n-1} + ar^n \ldots..(2)$$

Subtracting **(2)** from **(1)**,

$$S_n - rS_n = a - ar^n \quad \text{(all other terms cancel out)}$$

$$\Rightarrow \quad (1-r)S_n = a(1-r^n) \quad \Rightarrow \quad S_n = \frac{a(1-r^n)}{1-r}$$

If $r > 1$, it is best to express this result as $S_n = \dfrac{a(r^n - 1)}{r - 1}$.

$$\boxed{\text{The sum of a G.P. to } n \text{ terms, } S_n = \dfrac{a(1 - r^n)}{1 - r} = \dfrac{a(r^n - 1)}{r - 1}}$$

WORKED EXAMPLES

1 Find the 15th term and the sum of the first 10 terms of the geometrical progression $64 + 32 + 16 + 8 + \cdots$.

The first term a is 64 and the common ratio, r, is $\frac{1}{2}$.

The value of the nth term $= ar^{n-1}$

The value of the 15th term $= 64(\frac{1}{2})^{14} = \dfrac{2^6}{2^{14}} = \dfrac{1}{2^8} = \dfrac{1}{256}$

The sum of a G.P. to n terms $= \dfrac{a(1 - r^n)}{1 - r}$

$$\Rightarrow \quad S_{10} = \frac{64[1 - (\frac{1}{2})^{10}]}{1 - \frac{1}{2}} = 128[1 - (\frac{1}{2})^{10}]$$

$$= 128\left(1 - \frac{1}{1024}\right) = \frac{1023}{1024} \times 128 = \frac{1023}{8}$$

2 How many terms of the series $2 + 6 + 18 + 54 + \cdots$ are required to make a sum greater than 5000?

The first term, a, is 2 and the common ratio, $r = 3$.

Using the sum of a G.P. to n terms, $S_n = \dfrac{a(r^n - 1)}{r - 1}$

Consider a sum equal to 5000.

$$5000 = \frac{2(3^n - 1)}{3 - 1} \quad \Rightarrow \quad 3^n - 1 = 5000 \quad \Rightarrow \quad 3^n = 5001$$

In general, this type of equation is solved using logarithms.

Taking logarithms, base 10, of each side,

$$\log 3^n = \log 5001$$

$$\Rightarrow \quad n \log 3 = \log 5001 \quad \Rightarrow \quad n = \frac{\log 5001}{\log 3}$$

$$\Rightarrow \quad n = 7.75 \text{ to 2 dec. places (by calculator)}$$

Now, n represents the number of terms required and must be an integer. In order that the sum exceeds 5000, 8 terms will be needed.

3   The sum of the 2nd and 3rd terms of a G.P. is 6 and the sum of the first 4 terms is -15. Find the first term, the common ratio and the first 5 terms of the series.

Take each of the statements given in the question and translate them into a mathematical relation.

The sum of the 2nd and 3rd terms $= 6$ gives

$$ar + ar^2 = 6 \ldots\ldots\ldots\ldots\ldots \textbf{(1)}$$

The sum of the first 4 terms $= -15$. This implies that the sum of the 1st and 4th terms $= -21$, since the sum of the 2nd and 3rd terms is 6.

$$\therefore a + ar^3 = -21 \ldots\ldots\ldots \textbf{(2)}$$

Dividing equation **(2)** by equation **(1)**,

$$\frac{a + ar^3}{ar + ar^2} = -\frac{21}{6}$$

$$\Rightarrow \quad \frac{a(1 + r^3)}{a(r + r^2)} = -\frac{7}{2} \quad \text{or} \quad \frac{1 + r^3}{r + r^2} = -\frac{7}{2}$$

Multiplying by $2(r + r^2)$ gives

$$2(1 + r^3) = -7(r + r^2)$$

$$\Rightarrow \quad 2r^3 + 7r^2 + 7r + 2 = 0$$

Use the remainder theorem to find possible factors.

Let $\qquad\qquad\qquad\qquad f(r) = 2r^3 + 7r^2 + 7r + 2$

Since these terms of $f(r)$ form a sum, try negative values for r.

$$f(-1) = -2 + 7 - 7 + 2 = 0$$

Thus $(r + 1)$ is a factor of $f(r)$.

$$\therefore \ f(r) = (r + 1)(2r^2 + 5r + 2) \qquad \text{(by inspection)}$$

$$= (r + 1)(r + 2)(2r + 1)$$

Hence, $f(r) = 0$ provides 3 real solutions for r: $\qquad r = -\frac{1}{2}, -1$ or -2.

If $r = -\frac{1}{2}$, then from **(1)**, $\quad a = \dfrac{6}{r + r^2} \quad \Rightarrow \quad a = \dfrac{6}{-\frac{1}{2} + \frac{1}{4}} = -24$

If $r = -1$, then $a = \dfrac{6}{0}$, which is not finite.

If $r = -2$, then $a = \dfrac{6}{2} = 3$.

There are two possible series. The first 5 terms are either

$$-24 + 12 - 6 + 3 - 1\tfrac{1}{2} \quad \text{or} \quad 3 - 6 + 12 - 24 + 48$$

Exercise 14.4

1 Which of the following series are geometrical progressions?

    (a) $1 + 2 + 4 + 8 + \cdots + 256$

    (b) $-1 + 3 - 9 + 27 - \cdots$

    (c) $1 + \frac{1}{2} + \frac{1}{3} + \frac{1}{4} + \cdots$

    (d) $1 + 0.1 + 0.01 + 0.001 + \cdots$

    (e) $2 + 2\frac{1}{2} + 2\frac{1}{4} + 2\frac{1}{8} + \cdots$

    (f) $1 - 1 + 1 - 1 + \cdots$

    (g) $a + 2ax + 4ax^2 + \cdots$

    (h) $64 - 16 + 4 - 1 + \frac{1}{4}$

2 Find the value of the terms indicated in the following G.P.s. Simplify your answer where appropriate.

    (a) $4 + 12 + 36 + \cdots$ 9th, 20th

    (b) $\dfrac{1}{2} + \dfrac{1}{4} + \dfrac{1}{8} + \cdots$ 8th, 19th

    (c) $5 - 4 + \cdots$ 6th, 14th

    (d) $\dfrac{3}{7} + \dfrac{9}{14} + \dfrac{27}{28} + \cdots$ 10th, nth

    (e) $0.4 + 0.04 + 0.004 + \cdots$ 5th, nth

    (f) $81 - 27 + 9 - \cdots$ 7th, nth

    (g) $1 + \dfrac{2}{3} + \dfrac{4}{9} + \cdots$ $(2n)$th

3 Find the number of terms is the following G.P.s

    (a) $16 + 32 + 64 + \cdots + 1024 +$

    (b) $16 + 8 + 4 + \cdots + \dfrac{1}{128}$

    (c) $1 - 3 + 9 - 27 + \cdots - 2187$

4 Find the sums of the G.P.s, simplifying your answer where appropriate.

    (a) $6 + 12 + 24 + \cdots + 3072$

    (b) $100 + 50 + 25 + \cdots + \dfrac{25}{256}$

    (c) $2 - \dfrac{3}{2} + \dfrac{9}{8} + \cdots$ to 8 terms

    (d) $1 - 2 + 4 - 8 + \cdots$ to 50 terms

    (e) $4 + 16 + 64 + \cdots$ to 13 terms

    (f) $360 - 120 + 40 - \cdots$ to n terms

5 If the 4th term of a G.P. is 12 and the 5th term is 96, find the value of the common ratio and the value of the first term.

6 The 6th term of a G.P. is 4 times the 4th term. If the sum of the first 5 terms is $5\frac{1}{2}$ and $r < 0$, find the sum of the first 6 terms of the series.

7 If the first term of a G.P. is $\frac{2}{3}$ and the 4th term is $2\frac{1}{4}$, find the 10th term and the sum of the first 10 terms.

8 Find the number of terms of the series $3 + 12 + 48 + \cdots$ required for the sum to exceed $10\,000$.

9 The 5th term of a G.P. is 216 and the 2nd term is 27. Find the sum of the first 7 terms.

10 The numbers $x - 3$, $x + 3$, $2x + 1$, form a geometrical progression. Find the value of the common ratio.

11 The 2nd, 4th and 8th terms of an A.P. form a geometrical progression. If the sum of the first 4 terms of the A.P. is 60, find the sum of the first 8 terms.

12 The sum of the first two terms of a G.P. is 32. If the sum of the first 3 terms is 104, find the value of the common ratio if $r > 0$, and write down the first 4 terms of the series.

Infinite geometrical progression

You have already seen how the terms of a sequence can tend to a limit as a large number of terms is taken. In a similar way, the sum of a G.P. can tend to a finite limit.

The series $S = 1 + 2 + 4 + 8 + 16 + \cdots$ will give the following sums to the stated numbers of terms.

$$S_1 = 1, \quad S_2 = 3, \quad S_3 = 7, \quad S_4 = 15, \quad S_5 = 31, \quad S_6 = 63, \ldots$$

Clearly the sum is increasing without limit as more terms are taken.

The series $S = 1 + 0.1 + 0.01 + 0.001 + 0.0001 + \cdots$ is a G.P. whose common ratio is 0.1 and the sums to given numbers of terms will be

$$S_1 = 1, \quad S_2 = 1.1, \quad S_3 = 1.11, \quad S_4 = 1.111, \quad S_5 = 1.1111, \ldots$$

Clearly, although the sum increases, it will never reach a value of 1.12. We say that this series **converges** to a limit as n (the number of terms) becomes large.

The trend can be shown on a graph, as in the figure, right.

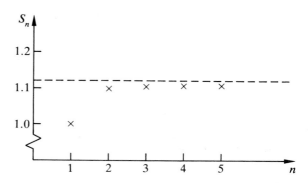

By considering the general series and its sum to n terms, you can find a result for the case when n is large.

$$S_n = a + ar + ar^2 + ar^3 + \cdots ar^{n-1} = \frac{a(1 - r^n)}{1 - r}$$

Now, if $-1 < r < 1$, then r^n will be very small if n is large. In this case, r^n can be neglected and S_n tends to a finite limit called the **sum to infinity**, and the series **converges**.

$$\boxed{\text{The sum to infinity, } S_\infty = \frac{a}{1 - r}}$$

For the series above, the limit is $S_\infty = \dfrac{1}{1 - 0.1} = \dfrac{1}{0.9} = 1.\dot{1}.$

If $r > 1$ r^n does not tend to zero as n becomes large. Thus, S_n does not tend to a finite limit and the series **diverges**.

If $r < -1$ r^n oscillates between positive and negative values and the sum oscillates infinitely.

WORKED EXAMPLES

1 Find the sum of the first n terms of the series $2 - 1 + \frac{1}{2} - \frac{1}{4} + \cdots$ and deduce its sum to infinity.

This is a G.P. with common ratio $r = -\frac{1}{2}$ and $a = 2$.

The sum to n terms $= \dfrac{a(1-r^n)}{1-r}$

$$= \frac{2[1-(-\frac{1}{2})^n]}{1-(-\frac{1}{2})} = \frac{4}{3}[1-(-\frac{1}{2})^n]$$

Since $(-\frac{1}{2})^n \to 0$ as $n \to \infty$, the sum to infinity $= \frac{4}{3}$.

2 Express the recurring decimal $0.\dot{2}\dot{7}$ as a fraction in its lowest terms.

Now, $0.\dot{2}\dot{7} = 0.272\,727\cdots$

$$= 0.27 + 0.0027 + 0.000\,027 + \cdots$$

This is an infinite G.P. with a first term of 0.27 and a common ratio of 0.01.

Using the sum to infinity, $S_\infty = \dfrac{a}{1-r} = \dfrac{0.27}{1-0.01}$

$$= \frac{0.27}{0.99} = \frac{27}{99} = \frac{3}{11}$$

Exercise 14.5

1 Write down the sum of the following G.P.s to n terms and, if the series possesses a sum to infinity, deduce its value.

    (a) $13 + 6\frac{1}{2} + 3\frac{1}{4} + \cdots$ **(c)** $400 - 100 + 25 - \cdots$ **(e)** $4 - 3 + 2\frac{1}{4} - \cdots$

    (b) $0.6 + 0.06 + 0.006 + \cdots$ **(d)** $0.2 + 0.4 + 0.8 + \cdots$

2 Find the sum to infinity of the geometrical progressions.

    (a) $4 + 2 + 1 + \frac{1}{2} + \cdots$ **(c)** $1 + \frac{2}{3} + \frac{4}{9} + \cdots$ **(e)** $1 + a + a^2 + \cdots$ where $0 < a < 1$

    (b) $1 - \frac{1}{4} + \frac{1}{16} - \frac{1}{64} + \cdots$ **(d)** $1 - \frac{1}{10} + \frac{1}{100} - \cdots$ **(f)** $1 + \cos\theta + \cos^2\theta + \cdots$

3 If the first term of a G.P. is 3 and its sum to infinity is $2\frac{1}{4}$, find its common ratio.

4 If the 3rd term of a G.P. is $-\frac{1}{3}$ and the 6th term is $\frac{1}{81}$, find the sum to infinity.

5 Find the possible values of x for which the following G.P.s possess a sum to infinity.

    (a) $x + \dfrac{3x^2}{2} + \dfrac{9x^3}{4} + \cdots$ **(c)** $2 - 6x + 18x^2 + \cdots$

    (b) $1 + 4x + 16x^2 + \cdots$ **(d)** $1 + 4x^2 + 16x^4 + \cdots$

6 Express the following recurring decimals as fractions in their lowest terms.

    (a) $0.\dot{3}\dot{2}$ **(b)** $0.\dot{7}$ **(c)** $0.\dot{0}6\dot{3}$

7 If the sum to infinity of a G.P. is 12 and the 2nd term is 3, find the sum of the first six terms.

8 If the sum of the first three terms of a G.P. is $\frac{3}{4}$ and its sum to infinity is $\frac{2}{3}$, find the first three terms of the series.

---

15 Laws of motion

Introduction

When we push or pull an object we apply a force to that object.

There are many different types of forces, including weight, tension, reaction and friction. In Chapter 8, we considered how these are applied to bodies.

If a force is applied to a body it can produce motion, such as moving a garden roller or pushing a fork into the ground.

Often more than one force will act on a body – the person pushing the fork into the ground will apply a downward force to the fork which will also be subject to an upward resistance of the ground.

The parachutist may well be subject to three forces simultaneously – his own weight, an upward resistance due to the air and a wind force sideways.

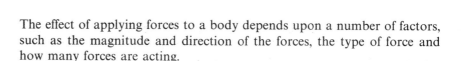

The effect of applying forces to a body depends upon a number of factors, such as the magnitude and direction of the forces, the type of force and how many forces are acting.

A force may be applied over a period of time, as in the case of the forward pulling force developed by the engine of a car. Indeed, the force may vary. When, for instance, starting from traffic lights, the pulling force will be greater than the resistance to motion and the car will accelerate, but when cruising on a motorway, the force developed will be just sufficient to balance the resistances and the car will move at a constant velocity.

Forces which act for a very short time interval are usually called **impulsive forces**, and their effect will be considered later.

Forces applied to a single body are called **external forces**. (e.g. the weight of a body, a resistance to motion). If two bodies are acted upon by mutual equal and opposite forces, they are called **internal forces** and are known as the **action** and **reaction** (e.g. the tension in the tow bar between a car and a caravan).

Newton's laws of motion

Most of the examples mentioned in the introduction are familiar to our experience and were known to early scientists and mathematicians such as Galileo, Huygens and Newton. In fact, it was Sir Isaac Newton (1642–1727) who formed the observations into **three laws** which he published in his *Principia* in 1686.

Newton's first law: Every body continues in its state of rest or of uniform motion in a straight line unless compelled to change that state by externally applied forces.

A book placed on a table remains at rest unless someone pushes it or tilts the table, i.e. unless a force is applied to it to cause it to move.

Remember that the book is subject to its weight and a reaction force upwards, but these balance.

A spacecraft that has escaped from the earth's gravitational pull will continue with constant speed, without the use of the rockets, until it comes into the influence of the moon's gravitation, under which it will accelerate.

In effect, Newton's first law gives a definition of force as that cause which changes or tends to change the state of rest or uniform motion of a body.

> **Newton's second law:** The rate of change of momentum of a body is proportional to the applied force and takes place in the direction of that force.

Now, the **momentum of a body** is defined as the product of its mass m and its velocity $\mathbf{v}$.

$$\therefore \text{Momentum} = m\mathbf{v} \qquad \text{(a vector quantity)}$$

If $\mathbf{F}$ is the resultant of all the external forces applied to a body of mass m, then by Newton's second law,

$$\frac{d}{dt}(m\mathbf{v}) \propto \mathbf{F}$$

Thus, if the mass m is constant and k is a constant of proportionality,

$$km\frac{d\mathbf{v}}{dt} = \mathbf{F} \quad \text{or} \quad \mathbf{F} = km\mathbf{a}$$

where $\mathbf{a}$ is the acceleration of the body.

The value of k depends upon the units chosen for $\mathbf{F}$, m and $\mathbf{a}$. We choose the unit of force to be the **Newton** (symbol N).

A Newton is the force required to give a mass of $1\,\text{kg}$ an acceleration of $1\,\text{m s}^{-2}$.

In this case, if the force is in newtons, mass is in kilogrammes and acceleration in m s^{-2}, then $k = 1$.

$$\text{Newton's 2nd law then gives} \qquad \boxed{\mathbf{F} = \frac{d}{dt}(m\mathbf{v})} \quad \text{or} \quad \boxed{\mathbf{F} = m\frac{d\mathbf{v}}{dt} = m\mathbf{a} \quad \text{if } m \text{ is constant}}$$

Note that the force $\mathbf{F}$ must denote the resultant (or effective force) in the direction of motion.

WORKED EXAMPLES

1 A body of mass $2\,\text{kg}$ resting on a table is moved by a force of $15\,\text{N}$ against resistances of $3\,\text{N}$ and $8\,\text{N}$. Find the acceleration of the body.

Since the mass will move horizontally along the table, we apply Newton's 2nd law in that direction.

Using $\mathbf{F} = m\mathbf{a}$,

$$15 - (3 + 8) = 2a \quad \Rightarrow \quad a = 2\,\text{m s}^{-2}$$

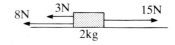

The acceleration of the body is $2\,\text{m s}^{-2}$ in the direction of the $15\,\text{N}$ force.

2 A 3 kg mass slides down a rough plane inclined at 30° to the horizontal, against a frictional resistance of 4 N. Find the acceleration of the mass and the reaction between the mass and the plane. Take $g = 9.8\,\mathrm{m\,s^{-2}}$.

Newton's 2nd law can be applied in any direction so long as we remember to take the component parts of the forces in the chosen direction.

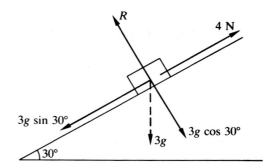

In this example, the weight $3g$ N acts vertically and is shown by the dotted line in the figure. This can be resolved into its components of $3g\sin 30°$ down the plane and $3g\cos 30°$ perpendicular to the plane, these being used instead of the single force of $3g$ acting vertically.

Consider motion parallel to the plane. By Newton's 2nd law, $\mathbf{F} = m\mathbf{a}$, we have,

$$3g\sin 30° - 4 = 3a$$

where a is the acceleration of the mass down the plane.

$$\therefore\ 3g(\tfrac{1}{2}) - 4 = 3a \quad\Leftrightarrow\quad 10.7 = 3a \quad\Leftrightarrow\quad a = 3.57\,\mathrm{m\,s^{-2}}$$

If R is the reaction between the mass and the plane, then using Newton's 2nd law perpendicular to the plane,

$$R - 3g\cos 30° = 3 \times 0 \quad\Rightarrow\quad R = 3g\cos 30° = 25.46\,\mathrm{N}$$

The acceleration in this direction is zero, since the mass does not leave the surface of the plane. Thus the acceleration is $3.57\,\mathrm{m\,s^{-2}}$ down the plane, and the normal reaction is 25.46 N.

3 A force given by $\mathbf{F} = 3\mathbf{i} + 2\mathbf{j} - \mathbf{k}$ acts upon a mass of 2 kg. Find the acceleration of the mass.

Using Newton's 2nd law, $\mathbf{F} = m\mathbf{a}$

$$3\mathbf{i} + 2\mathbf{j} - \mathbf{k} = 2\mathbf{a}$$

Hence, $\quad \mathbf{a} = \tfrac{3}{2}\mathbf{i} + \mathbf{j} - \tfrac{1}{2}\mathbf{k}$

4 A girl pulls a sledge of mass 6 kg up a hill inclined at 30° to the horizontal, by a rope inclined at 30° to the face of the hill. If the tension in the rope is $32\sqrt{3}$ N and there is a resistance to motion of 10 N, find the acceleration of the sledge. Find also the normal reaction between the sledge and the face of the hill.

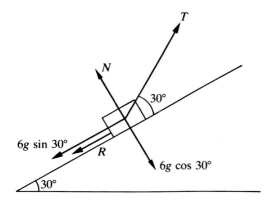

Let R N be the resistance to motion acting down the hill and N N be the normal reaction.

The weight of $6g$ N is resolved into two components, $6g\cos 30°$ perpendicular to the face of the hill and $6g\sin 30°$ down the slope.

The tension in the rope, $T = 32\sqrt{3}$ N.

Using Newton's 2nd law up the face of the hill where $a\,\mathrm{m\,s^{-2}}$ is the acceleration in that direction,

$$T \cos 30° - R - 6g \sin 30° = 6a$$

$$\Rightarrow \quad 32\sqrt{3}\left(\frac{\sqrt{3}}{2}\right) - 10 - 6g\left(\frac{1}{2}\right) = 6a$$

$$\Leftrightarrow \quad 48 - 10 - 29.4 = 6a \quad \Leftrightarrow \quad a = \frac{8.6}{6} = 1.43$$

Since there is no acceleration perpendicular to the face of the hill, we have, using Newton's 2nd law in the direction,

$$N + T \sin 30° - 6g \cos 30° = 0$$

$$\Rightarrow \quad N + 32\sqrt{3}\left(\frac{1}{2}\right) - 6g\left(\frac{\sqrt{3}}{2}\right) = 0$$

$$\Leftrightarrow \quad N = 50.92 - 27.71 = 23.21$$

Thus the acceleration of the mass is $1.43\,\mathrm{m\,s^{-2}}$ up the slope and the normal reaction is 23.21 N.

Newton's third law: To every action there is an equal and opposite reaction.

This means that if a body A exerts a force on a body B, then B exerts an equal and opposite force on A.

A man standing in a lift experiences an action, R, from the floor and applies an equal and opposite force, R, to the floor of the lift.

The diagram shows the man subject to his weight mg and the upward action R, whereas the lift is subject to the tension in the cable, its weight Mg and a downward reaction R.

Forces on lift

Forces on man

Investigation 1

Explain why:

(a) when a balloon is blown up and released so that the air escapes, the balloon flies across the room;

(b) an astronaut could manoeuvre himself in space with an aerosol;

(c) a sprinter uses starting blocks;

(d) a gun recoils when it is fired;

(e) a particle sliding down the face of a wedge which lies on a horizontal plane, causes the wedge to move.

Can you draw diagrams to show the forces which act in these situations?

WORKED EXAMPLE ▶

A man of mass 70 kg stands in a lift. Find the reaction between the man and the floor of the lift when it is moving **(a)** with constant speed, **(b)** upwards with an acceleration of $1.5\,\mathrm{m\,s^{-2}}$ and **(c)** downwards with an acceleration of $2\,\mathrm{m\,s^{-2}}$. Find the tension in the cable in part **(b)** if the lift has a mass of 1000 kg.

Refer to the diagram opposite, showing the forces on the man – his weight and the reaction.

(a) Let the reaction be R_1 N when the lift has constant velocity. Using Newton's 2nd law in the direction of motion, $\mathbf{F} = m\mathbf{a}$

$$R_1 - 70g = 70 \times 0 \quad \Leftrightarrow \quad R_1 = 70g = 686$$

The reaction at constant velocity is 686 N.

(b) Let the reaction be R_2 N when the lift is accelerating upward at $1.5\,\mathrm{m\,s^{-2}}$. Using Newton's 2nd law in the direction of motion,

$$R_2 - 70g = 70 \times 1.5 = 105$$
$$\therefore R_2 = 70g + 105 = 791$$

The reaction when accelerating upwards at $1.5\,\mathrm{m\,s^{-2}}$ is 791 N.

(c) Let the reaction be R_3 N when the lift is accelerating downwards at $2\,\mathrm{m\,s^{-2}}$. Using Newton's 2nd law in the direction of motion,

$$70g - R_3 = 70 \times 2$$
$$\Rightarrow R_3 = 70g - 140 = 546$$

The reaction when the lift is accelerating downwards at $2\,\mathrm{m\,s^{-2}}$ is 546 N.

Now, by Newton's 3rd law, the force exerted by the man on the floor of the lift is equal and opposite to that exerted by the floor on the man. Thus, in case **(b)**, $R_2 = 791$ and if T is the tension in the cable, we have, from Newton's 2nd law applied to the lift,

$$T - 1000g - R_2 = 1000 \times 1.5$$
$$\Rightarrow T = 1500 + 1000g + 791 = 12\,091\ \mathrm{N}$$

The tension in the cable is 12 091 N.

Newton's law of gravitation

Newton also deduced, from observations of astronomical bodies, that every body attracts every other body with a force called gravitational attraction.

> **Newton's law of gravitation:** Every body attracts every other body with a force which varies directly as the product of the masses and inversely as the square of the distance between them.

Thus, the gravitational force between two masses m_1 and m_2, at a distance r apart, is

$$F = \frac{Gm_1m_2}{r^2}$$

where G is the universal gravitational constant, which has a value of 6.67×10^{-11} in S.I. units.

For a body near the earth's surface, $\frac{Gm_2}{r^2}$ is approximately constant, since m_2 is the mass of the earth and r is the radius of the earth.

Thus, the gravitational force (the weight) of a mass, $m\,\text{kg} = mg$ N, where $g = \frac{Gm_2}{r^2} = 9.8\,\text{m s}^{-2}$.

By Newton's 2nd law, $\mathbf{F} = m\mathbf{a}$ and if a body falls freely under its own weight, it has an acceleration of $g = 9.8\,\text{m s}^{-2}$.

This value of g varies slightly at different points on the earth's surface and hence the weight of a body varies. On the moon, the gravitational pull is only about one sixth of the gravity on earth. In outer space, with no large bodies close by, objects become weightless.

Remember, however, that the **mass** of a body **does not change**.

It is important to remember that Newton's laws are basic assumptions that cannot be proved and they are only justified because they produce results which agree with observations. In certain astronomical observations and in the motion of atomic particles when velocities are close to the velocity of light, errors are evident when using Newton's laws, which can only be resolved by the theory of relativity introduced by Einstein in the first decade of the twentieth century. However, for ordinary velocities, the two theories agree closely and Newton's laws describe the motion of moving bodies to sufficient accuracy for normal use.

Exercise 15.1   (Take $g = 9.8\,\text{m s}^{-2}$, if necessary.)

1 Find the weight in Newtons of a mass of

(a) 5 kg   (b) 2.4 kg   (c) 0.75 kg

2 A particle of mass $\frac{1}{2}$ kg has a force $\mathbf{F} = 5\mathbf{i} + 4\mathbf{j}$ applied to it. Find its acceleration.

3 A particle of mass 3 kg is subject to a force of $\mathbf{F} = 3\mathbf{i} + 6\mathbf{j} - 2\mathbf{k}$. Find its acceleration.

4 Find the force necessary to give a particle of mass 2 kg an acceleration of $2\mathbf{i} - \mathbf{j} - 4\mathbf{k}$.

5 Find the acceleration produced when (a) a force of 2 N is applied to a mass of 6 kg and (b) a force of 10 N is applied to a mass of 0.5 kg.

6 What force is required to give a 4 kg mass an acceleration of $3\,\text{m s}^{-2}$?

7 Three forces, $\mathbf{F}_1 = 2\mathbf{i} - \mathbf{j} + \mathbf{k}$, $\mathbf{F}_2 = 3\mathbf{i} + 2\mathbf{j} + 2\mathbf{k}$ and $\mathbf{F}_3 = -\mathbf{i} + \mathbf{j} - \mathbf{k}$, are applied to a mass of 4 kg. Find the resultant force and the acceleration of the particle.

8 The engine of a car produces a pulling force of 2450 N. If the mass of the car is 1000 kg, find the acceleration.

9 A parachutist of mass 150 kg falls from an aircraft and, at one moment, experiences a resistance of 320 N. Find his acceleration in free fall at that moment. Later in the descent he falls with constant speed with the parachute open. Find the total retarding force of the parachute and the air.

10 A train driver sees a red signal and applies the brakes of a train causing a retarding force of 200 kN to act. If the mass of the train is 150 000 kg, find the deceleration of the train.

11 A crate of mass 28 kg is lifted by a crane with an acceleration of $0.7\,\text{m s}^{-2}$. Find the tension in the cable.

12 A particle of mass 5 kg is pulled along a horizontal table by a force of 10 N against a resistance of 8 N. Find the acceleration of the particle.

13 A block of wood of mass 4 kg is pulled up a slope of inclination 20° to the horizontal by a force of 25 N acting parallel to the slope, against a resistance of 8 N. Find the acceleration of the block.

14 A mass of 2 kg slides down the face of a desk top which is inclined at 40° to the horizontal. Find the acceleration of the mass if there is a frictional force of 6 N acting.

15 A car of mass 800 kg moves with an acceleration of 1.2 m s^{-2} on level ground. If there is a constant resistance of 700 N, find the pulling force of the engine.

16 A parcel of mass 30 kg is placed in a lift. Find the reaction between the parcel and the floor, if the lift moves **(a)** with constant speed, **(b)** with an acceleration of 1.2 m s^{-2} upwards, **(c)** with an acceleration of 1.8 m s^{-2} downwards. Find the tension in the lift cable in each case, if the lift has a mass of 800 kg.

17 A man of mass 60 kg holds a case of mass 10 kg. If he stands in a lift which accelerates upwards at 1.8 m s^{-2}, find the reaction between the floor and the man and the tension in the lift cable, if the lift has a mass of 750 kg.

18 A body of mass 3 kg is moved on a rough slope inclined at 40° to the horizontal by a string inclined at 20° to the line of greatest slope. If the resistance due to friction is 12 N, find the tension in the string if the body moves **(a)** up the slope with constant speed, **(b)** down the slope at constant speed, **(c)** up the slope with an acceleration of 0.6 m s^{-2}.

19 A body of mass 5 kg is attached to a spring balance and placed in a moving lift. If the body appears to weigh 39 N, find the acceleration of the lift.

Further applications

Newton's second law implies that if the effective force applied to a body is constant, then the acceleration produced will be constant. Hence, many questions also include the use of the equations of motion for **constant acceleration**, namely,

$$v^2 = u^2 + 2as; \quad v = u + at; \quad s = ut + \tfrac{1}{2}at^2; \quad s = \left(\frac{u+v}{2}\right)t$$

WORKED EXAMPLES

1 A force of 6 N acts on a particle of mass 2 kg for 3 seconds. Find the acceleration of the particle and the speed it attains if its initial speed is 4 m s^{-1}.

Using Newton's 2nd law, **F** = m**a**, in the direction of the force.

$$6 = 2a \quad \Leftrightarrow \quad a = 3 \, \text{m s}^{-2}$$

So the particle starts with a speed of 4 m s^{-1} and moves with an acceleration of 3 m s^{-2} for 3 seconds.

Using $v = u + at$, gives $v = 4 + 3(3) = 13 \, \text{m s}^{-1}$

Thus the final speed is 13 m s^{-1}.

2 A boy of mass 40 kg moves down a straight slide inclined at 35° to the horizontal. If he experiences a resistance of 160 N, find his acceleration. If he starts from rest and the slide is 16 m long, find the time taken to reach the bottom of the slide and his speed at that point.

Consider the diagram, right, on which the forces acting on the boy are shown. His weight of $40g$ N is resolved into components along and perpendicular to the slide.

Let a m s^{-2} be the acceleration of the boy and R N the normal reaction between the boy and the slide.

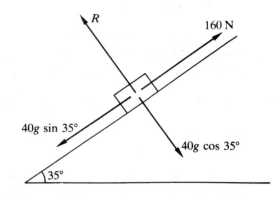

Using Newton's 2nd law, $\mathbf{F} = m\mathbf{a}$, in the direction of motion.

$$40g \sin 35° - 160 = 40a$$

$$\Leftrightarrow \quad 40a = 64.84 \quad \Leftrightarrow \quad a = 1.62$$

The acceleration of the boy is 1.62 m s^{-2}.

The boy starts from rest and moves with an acceleration of 1.62 m s^{-2} for 16 m.

Using $s = ut + \frac{1}{2}at^2$, we have,

$$16 = 0 + \tfrac{1}{2}(1.62)t^2$$

$$\Rightarrow \quad t^2 = \frac{32}{1.62} = 19.75 \quad \Rightarrow \quad t = 4.44 \text{ seconds, since } t > 0.$$

If his speed at the foot of the slide is v m s^{-1}, then using $v^2 = u^2 + 2as$,

$$v^2 = 0 + 2(1.62)16 = 51.84 \qquad \therefore \ v = 7.2$$

His speed at the foot of the slide is 7.2 m s^{-1}.

Exercise 15.2

1 A force of 1.4 N acts on a particle of mass 0.2 kg for 5 s. Find the acceleration of the particle and its speed after 5 s, if it starts from rest.

2 A particle of mass 10 kg increases its speed from 6 m s^{-1} to 10 m s^{-1} in a distance of 20 m. Find the force which is applied to the particle to cause this change.

3 A body of mass 80 kg is acted on by a force of 15 N. If the body starts from rest, how long will it be before its speed is 18 km h^{-1}?

4 A particle of mass 1.5 kg falls from rest from the top of a tower 300 m high. If it experiences an air resistance of 7 N, find the time taken to reach the ground and the speed with which it hits the ground.

5 A car of mass 1000 kg is moving with a speed of 15 m s^{-1} when the brakes are applied causing a constant force of 1500 N to act on the car. Find the distance the car covers before coming to rest.

6 A mass of 0.6 kg slides from rest down a plane, inclined at 30° to the horizontal, against a resistance of 1.6 N. Find the time taken to slide a distance of 5 m.

7　A balloon of mass 200 kg is stationary 300 m above the ground when it begins to descend. If there is an air resistance of 1800 N, find the acceleration of the balloon and its speed after 2 seconds. At this point, 50 kg of ballast is thrown out. Find the total height lost before the balloon comes instantaneously to rest.

8　A bullet of mass 20 g is fired into a fixed block of wood with a velocity of $300\,\mathrm{m\,s^{-1}}$ and comes to rest in 0.01 s. Find the resistance of the wood, assuming this to be constant.

9　An arrow of mass 0.2 kg strikes a fixed target when moving at $50\,\mathrm{m\,s^{-1}}$. If it comes to rest in 8 cm, find the resistance of the target, assuming this to be constant. If the target is only 5 cm thick instead, find the speed with which the arrow would emerge, if the resistance is unaltered.

10　Some trucks of mass 40 tonnes start from rest and roll down an incline of $\sin^{-1}(\frac{1}{400})$ against resistances of 10 N per tonne. If the slope is 50 m long, find the speed of the trucks at the bottom. After this, the track becomes level. Find the further distance moved by the trucks before they come to rest, if the resistance remains the same.

11　A train is travelling on level track at $144\,\mathrm{km\,h^{-1}}$ when it begins to climb a hill of $\sin^{-1}(\frac{1}{80})$. If the tractive force of the engine is constant at $\frac{1}{100}$ of the weight of the train, and the resistances are constant at $\frac{1}{40}$ of the weight of the train, find the distance covered before the train comes to rest.

Connected bodies

When objects are connected in some way by a towbar, coupling or piece of string, such that both objects move with the same velocity and acceleration, then it is possible to regard them **(a)** as a single combined object, or to **(b)** consider the motion of each separately under the action of the forces applied to them.

Consider the car of mass M kg towing a caravan of mass m kg mentioned in the introduction to this chapter (and see figure, right).

(a) We can use Newton's 2nd law on a combined mass of $(M + m)$ kg, under the action of forces P N and R N.

(b) It is possible to consider the separate masses moving with common velocity and acceleration. Mass M kg acted upon by forces P N, T N and R_1 N and mass m kg by forces of T N and R_2 N where R_1 and R_2 are the proportional resistances.

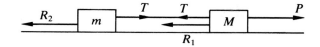

WORKED EXAMPLES

1　A car of mass 1000 kg tows a caravan of mass 800 kg. If the pull of the engine of the car is 3600 N and the resistances are 1 N per 2 kg, find the acceleration of the car and the caravan and the tension in the tow bar.

Consider the car and the caravan as a single object of mass 1800 kg. Using Newton's 2nd law in the direction of motion.

$$P - R = (M + m)a$$

$$\Rightarrow\quad 3600 - (900) = 1800a \quad\Leftrightarrow\quad a = \frac{2700}{1800} = 1.5\,\mathrm{m\,s^{-2}}$$

Hence, the acceleration of the vehicles is $1.5\,\mathrm{m\,s^{-2}}$.

Using the second diagram on the previous page, the tension in the towbar, T N, acts in opposite directions on the car and the caravan. The total resistance of 900 N is divided in the ratio of the masses between the car and the caravan.

Hence, the resistance on the car, $R_1 = 500$ N and that on the caravan, $R_2 = 400$ N.

Using Newton's 2nd law applied to the car only, $\mathbf{F} = m\mathbf{a}$, gives

$$P - R_1 - T = Ma$$
$$\Rightarrow \quad 3600 - 500 - T = 1000 \times 1.5$$
$$\Rightarrow \quad T = 3600 - 500 - 1500 = 1600 \text{ N}$$

Thus, the tension in the tow bar is 1600 N.

Note that, in this last part, we could have applied Newton's 2nd law to the caravan only, in which case, $\mathbf{F} = m\mathbf{a}$, gives

$$T - R_2 = ma$$
$$\Rightarrow \quad T - 400 = 800 \times 1.5$$
$$\Rightarrow \quad T = 1600 \text{ N, as before.}$$

2 Two particles of masses 3 kg and 5 kg are connected by a light inextensible string passing over a smooth fixed pulley. Find the acceleration of the system after it is released from rest and the tension in the string.

When the system is in motion, let the acceleration be a m s^{-2} and the tension in the string be T N. Since the pulley is smooth, the tension will be constant throughout the length of the string.

Using Newton's 2nd law in the direction of motion for each mass, we have,

on the 3 kg mass, moving upwards,

$$T - 3g = 3a \cdots\cdots (1)$$

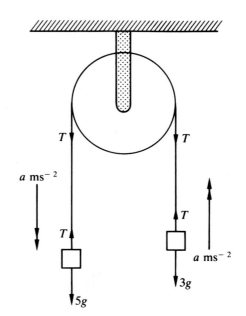

on the 5 kg mass, moving downwards,

$$5g - T = 5a \cdots\cdots (2)$$

Adding equations (1) and (2),

$$2g = 8a \quad \Rightarrow \quad a = \frac{g}{4}$$

Using equation (1) with $a = \frac{g}{4}$,

$$T - 3g = 3\left(\frac{g}{4}\right)$$
$$\Rightarrow \quad T = 3g + \frac{3g}{4} = \frac{15g}{4} = 36.75 \text{ N}$$

The acceleration of the masses will be $\frac{g}{4} = 2.45$ m s^{-2} and the tension in the string is 36.75 N.

Notice that the effect of the tension in the string on the pulley is a downward force of $(T + T) = 2T$ N.

3 A particle, of mass 7 kg, rests on a rough horizontal table and is connected to a particle, of mass 3 kg, hanging freely, by a light inextensible string passing over a smooth pulley at the edge of the table. If the system is released from rest with the string taut, find the acceleration of the masses and the force exerted on the pulley, given that a frictional force of $2g$ N acts on the 7 kg mass.

If the 3 kg mass is initially 1.2 m above the floor, find the speed of the 7 kg mass at the moment when the 3 kg mass strikes the floor. Find the further time that elapses before the 7 kg mass comes to rest, assuming that it does not reach the edge of the table.

When the system is in motion, let the acceleration of the masses be $a\,\mathrm{m\,s^{-2}}$ and the tension in the string be T N. Since the 7 kg mass rests on the table, the normal reaction N will equal $7g$.

Using Newton's 2nd law in the direction of motion for each mass, we have,

on the 7 kg mass, moving horizontally,

$$T - 2g = 7a \cdots\cdots (1)$$

on the 3 kg mass moving downwards,

$$3g - T = 3a \cdots\cdots (2)$$

adding equations **(1)** and **(2)**,

$$g = 10a \quad \Rightarrow \quad a = \frac{g}{10} = 0.98$$

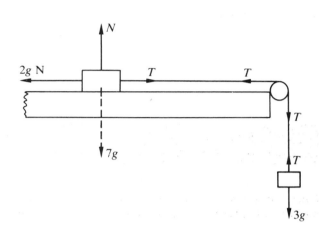

Hence, the acceleration of the system is $0.98\,\mathrm{m\,s^{-2}}$.

Substituting into equation **(1)**

$$T - 2g = \frac{7g}{10}$$

$$\Rightarrow \quad T = 2g + \frac{7g}{10} = \frac{27g}{10} = 26.46\ \text{N}$$

Thus, the tension in the string is 26.46 N and two such forces are applied to the pulley, one vertically and one horizontally.

Since the tensions are at right angles, the resultant can be found using Pythagoras' theorem.

$$\therefore\ R^2 = T^2 + T^2$$

$$\Rightarrow\ R = \sqrt{2}T$$

$$\therefore\ R = \sqrt{2} \times 26.46 = 37.42\ \text{N}$$

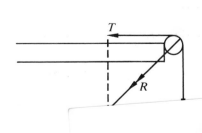

The force exerted on the pulley will be 37.42 N at 45° to the vertical.

The masses move from rest with an acceleration of $\frac{g}{10} = 0.98\,\mathrm{m\,s^{-2}}$. If $v\,\mathrm{m\,s^{-1}}$ is the speed of both masses after 1.2 m (i.e. when the 3 kg mass hits the floor), then,

using $v^2 = u^2 + 2as$,

$$v^2 = 0 + 2\left(\frac{g}{10}\right)(1.2) = 2.35$$

$$\Rightarrow \quad v = 1.53$$

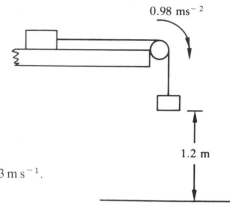

0.98 ms^{-2}

1.2 m

The speed of the 7 kg mass when the 3 kg mass strikes the floor is $1.53\,\mathrm{m\,s^{-1}}$.

At this point, the 3 kg mass comes to rest on the floor and the string becomes slack, so the tension, $T = 0$. The 7 kg mass now moves under the action of the $2g$ N frictional force.

Using Newton's 2nd law, in the direction of motion,

$$-2g = 7a_1 \quad \Rightarrow \quad a_1 = -\frac{2g}{7} = -2.8\,\mathrm{m\,s^{-2}}$$

i.e. the particle decelerates at $2.8\,\mathrm{m\,s^{-2}}$ from a speed of $1.53\,\mathrm{m\,s^{-1}}$.

If t s is the time taken to come to rest, then using $v = u + at$,

$$0 = 1.53 - 2.8t \quad \Rightarrow \quad t = \frac{1.53}{2.8} = 0.55$$

The 7 kg mass comes to rest in 0.55 seconds.

Exercise 15.3

1   Two particles of masses 6 kg and 4 kg are connected by a light inextensible string which passes over a fixed smooth pulley. Find the acceleration of the system, the tension in the string and the force on the pulley.

2   Repeat Question **1** for masses of 4 kg and 3 kg.

3   A car of mass 1000 kg tows a caravan of mass 600 kg along a level road, and accelerates at $0.5\,\mathrm{m\,s^{-2}}$. If the total resistance of the car and the caravan is 1200 N, find the pulling force developed by the engine of the car. If the resistance of the caravan is $\frac{3}{8}$ of the total resistance, find the tension in the tow bar.

4   A lorry of mass 2000 kg pulls a trailer of mass 1000 kg up a slope of $\sin^{-1}\left(\frac{1}{20}\right)$, against resistances of $180g$ N. Find the acceleration of the lorry and the tension in the tow bar, if the resistances of the lorry and its trailer are proportional to their masses and the lorry develops a pulling force of $1680g$ N.

5   A particle of mass 4 kg lies on a smooth plane inclined at $60°$ to the horizontal. It is connected by a light inextensible string, passing over a smooth pulley at the top of the plane, to mass of 4 kg hanging freely. Find the acceleration of the system, the tension in the string and the force exerted on the pulley.

6   A particle of mass 3 kg lies on a smooth horizontal table and is connected by a light inextensible string passing over a smooth pulley, at the edge of the table, to a mass of 2 kg hanging freely. Find the acceleration of the masses, the tension in the string and the force exerted on the pulley.

7 A particle of mass 2 kg rests on a smooth horizontal table and is connected by two light inextensible strings to masses of 2 kg and 1 kg, which hang freely over opposite edges of the table. Find the acceleration of the system and the tensions in the strings.

8 A double inclined plane in the form of a fixed wedge has two smooth faces inclined at 30° and 45° to the horizontal. Two particles of masses 5 kg and 7 kg are placed on the planes at 30° and 45°, respectively, and are connected by a light inextensible string passing over a smooth pulley at the top of the wedge. Find the acceleration of the masses and the tension in the string.

9 Two particles of masses 3 kg and 4 kg are connected by a light inextensible string passing over a fixed smooth pulley. If the system is released from rest with the 4 kg mass 2 m above the floor, find the time that elapses before this mass hits the floor and its speed at that moment.

10 Two particles of masses 1.5 kg and 2.5 kg are connected by a light inextensible string over a smooth fixed pulley. If the 2.5 kg is initially 4 m above the ground, find its speed when it hits the ground if it is released from rest. Find the further time that elapses before the 1.5 kg mass comes to instantaneous rest.

11 An engine of mass 100 tonnes pulls six coaches, each of mass 30 tonnes, up a slope of $\sin^{-1}(\frac{1}{280})$ when the engine develops a tractive force of 30 kN. If the resistances are 40 N per tonne, find the acceleration up the slope. Find the tension in the coupling between **(a)** the engine and the first coach, **(b)** the last two coaches.

12 A goods train travels up a slope of $\sin^{-1}(\frac{1}{98})$ at 54 km h^{-1}, when the coupling in front of the last two trucks breaks. If these trucks have masses of 60 kg and 40 kg, and the resistance to motion is 0.5 N per kg, find the deceleration of the trucks and the time before they come to rest.
   Find the tension in the coupling if **(a)** the 60 kg trucks leads, and **(b)** the 40 kg truck leads.

13 A particle of mass 18 kg is placed on a smooth horizontal table, 6 m from the edge. It is connected to a particle of mass of 2 kg, hanging freely, by a light inextensible string passing over a pulley at the edge of the table. If the 2 kg mass falls for 2 s before reaching the floor, find the total time that elapses before the 18 kg mass reaches the edge of the table.

16 Friction

Introduction

Why do skiers wax their skis?

In curling, why is the ice brushed?

Why is machinery oiled?

Why do contact lens wearers wet the lenses?

Why do cars need chains in icy conditions?

All these questions lead to the answer, **friction**.

Investigations

1 Tip the desk or table-top on which this book is resting until the book begins to slide. At what angle did the book begin to move?

2 Repeat the experiment with several articles on the table e.g. book, single sheet of paper, rubber, ruler, pencil, calculator. Which slides first and why?

3 Repeat the experiment with one book on top of another. Will two books move before one book moves?

4 My pen and ruler weigh about the same, but why does the pen always slide first?

5 Why do you rub your hands together to make them warm? Why press harder?

Laws of friction

What does 'friction' depend on?

Friction is a force which opposes motion or the tendency for a body to slide over a surface.

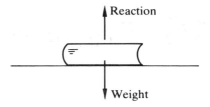

When our book lies on a flat surface its weight (downwards) is balanced by the reaction (support) force upwards.

When the surface is angled and the book remains stationary, the total reaction force, R, upwards still balances the weight, W, but we can regard the reaction as made up of the friction force, F, and the normal reaction force, N, which is normal (at right angles) to the surface.

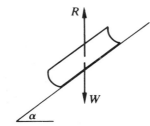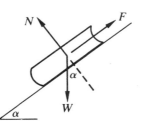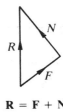

$$\mathbf{R} = \mathbf{F} + \mathbf{N} = \mathbf{W} = m\mathbf{g}$$

Resolving $\qquad \begin{array}{l} N = mg\cos\alpha \\ F = mg\sin\alpha \end{array} \Bigg\} \quad \Rightarrow \quad F = N\tan\alpha$

When the book slides down the plane surface, the overall force down the slope, $mg\sin\alpha$, must exceed the friction force causing the book to accelerate down the slope.

Thus $\qquad mg\sin\alpha - F = ma$ but $\qquad N = mg\cos\alpha$ (no motion $\Rightarrow$ forces balance)

We can see that,

> friction opposes the direction of motion (or potential motion);
>
> friction depends on the nature of the surfaces in contact;
>
> friction is proportional to the contact force between the objects;
>
> friction will increase until motion is about to begin (limiting equilibrium).

Coefficient of friction

In the figure, right, the book is about to move and F has reached its largest (limiting) value when the angle $\alpha = \lambda$.

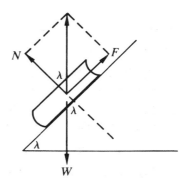

At this angle λ, $\quad \dfrac{F}{N} = \tan\lambda = \mu$ and $F = \mu N$

μ is called the coefficient of friction and λ the angle of friction.

In general, $F \leqslant \mu N$

When the particle is moving, the friction (**dynamic**) is still acting and we take this to be equal to the (**static**) limiting friction when the particle is about to move.

Friction

If the book is flat on the table, being pulled by an increasing force, P, (figure, right) then F increases until the book moves. At this time, F has reached its limiting value (μN) and $\mathbf{R} = \mathbf{N} + \mathbf{F}$ (the total reaction) makes an angle of λ with the normal reaction, N.

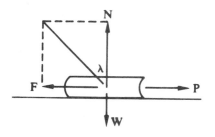

$F = \mu N$ and $F = N \tan \lambda \quad \Rightarrow \quad \mu = \tan \lambda$

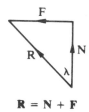

$\mathbf{R} = \mathbf{N} + \mathbf{F}$

WORKED EXAMPLES

---

1 A particle on an inclined plane is about to slip when the plane slopes at $20°$. Find the coefficient of friction between the particle and the plane.

$\diagdown \quad N = W \cos 20° \quad \nearrow \quad F = W \sin 20°$

$\dfrac{F}{N} = \tan 20°$ and $\dfrac{F}{N} = \mu \quad \Rightarrow \quad \mu = \tan 20° = 0.36$

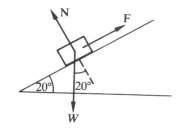

The coefficient of friction is 0.36.

---

2 A particle of mass 3 kg rests on a horizontal table with which the coefficient of friction is μ. A force P, inclined at angle θ to the horizontal, is applied to the particle until it moves. Find the value of θ for which P is least.

Method 1

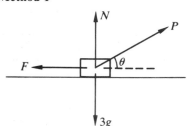

$\uparrow \quad N + P \sin \theta = 3g \quad \longleftarrow F = P \cos \theta \quad F = \mu N \text{ (on the point of moving)}$

$\Rightarrow \quad N = 3g - P \sin \theta$ and $F = \mu N = P \cos \theta$

$\Rightarrow \quad \mu N = 3\mu g - P \mu \sin \theta = P \cos \theta$

$\Rightarrow \quad P \cos \theta + P \mu \sin \theta = 3 \mu g \quad \Rightarrow \quad P = \dfrac{3 \mu g}{\cos \theta + \mu \sin \theta}$

Writing $\mu = \tan \lambda \quad \Rightarrow \quad P = \dfrac{3g \tan \lambda}{\cos \theta + \tan \lambda \sin \theta} = \dfrac{3g \sin \lambda}{\cos \theta \cos \lambda + \sin \lambda \sin \theta}$ (multiplying top and bottom by $\cos \lambda$)

$\Rightarrow \quad P = \dfrac{3g \sin \lambda}{\cos(\theta - \lambda)}$

which is least when $\cos(\theta - \lambda)$ is greatest, i.e. when $\theta = \lambda$ (the angle of friction) and P (minimum) $= 3g \sin \lambda$.

Method 2 Replace $\mathbf{F} + \mathbf{N}$ by $\mathbf{R}$, the total reaction, which acts at an angle

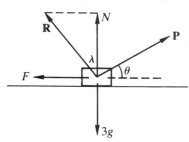

λ to N when F is limiting, because

$\dfrac{F}{N} = \tan \lambda = \mu$ (figure, left).

The weight $3\mathbf{g}$, $\mathbf{R}$ and $\mathbf{P}$ are in equilibrium so the triangle of forces ABC has AB representing $3\mathbf{g}$. $BC \equiv \mathbf{R}$ and $CA \equiv \mathbf{P}$ (figure, right).

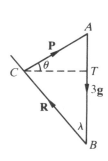

Angle $ABC = \lambda$ and the least value of **P** occurs when **P** is perpendicular to **R**   i.e. $A\hat{C}B = 90°$.

Draw $CT \perp AB$ $\Rightarrow$ $A\hat{C}T = \theta$ $\Rightarrow$ $\theta + T\hat{C}B = 90°$, but $T\hat{C}B + \lambda = 90°$ $\Rightarrow$ $\theta = \lambda$.

So $P\,(\text{minimum}) = 3g \sin \lambda$, from $\triangle ABC$.

Method 3   is a variation of Method 2, but using **Lami's Theorem** (figure, right).

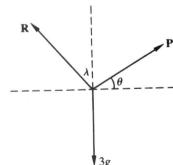

$\mathbf{R} = \mathbf{F} + \mathbf{N}$ is the total reaction which acts at λ to the vertical when the friction is limiting $(F = \mu N)$.

The 3 forces act at a point, so using Lami's theorem,

$$\frac{P}{\sin(180 - \lambda)} = \frac{R}{\sin(90 + \theta)} = \frac{3g}{\sin(90 - \theta + \lambda)}$$

$$\Rightarrow \quad \frac{P}{\sin \lambda} = \frac{3g}{\cos(\theta - \lambda)} \quad \Rightarrow \quad P = \frac{3g \sin \lambda}{\cos(\theta - \lambda)} \quad (\text{since } \sin(90 - \theta + \lambda) = \sin[90 - (\theta - \lambda)] = \cos(\theta - \lambda))$$

which is least when $\cos(\theta - \lambda)$ is greatest, i.e. when $\theta = \lambda$.

$$P\,(\text{minimum}) = \frac{3g \sin \lambda}{\cos 0°} = 3g \sin \lambda$$

Exercise 16.1

1   A block of mass 5 kg rests on a rough table, the coefficient of friction being 0.5. A force of 30 N acts on the block at an angle of $\theta°$ above the horizontal. Find the acceleration of the block when **(a)** $\theta = 40°$,   **(b)** $\theta = 30°$,   **(c)** $\theta = 20°$,   **(d)** $\theta = 0°$.

2   A particle of weight 10 N rests on a rough plane $(\mu = \frac{1}{2})$ inclined at $30°$ to the horizontal. Find the horizontal force required to **(a)** prevent the particle slipping down the plane and **(b)** make the particle just slide up the plane.

3   A block rests on a rough plane inclined at angle θ to the horizontal and is about to slide down the plane.

   (a)   Find μ, the coefficient of friction, if $\theta = 35°$.

   (b)   Find θ if $\mu = \frac{1}{2}$.

   (c)   Find the friction force, F, if $\mu = \frac{1}{2}$ and the block weighs 10 kg.

   (d)   Find the normal reaction, if $\mu = \frac{1}{2}$ and the block weighs 10 kg.

   (e)   Find the total reaction, if $\mu = \frac{1}{2}$ and the block weighs 20 kg.

4   My desk is inclined at $20°$ to the horizontal. I have a glass paperweight which slides down the desk at constant velocity.

   (a)   Find the coefficient of friction between the desk top and the paperweight.

   (b)   What will happen if I increase the angle of slope to $30°$?

   (c)   At an angle of $15°$ the paperweight (shaped like a bread roll) is still, but when I turn it upside down it slides down. Why?

5   A block of weight 20 N rests on a horizontal plane. The coefficient of friction between the block and the plane is 0.25. A force, P, is applied to the block so that the block is about to slide. Find P given that **(a)** P acts horizontally, **(b)** P acts upwards at 45° to the horizontal, **(c)** P acts downwards at 45° to the horizontal.

6   A particle of mass 10 kg slides down a plane inclined at 30° to the horizontal. If the coefficient of friction is 0.3, find the acceleration of the particle.

7   A skier slides down a nursery slope of 1 in 10 at constant velocity. If the mass of the skier is 60 kg, find the tension in the tow bar (parallel to the slope) when the skier is pulled up the lift on the same slope at constant speed. What is the tension in the tow bar if it is inclined at 45° to the horizontal?

8   What is the least value for a coefficient of friction? What is the greatest, theoretically? What is the greatest (practically) you can achieve in the classroom? (e.g. book on the desk.) What is the least coefficient of friction you can think of in practice? For example, skis on snow or ice, or curling stones on ice?

9   Analyse and identify the forces acting on a ladder leaning against a wall. How do the forces vary as someone climbs the ladder?

10 A girl on a sledge accelerates down a slope of 45°, the weight of the sledge and the girl being 30 kg and the coefficient of friction being 0.5. Find the acceleration when the girl **(a)** slides on her own, and **(b)** slides with her father also on the sledge (her father's mass is 70 kg).

11 A particle of mass 5 kg is projected across a horizontal plane with a speed of $8\,\mathrm{m\,s}^{-1}$. If it comes to rest in a distance of 12 m, find the magnitude of the frictional force acting on the particle and the coefficient of friction between the particle and the plane.

12 A block of weight 10 N rests on a horizontal plane, the coefficient of friction between the block and the plane being 0.25. Find the least force required to move the block and the direction in which it acts.

13 A particle of weight 10 N rests on a rough horizontal plane, the coefficient of friction being $\frac{1}{3}$. The particle is connected by a light inextensible string, passing over a smooth pulley at the edge of the plane, to a particle of mass m hanging vertically. Find the largest value of m for the system to remain at rest. If $mg = 10$ N, find the acceleration of the system.

14 Analyse the forces acting on a step ladder consisting of two 'legs' of equal length. Is there any difference if the step side of the ladder is twice the weight of the other side? What happens when someone steps on the ladder?

17 The Binomial Theorem and other series

Pascal's triangle

Have you ever wondered why it is that, when marbles are dropped through a framework of pegs set out as in the diagram, right, the greatest number of marbles accumulate towards the centre of the last row rather than at the ends?

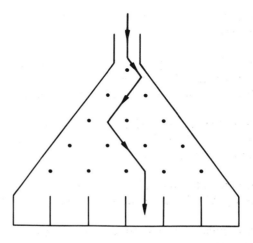

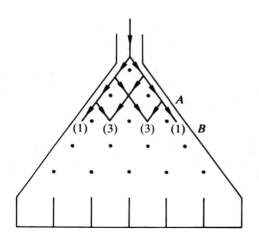

If you investigate the number of possible routes a single marble can take to reach a particular point, you will begin to understand.

For example, in the diagram, left, at level *B*, there are 3 pegs and 4 spaces through which the marbles can fall.

The numbers in brackets indicate the number of possible routes the marble could take to reach that point.

Investigation 1

Draw a diagram, similar to those shown, but with six rows of pegs. Record in the spaces between the pegs on each row, the total number of routes by which a marble can reach that point.

The first few entries have been made for you in the figure, right.

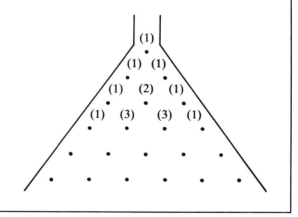

The pattern you should obtain in Investigation 1 can be achieved in a different way. If we consider successive powers of the binomial (two terms) expression $(a + b)$.

$$(a + b)^0 = 1$$
$$(a + b)^1 = a + b$$
$$(a + b)^2 = a^2 + 2ab + b^2$$

Investigation 2

Expand the following binomial expressions and complete the pattern started above.

(a) $(a + b)^3$ **(b)** $(a + b)^4$ **(c)** $(a + b)^5$ **(d)** $(a + b)^6$

If we now write these expansions including only the coefficients of each term, we obtain an array of numbers called **Pascal's Triangle**, after the French Mathematician, Blaise Pascal (1623–1662).

```
              1
           1     1
        1     2     1
     1     3     3     1
  1     4     6     4     1
1     5    10    10    5     1
1  6   15    20    15   6    1
```

This pattern should be the same as the result of Investigation 1. Can you see how to form each new line? Apart from the 1 at each end, an entry consists of the sum of the two entries on either side of it on the line above.

This enables us to write out an expansion in the form of $(a + b)^n$, since Pascal's triangle will identify the coefficients and the terms will be successively

$$a^n, \ a^{n-1}b, \ a^{n-2}b^2, \ \ldots, \ ab^{n-1}, \ b^n$$

Notice that the powers of a decrease by one, and the powers of b increase by one, for each term. Thus,

$$(a + b)^7 = a^7 + 7a^6b + 21a^5b^2 + 35a^4b^3 + 35a^3b^4 + 21a^2b^5 + 7ab^6 + b^7$$

Notes

(i) This is only valid for positive integral values of n.

(ii) The expansion of $(a + b)^n$ has $(n + 1)$ terms.

(iii) Each term has degree n. i.e. in $(a + b)^7$, the sum of the powers of a and b in each term is 7. $(a^5b^2, a^4b^3, \ldots)$.

(iv) Each line of Pascal's triangle has symmetry about the centre.

WORKED EXAMPLE

Expand, **(a)** $(x + 2y)^5$ and **(b)** $\left(2x - \dfrac{1}{x}\right)^6$, using Pascal's triangle.

(a) Since the index is 5, we need the row   1   5   10   10   5   1   of Pascal's triangle.

Hence, $(a + b)^5 = a^5 + 5a^4b + 10a^3b^2 + 10a^2b^3 + 5ab^4 + b^5$

$\therefore \ (x + 2y)^5 = x^5 + 5x^4(2y) + 10x^3(2y)^2 + 10x^2(2y)^3 + 5x(2y)^4 + (2y)^5$

$\qquad\qquad = x^5 + 10x^4y + 40x^3y^2 + 80x^2y^3 + 80xy^4 + 32y^5$

(b) Since the index is 6, we need the row   1   6   15   20   15   6   1 of Pascal's triangle.

Hence, $(a + b)^6 = a^6 + 6a^5b + 15a^4b^2 + 20a^3b^3 + 15a^2b^4 + 6ab^5 + b^6$

$\therefore \left(2x - \dfrac{1}{x}\right)^6 = (2x)^6 + 6(2x)^5\left(-\dfrac{1}{x}\right) + 15(2x)^4\left(-\dfrac{1}{x}\right)^2 + 20(2x)^3\left(-\dfrac{1}{x}\right)^3 + 15(2x)^2\left(-\dfrac{1}{x}\right)^4 + 6(2x)\left(-\dfrac{1}{x}\right)^5 + \left(-\dfrac{1}{x}\right)^6$

$\qquad\qquad = 64x^6 - 192x^4 + 240x^2 - 160 + \dfrac{60}{x^2} - \dfrac{12}{x^4} + \dfrac{1}{x^6}$

Watch that you take powers of $(2x)$ and $\left(-\dfrac{1}{x}\right)$; remember the signs!

Exercise 17.1

1 Use Pascal's triangle to expand the following.

(a) $(a + b)^4$

(b) $(x + y)^6$

(c) $(1 - x)^3$

(d) $(2a + b)^5$

(e) $(x - 2y)^4$

(f) $(x - 3)^3$

(g) $(a - b)^2$

(h) $(2x - 1)^7$

(i) $\left(x + \dfrac{1}{x}\right)^4$

(j) $(x^2 + 2)^5$

(k) $\left(x^2 - \dfrac{2}{x}\right)^6$

(l) $\left(2x^2 + \dfrac{1}{x^2}\right)^3$

2 Expand $\left(1 + \dfrac{x}{2}\right)^4$. By letting $x = 0.2$, find the value of $(1.1)^4$.

3 Expand $(2 + x)^5$. By taking the first three terms of the expansion with $x = 0.01$, find an approximate value of $(2.01)^5$.

4 Expand $(1 - x)^6$. By choosing a suitable value of x, find the value of $(0.995)^6$ correct to three decimal places.

5 Expand $(1 + 3x)^4$ and $(1 - 2x)^5$. Hence find the first four terms of the expansion of $(1 + 3x)^4(1 - 2x)^5$ in ascending powers of x.

6 Expand $(x - 2)^4$ and $(2x - 3)^5$. Hence find the first three terms of the expansion of $(x - 2)^4(2x - 3)^5$ in descending powers of x.

7 Expand $(1 - x - x^2)^5$ in ascending powers of x as far as the term in x^3, by considering $(1 - x - x^2)^5$ as $[1 - (x + x^2)]^5$.

The Binomial Theorem for a positive integral index

Although Pascal's triangle can be used to find the expansion of $(a + b)^n$, this method is very tedious if n is large. An alternative approach can be used which considers $(a + b)^n$ to be a product of n factors, each of the form $(a + b)$.

$$\text{i.e.} \quad (a + b)^n = (a + b)(a + b)(a + b) \cdots\cdots (a + b)(a + b)$$

Now, we know that the terms of the final expansion will be

$$a^n, \ a^{n-1}b, \ a^{n-2}b^2, \ a^{n-3}b^3, \ \ldots\ldots, \ ab^{n-1}, \ b^n$$

In fact, each term is formed by choosing one letter from each bracket and multiplying them together.

So the term $a^{n-3}b^3$ is formed by choosing a from $n - 3$ of the brackets, and b from three of the brackets.

Clearly we can choose the three bs in several ways from the n brackets – in fact in $^nC_3$ ways. Thus, the coefficient of each term will be given by the number of ways the bs can be chosen from n, i.e.

$$\text{the term } a^n \text{ chooses no } bs = {}^nC_0 = 1 \text{ way}$$

$$\text{the term } a^{n-1}b \text{ chooses } 1 \ b = {}^nC_1 = n \text{ ways}$$

$$\text{the term } a^{n-2}b^2 \text{ chooses } 2 \ bs = {}^nC_2 = \frac{n!}{(n-2)!2!} = \frac{n(n-1)}{2!} \text{ ways}$$

$$\text{the term } a^{n-3}b^3 \text{ chooses } 3 \ bs = {}^nC_3 = \frac{n!}{(n-3)!3!} = \frac{n(n-1)(n-2)}{3!} \text{ ways}$$

Hence, $(a + b)^n = a^n + {}^nC_1 a^{n-1}b + {}^nC_2 a^{n-2}b^2 + {}^nC_3 a^{n-3}b^3 + \cdots\cdots + {}^nC_n b^n$

Clearly, $^nC_n = \dfrac{n!}{n!0!} = 1$

> The general result, called the binomial theorem is,
>
> $$(a + b)^n = a^n + {}^nC_1 a^{n-1}b + {}^nC_2 a^{n-2}b^2 + \cdots\cdots + {}^nC_r a^{n-r}b^r + \cdots\cdots + b^n$$

Note that this form only holds if n is a positive integer. It is sometimes useful to replace the coefficients $^nC_1, {}^nC_2, {}^nC_3$, etc. with their values, and hence,

> $$(a + b)^n = a^n + na^{n-1}b + \frac{n(n-1)}{2!}a^{n-2}b^2 + \frac{n(n-1)(n-2)}{3!}a^{n-3}b^3 + \cdots + b^n$$

In this case, the general term $^nC_r a^{n-r}b^r$ is given by

$${}^nC_r a^{n-r}b^r = \frac{n(n-1)(n-2)(n-3)\cdots(n-r+1)}{r!}a^{n-r}b^r$$

The binomial theorem will be proved later, by the method of induction.

WORKED EXAMPLES

1 Expand $(1 - 2x)^4$ in ascending powers of x.

Using the binomial theorem with $n = 4$,

$$(1 - 2x)^4 = 1 + {}^4C_1(-2x) + {}^4C_2(-2x)^2 + {}^4C_3(-2x)^3 + (-2x)^4$$

Notice that, since $a = 1$ and any power of a will equal 1, the working is considerably simplified.

$$\therefore (1 - 2x)^4 = 1 + 4(-2x) + 6(-2x)^2 + 4(-2x)^3 + (-2x)^4$$
$$= 1 - 8x + 24x^2 - 32x^3 + 16x^4$$

2 Express $(3 + 2x)^6$ in ascending powers of x, using the binomial theorem.

Using the general expansion,

$$(a + b)^6 = a^6 + {}^6C_1a^5b + {}^6C_2a^4b^2 + {}^6C_3a^3b^3 + {}^6C_4a^2b^4 + {}^6C_5ab^5 + b^6$$

Hence,

$$(3 + 2x)^6 = 3^6 + {}^6C_1(3)^5(2x) + {}^6C_2(3^4)(2x)^2 + {}^6C_3(3^3)(2x)^3 + {}^6C_4(3^2)(2x)^4 + {}^6C_5(3)(2x)^5 + (2x)^6$$

$$= 729 + 6(3^5)(2x) + \frac{6!}{4!2!}(3^4)(2x)^2 + \frac{6!}{3!3!}(3^3)(2x)^3 + \frac{6!}{4!2!}(3^2)(2x)^4 + \frac{6!}{5!1!}(3)(2x)^5 + (2x)^6$$

$$= 729 + 2916x + 15(3^4)(4x^2) + 20(3^3)(8x^3) + 15(3^2)(16x^4) + 6(3)(32x^5) + 64x^6$$

$$= 729 + 2916x + 4860x^2 + 4320x^3 + 2160x^4 + 576x^5 + 64x^6$$

3 Expand $(2 + x)^5$ in ascending powers of x.

The working can be simplified if the value of the first term inside the bracket can be made equal to 1 by extracting a factor from the bracket.

In this case remove a factor of 2^5.

$$(2 + x)^5 = 2^5\left(1 + \frac{x}{2}\right)^5$$

$$= 2^5\left[1 + {}^5C_1\left(\frac{x}{2}\right) + {}^5C_2\left(\frac{x}{2}\right)^2 + {}^5C_3\left(\frac{x}{2}\right)^3 + {}^5C_4\left(\frac{x}{2}\right)^4 + \left(\frac{x}{2}\right)^5\right]$$

$$= 32\left[1 + \frac{5}{2}x + 10\left(\frac{x}{2}\right)^2 + 10\left(\frac{x}{2}\right)^3 + 5\left(\frac{x}{2}\right)^4 + \left(\frac{x}{2}\right)^5\right]$$

$$= 32 + 80x + 80x^2 + 40x^3 + 10x^4 + x^5$$

Note

To express $(2 + x)^5$ in **descending** powers of x, write $(2 + x)^5$ as $(x + 2)^5$ i.e. $x^5\left(1 + \frac{2}{x}\right)^5$

$$x^5\left(1 + \frac{2}{x}\right)^5 = x^5\left[1 + {}^5C_1\left(\frac{2}{x}\right) + {}^5C_2\left(\frac{2}{x}\right)^2 + {}^5C_3\left(\frac{2}{x}\right)^3 + {}^5C_4\left(\frac{2}{x}\right)^4 + \left(\frac{2}{x}\right)^5\right]$$

$$= x^5\left[1 + 5\left(\frac{2}{x}\right) + 10\left(\frac{4}{x^2}\right) + 10\left(\frac{8}{x^3}\right) + 5\left(\frac{16}{x^4}\right) + \frac{32}{x^5}\right]$$

$$= x^5 + 10x^4 + 40x^3 + 80x^2 + 80x + 32$$

4 Find the coefficient of x^5 in the expansion of **(a)** $\left(2 - \dfrac{x}{3}\right)^9$, **(b)** $\left(x - \dfrac{2}{x}\right)^{11}$

It is not necessary to write out all the terms of the expansion, because we can use the general term expression.

The term involving b^r in $(a + b)^n$ is ${}^nC_r a^{n-r} b^r$.

(a) In $\left(2 - \dfrac{x}{3}\right)^9$, the term in x^5 is ${}^9C_5 (2)^4 \left(-\dfrac{x}{3}\right)^5$

$$= -\frac{9!}{5!4!} \times \frac{16}{1} \times \frac{x^5}{243} = -\frac{9.8.7.6}{4.3.2.1} \times \frac{16x^5}{243} = -\frac{224}{27}x^5$$

The coefficient of x^5 in $\left(2 - \dfrac{x}{3}\right)^9$ is $-\dfrac{224}{27}$.

(b) In this case, where $n = 11$, the general term is given by ${}^{11}C_r x^{11-r} \left(-\dfrac{2}{x}\right)^r$.

Now, we need the final power of x to be 5, i.e. $11 - r - r = 5$

$$\Leftrightarrow \quad 11 - 2r = 5 \quad \Leftrightarrow \quad 2r = 6 \quad \Leftrightarrow \quad r = 3$$

The term in x^5 is ${}^{11}C_3 x^8 \left(-\dfrac{2}{x}\right)^3 = {}^{11}C_3 (-2)^3 x^5 = -\dfrac{11!}{3!8!} \times 8x^5$

The coefficient of x^5 is $-\dfrac{11.10.9.8}{3.2} = -1320$.

5 Expand $(1 + x - x^2)^7$ as far as the term in x^3.

Now, $(1 + x - x^2)^7 = [1 + (x - x^2)]^7$

$$= 1 + 7(x - x^2) + \frac{7.6}{2!}(x - x^2)^2 + \frac{7.6.5}{3!}(x - x^2)^3 + \cdots$$

We have used the values of ${}^7C_1, {}^7C_2, {}^7C_3, \ldots$. No more terms are needed since all powers of x in subsequent terms will be greater than 3.

$$\therefore (1 + x - x^2)^7 = 1 + 7(x - x^2) + 21(x - x^2)^2 + 35(x - x^2)^3 + \cdots$$

Now evaluate the powers of $(x - x^2)$ including only those powers of x which are $\leqslant 3$.

$$\therefore (1 + x - x^2)^7 = 1 + 7x - 7x^2 + 21(x^2 - 2x^3 + \cdots) + 35(x^3 + \cdots) + \cdots$$
$$= 1 + 7x + 14x^2 - 7x^3 + \cdots$$

Exercise 17.2

1 Expand the following in ascending powers of x, using the binomial theorem.

(a) $(1 + x)^5$ (c) $(1 + \frac{1}{2}x)^8$ (e) $(4 - 3x)^3$

(b) $(2 - x)^6$ (d) $\left(\dfrac{1}{2x} + x\right)^7$ (f) $(1 - x^2)^{10}$

2 Expand in descending powers of x,

(a) $\left(x^2 + \dfrac{1}{x}\right)^3$ (b) $\left(1 - \dfrac{2}{x^2}\right)^4$ (c) $(1 + x^3)^5$

3 Expand the following in ascending powers of x, as far as the term indicated.

(a) $(1 + 3x)^{12}$; term in x^4 (d) $(1 - x)^9$; term in x^5

(b) $\left(2 - \dfrac{x}{4}\right)^{10}$; terms in x^5 (e) $\left(5 - \dfrac{1}{5}x\right)^5$; term in x^3

(c) $(1 + x^2)^{20}$; term in x^8 (f) $\left(\dfrac{1}{x} + 2x\right)^6$; constant term

4 Find the term indicated in the following expansions.

(a) $(4 - x)^7$; term in x^5 (c) $\left(x + \dfrac{1}{x}\right)^6$; term in x^2

(b) $\left(1 - \dfrac{x}{4}\right)^{10}$; term in x^3 (d) $\left(x^2 - \dfrac{1}{2x}\right)^6$; constant term

5 Expand the following in descending powers of x.

(a) $(x - 1)^4$ (b) $(2x - 1)^7$ (c) $(2 - 5x)^5$

6 Expand the following in ascending powers of x, as far as the term in x^4.

(a) $(1 + x + x^2)^6$ (b) $(1 - 2x - x^2)^5$ (c) $(3 - 2x + x^2)^4$ (d) $(2 + x - x^3)^7$

7 If the first three terms of $(1 + ax)^n$ are $1 - 10x + 40x^2$, find the values of a and n.

8 Find the ratio of the term in x^6 to the term in x^4 in the expansion of $(3 + 2x)^{18}$.

9 If the coefficient of x in the expansion of $(1 + ax)^8$ is three times the coefficient of x^2 in $(1 + bx)^4$, and $a:b = 1:2$, find the values of a and b.

10 Expand $(1 + x)^{10}$ as far as the term in x^4, and use this result to find the value of $(1.01)^{10}$ correct to four decimal places.

11 Expand $(2 - x)^{12}$, giving the first four terms, and hence find the value of $(1.997)^{12}$ to two decimal places.

12 Find the first four terms in the expansion of $(1 + 2x)^7(1 - 3x)^6$ in ascending powers of x.

13 Without writing out the whole expansion of $(2 - x)^8(1 + x)$, find the term in x^7.

14 Find the term in x^4 in the expansion of $(1 - x + x^2)(1 - \frac{1}{2}x)^6$.

The Binomial Theorem for any index

We have seen that, when n is a positive integer, the expansion of $(a + b)^n$ has exactly $(n + 1)$ terms, but this is not the case if n is negative or fractional. It is also not possible to use the ^nC_r notation since this has no meaning unless n is an integer.

Now, if n is a positive integer,

$$(a + b)^n = a^n\left[1 + \frac{b}{a}\right]^n = a^n\left[1 + n\left(\frac{b}{a}\right) + \frac{n(n - 1)}{2!}\left(\frac{b}{a}\right)^2 + \frac{n(n - 1)(n - 2)}{3!}\left(\frac{b}{a}\right)^3 + \cdots\right]$$

Clearly, if n is not an integer, the numerators n, $n(n - 1)$, $n(n - 1)(n - 2)$, etc. never become zero and the series will contain an infinite number of terms.

Writing $a = 1$ and $b = x$ gives an easier format.

$$(1 + x)^n = 1 + nx + \frac{n(n - 1)}{2!}x^2 + \frac{n(n - 1)(n - 2)}{3!}x^3 + \cdots\cdots + \frac{n(n - 1)(n - 2)\cdots(n - r + 1)}{r!}x^r + \cdots$$

This gives the binomial expansion for any index. Although the proof is beyond the scope of this book, the expansion is only valid if $-1 < x < 1$, $(|x| < 1)$. This is because the series must converge to a definite limit otherwise the sum will increase without bound.

Notes

(a) The series has an infinite number of terms.

(b) The expansion is only valid for $-1 < x < 1$.

(c) The binomial expansion should contain 1 as the first term. If this is not the case, it will be necessary to remove a from $(a + b)^n$ to give $a^n\left(1 + \frac{b}{a}\right)^n$.

WORKED EXAMPLES

1 Expand $(1 - x)^{-1}$ in ascending powers of x.

Using the general expansion with $n = -1$ and x replaced by $-x$,

$$(1 - x)^{-1} = 1 + (-1)(-x) + \frac{(-1)(-2)}{2!}(-x)^2 + \frac{(-1)(-2)(-3)}{3!}(-x)^3 + \cdots\cdots$$

$$= 1 + x + x^2 + x^3 + \cdots\cdots$$

Notice that $1 + x + x^2 + x^3 + \cdots\cdots$ is a geometrical progression with common ratio x and first term 1.

Its sum to infinity $= \dfrac{a}{1 - r} = \dfrac{1}{1 - x} = (1 - x)^{-1}$

2 Expand $\sqrt{(1+2x)}$ in ascending powers of x as far as the term in x^3 and state the range of values of x for which the expansion is valid.

Using the binomial theorem for a fractional index,

$$\sqrt{(1+2x)} = (1+2x)^{\frac{1}{2}} = 1 + \tfrac{1}{2}(2x) + \frac{\frac{1}{2}(-\frac{1}{2})}{2!}(2x)^2 + \frac{\frac{1}{2}(-\frac{1}{2})(-\frac{3}{2})}{3!}(2x)^3 + \cdots\cdots$$

$$= 1 + x - \tfrac{1}{2}x^2 + \tfrac{1}{2}x^3 - \cdots\cdots$$

Be careful to take powers of $(2x)$ in the initial stage of the expansion and, when simplifying, it is easier to deal with the signs before calculating the numerical value of each coefficient.

The expansion is valid for $-1 < 2x < 1 \quad \Leftrightarrow \quad -\tfrac{1}{2} < x < \tfrac{1}{2}$.

3 Show that $\dfrac{1-x}{\sqrt{(1+x)}} = 1 - \dfrac{3}{2}x + \dfrac{7}{8}x^2$, if x is sufficiently small so that x^3 and higher powers of x may be neglected.

Now, $\dfrac{1-x}{\sqrt{(1+x)}} = (1-x)(1+x)^{-\frac{1}{2}}$

Hence, $(1-x)(1+x)^{-\frac{1}{2}} = (1-x)\left[1 - \dfrac{1}{2}x + \dfrac{(-\frac{1}{2})(-\frac{3}{2})}{2!}x^2 + \cdots \right]$ (neglecting x^3 and higher powers)

$$= (1-x)\left(1 - \dfrac{1}{2}x + \dfrac{3}{8}x^2 + \cdots \right)$$

Multiply the brackets, neglecting powers higher than x^2.

$$(1-x)(1+x)^{-\frac{1}{2}} = 1 - \dfrac{1}{2}x + \dfrac{3}{8}x^2 - x + \dfrac{1}{2}x^2 \cdots\cdots$$

$$= 1 - \dfrac{3}{2}x + \dfrac{7}{8}x^2$$

4 Expand $\dfrac{1}{(2+5x)^3}$ in ascending powers of x as far as the term in x^3.

$$\dfrac{1}{(2+5x)^3} = (2+5x)^{-3} = 2^{-3}\left(1 + \dfrac{5x}{2} \right)^{-3}$$

Using the binomial theorem for a negative index,

$$2^{-3}\left(1 + \dfrac{5x}{2} \right)^{-3} = \dfrac{1}{2^3}\left[1 + (-3)\left(\dfrac{5x}{2}\right) + \dfrac{(-3)(-4)}{2!}\left(\dfrac{5x}{2}\right)^2 + \dfrac{(-3)(-4)(-5)}{3!}\left(\dfrac{5x}{2}\right)^3 + \cdots\cdots \right]$$

$$= \dfrac{1}{8}\left[1 - \dfrac{15}{2}x + \dfrac{75}{2}x^2 - \dfrac{625}{4}x^3 + \cdots\cdots \right]$$

$$= \dfrac{1}{8} - \dfrac{15}{16}x + \dfrac{75}{16}x^2 - \dfrac{625}{32}x^3 + \cdots\cdots$$

5 Find the sum to infinity of the series,

$$1 - \frac{3}{8} - \frac{1}{2!2^2}\left(\frac{3}{4}\right)^2 - \frac{1.3}{3!2^3}\left(\frac{3}{4}\right)^3 - \frac{1.3.5}{4!2^4}\left(\frac{3}{4}\right)^4 - \cdots$$

In the general expansion,

$$(1 + x)^n = 1 + nx + \frac{n(n-1)}{2!}x^2 + \frac{n(n-1)(n-2)}{3!}x^3 + \cdots$$

The given series can be written,

$$1 + \frac{1}{2}\left(-\frac{3}{4}\right) + \frac{\frac{1}{2}(-\frac{1}{2})}{2!}\left(-\frac{3}{4}\right)^2 - \frac{\frac{1}{2}(-\frac{1}{2})(-\frac{3}{2})}{3!}\left(-\frac{3}{4}\right)^3 - \frac{\frac{1}{2}(-\frac{1}{2})(-\frac{3}{2})(-\frac{5}{2})}{4!}\left(-\frac{3}{4}\right)^4$$

i.e. $n = \dfrac{1}{2}$ and $x = -\dfrac{3}{4}$

Hence, the sum to infinity $= \left(1 - \dfrac{3}{4}\right)^{\frac{1}{2}} = \sqrt{\dfrac{1}{4}} = \dfrac{1}{2}$

Exercise 17.3

Expand **1** to **9** in ascending powers of x as far as the term in x^3. State the range of validity for each expansion.

1 $(1 + x)^{-1}$ **4** $(1 - 2x)^{\frac{3}{2}}$ **7** $\dfrac{1}{\sqrt{1 - x^2}}$

2 $(1 + 2x)^{-2}$ **5** $\sqrt{2 + x}$ **8** $\dfrac{1}{(2 + x)^2}$

3 $(1 - x)^{\frac{1}{2}}$ **6** $\sqrt[3]{(1 - 3x)}$ **9** $\dfrac{1}{4 + x}$

10 Expand $(1 + x)^{\frac{1}{2}} + (1 - x)^{\frac{1}{2}}$ giving the first three terms.

11 Expand $(1 + x)(1 - x)^{-\frac{1}{2}}$ as a series in ascending powers of x as far as the term in x^4.

12 Find the first four terms in the expansions of the following in ascending powers of x.

     (a) $\dfrac{1 + x}{1 - x}$ **(b)** $\dfrac{2 - x}{\sqrt{(1 + 2x)}}$ **(c)** $\dfrac{2x - 1}{(x + 2)^2}$

13 Expand $(4 - x)^{\frac{3}{2}}$ as a series in ascending powers of x as far as the term in x^3. Deduce the value of $\sqrt{(3.99)^3}$ to three significant figures.

14 Expand the following in ascending powers of x, giving the first four terms.

     (a) $\dfrac{1}{(1 - x + x^2)}$ **(b)** $(1 - 2x - x^2)^{\frac{1}{2}}$

15 Expand $\dfrac{1 - 5x}{1 + 3x}$ in ascending powers of x as far as the term in x^3, stating the range for which it is valid.

16 Multiply the numerator and denominator of $\sqrt{\dfrac{1 + x}{1 - x}}$ by $\sqrt{1 + x}$ and hence expand the result as a series in ascending powers of x. Use this to find the value of $\sqrt{3}$ to four significant figures.

17 Expand the following in ascending powers of x giving the first four terms.

(a) $(2+x)^{-1}(1+x)$ **(b)** $\dfrac{3}{(4-x)(2+x)}$ **(c)** $\dfrac{2x+1}{(1-x^2)(1+x)^2}$

18 Expand the following in ascending powers of $\dfrac{1}{x}$, giving the first three terms.

(a) $(x+2)^{-1}$ **(b)** $\dfrac{x+1}{x+2}$ **(c)** $\dfrac{x}{1-x^2}$ **(d)** $\dfrac{1}{x^2-5x-6}$

Other series

There are many other finite and infinite series and special methods for calculating the sums of these series.

WORKED EXAMPLE ▷

Find the sum of the series $1 + 6 + 8 + 13 + 15 + \cdots$ to $2n$ terms.

This series is made up of two arithmetical progressions.

Let $\quad S = 1 + 6 + 8 + 13 + 15 + 20 + 22 + \cdots$

$\qquad = (1 + 8 + 15 + 22 + 29 + \cdots) + (6 + 13 + 20 + 27 + \cdots)$

The terms in both series form A.P.s with a common difference of 7, and first term 1 and 6, respectively. As there are $2n$ terms in total, each bracket will contain n terms.

Using sum of an A.P. $= \dfrac{n}{2}[2a + (n-1)d]$,

$$S = \frac{n}{2}[2 + (n-1)7] + \frac{n}{2}[12 + (n-1)7]$$

$$= \frac{n}{2}[2 + 7n - 7 + 12 + 7n - 7] = 7n^2$$

Investigation 3

The series, $3 + 7x + 11x^2 + 15x^3 + \cdots + 39x^9$ has the coefficient of each term forming an arithmetical progression while the variable x increases its index by one. This is called the **arithmetico-geometrical series**. Its sum, S, can be found by multiplying the series by x and subtracting the result from the original series.

Let $\quad S = 3 + 7x + 11x^2 + 15x^3 + \cdots + 39x^9$

Show that, $\quad (1-x)S = 3 + 4x(1 + x + x^2 + x^3 + \cdots + x^8) - 39x^{10}$.

Since the terms in the bracket on the right-hand side form a G.P., they can be summed using the result in Chapter 14.

Show that, $\quad (1-x)S = 3 + \dfrac{4x(1-x^9)}{1-x} - 39x^{10}$ and hence, $\quad S = \dfrac{3 - 39x^{10}}{1-x} + \dfrac{4x(1-x^9)}{(1-x)^2}$

WORKED EXAMPLE

Find the sum of the series $\dfrac{1}{1 \times 3} + \dfrac{1}{2 \times 4} + \dfrac{1}{3 \times 5} + \dfrac{1}{4 \times 6} + \cdots$, to n terms.

The general term of this series can be written as $\dfrac{1}{r(r + 2)}$.

Thus, the whole series is, $\displaystyle\sum_{r=1}^{n} \dfrac{1}{r(r + 2)}$.

In series of this type, the general term can be split into two terms using partial fractions.

Let $\dfrac{1}{r(r + 2)} \equiv \dfrac{A}{r} + \dfrac{B}{r + 2} \quad \Rightarrow \quad 1 \equiv A(r + 2) + Br$

Let $r = 0 \quad \Rightarrow \quad 2A = 1 \quad \Rightarrow \quad A = \tfrac{1}{2}$

Let $r = -2 \quad \Rightarrow \quad -2B = 1 \quad \Rightarrow \quad B = -\tfrac{1}{2}$

The series can now be written $\displaystyle\sum_{r=1}^{n} \left(\dfrac{1}{2r} - \dfrac{1}{2(r + 2)} \right) = \dfrac{1}{2} \sum_{r=1}^{n} \left(\dfrac{1}{r} - \dfrac{1}{(r + 2)} \right)$

If we now substitute successive values of r (i.e. $1, 2, 3, \ldots n$) into the expression for the general term and the full series is written out, most terms will cancel, leaving a result for the sum.

$$\therefore \; \dfrac{1}{2} \sum_{r=1}^{n} \left(\dfrac{1}{r} - \dfrac{1}{r + 2} \right) = \dfrac{1}{2} \left[\; 1 \qquad\quad -\dfrac{1}{3} \right.$$

$$+\dfrac{1}{2} \qquad\quad -\dfrac{1}{4}$$

$$+\dfrac{1}{3} \qquad\quad -\dfrac{1}{5}$$

$$+\dfrac{1}{4} \qquad\quad -\dfrac{1}{6}$$

$$\cdots\cdots\cdots\cdots$$

$$\dfrac{1}{n-2} \qquad\quad -\dfrac{1}{n}$$

$$\dfrac{1}{n-1} \qquad\quad -\dfrac{1}{n+1}$$

$$\left. \dfrac{1}{n} \qquad\quad -\dfrac{1}{n+2} \right]$$

Thus, $\dfrac{1}{2} \displaystyle\sum_{r=1}^{n} \left(\dfrac{1}{r} - \dfrac{1}{r + 2} \right) = \dfrac{1}{2} \left[1 + \dfrac{1}{2} - \dfrac{1}{n+1} - \dfrac{1}{n+2} \right] = \dfrac{3}{4} - \dfrac{1}{2} \left(\dfrac{1}{n+1} + \dfrac{1}{n+2} \right)$

$$= \dfrac{3}{4} - \dfrac{2n + 3}{2(n + 1)(n + 2)}$$

Hence, the sum of the series $\dfrac{1}{1 \times 3} + \dfrac{1}{2 \times 4} + \dfrac{1}{3 \times 5} + \cdots + \dfrac{1}{n(n + 3)} = \dfrac{3}{4} - \dfrac{2n + 3}{2(n + 1)(n + 2)}$

Notice that we can deduce the sum to infinity since, if $n \to \infty$, the term $\dfrac{2n + 3}{2(n + 1)(n + 2)} \to 0$. Hence, $\displaystyle\sum_{r=1}^{\infty} \dfrac{1}{r(r + 2)} = \dfrac{3}{4}$.

The above method is known as the **method of differences**.

Exercise 17.4

1 Find the sums of the following series to n terms.

    (a) $1 + 2x + 3x^2 + 4x^3 + \cdots$ **(b)** $1 + 5x + 9x^2 + 13x^3 + \cdots$

2 Find the sum of the series $3 + 5x + 7x^2 + 9x^3 + \cdots$ to n terms and use the result to find the sum of,

    (a) $3 + 10 + 28 + \cdots$ to 8 terms   **(b)** $3 - \dfrac{5}{2} + \dfrac{7}{4} - \dfrac{9}{8} + \cdots$ to 10 terms

3 Find the sums of the following series to the number of terms indicated, by splitting them into two separate progressions.

    (a) $1 + 4 + 11 + 14 + 21 + 24 + \cdots$ to 18 terms

    (b) $2 + 5 + 6 + 9 + 10 + 13 + \cdots$ to 21 terms

    (c) $1 - 2 + 2 - 1 + 3 + 0 + 4 + \cdots$ to $(2n + 1)$ terms

4 Find the sums of the series to n terms by using the method of differences. Deduce the sums to infinity.

    (a) $\dfrac{1}{1 \times 4} + \dfrac{1}{4 \times 7} + \dfrac{1}{7 \times 10} + \cdots$ **(d)** $\dfrac{1}{1.3.5} + \dfrac{1}{3.5.7} + \dfrac{1}{5.7.9}$

    (b) $\dfrac{1}{2 \times 3} + \dfrac{1}{3 \times 4} + \dfrac{1}{4 \times 5} + \cdots$ **(e)** $\dfrac{1}{1.3.5} + \dfrac{2}{3.5.7} + \dfrac{3}{5.7.9}$

    (c) $\dfrac{1}{1.4.7} + \dfrac{1}{4.7.10} + \dfrac{1}{7.10.13} + \cdots$ **(f)** $\dfrac{1}{2.4.6} + \dfrac{2}{3.5.7} + \dfrac{3}{4.6.8}$

5 Express $\dfrac{r + 3}{(r - 1)r(r + 1)}$ in partial fractions, and hence find $\displaystyle\sum_{r=2}^{n} \dfrac{r + 3}{(r - 1)r(r + 1)}$.

Sums of powers of natural numbers

The arithmetical progression whose first term and common difference are both 1 produces the **series of natural numbers**.

$$S = 1 + 2 + 3 + 4 + 5 + \cdots\cdots\cdots + n = \sum_{r=1}^{n} r$$

The sum to n terms, $= S_n = \dfrac{n}{2}[2a + (n - 1)d]$

$$= \dfrac{n}{2}[2 + n - 1] = \tfrac{1}{2}n(n + 1)$$

Hence,
$$\boxed{\sum_{r=1}^{n} r = 1 + 2 + 3 + 4 + \cdots + n = \tfrac{1}{2}n(n + 1)}$$

The series of the **squares of the natural numbers** gives,

$$\sum_{r=1}^{n} r^2 = 1^2 + 2^2 + 3^2 + 4^2 + \cdots + (n-1)^2 + n^2$$

To find the sum of this series we use an identity together with the difference method.

$$\text{Consider,} \quad (r+1)^3 - r^3 = (r^2 + 2r + 1)(r+1) - r^3$$

$$= r^3 + 3r^2 + 3r + 1 - r^3$$

$$\therefore (r+1)^3 - r^3 = 3r^2 + 3r + 1$$

Hence, $\quad \sum_{r=1}^{n} 3r^2 + \sum_{r=1}^{n} 3r + \sum_{r=1}^{n} 1 = \sum_{r=1}^{n} [(r+1)^3 - r^3]$

Now, $\quad \sum_{r=1}^{n} 3r = 3 \sum_{r=1}^{n} r$, is an A.P. whose sum is $\dfrac{3n}{2}(n+1)$ and $\sum_{r=1}^{n} 1 = n$

Hence, $\quad \sum_{r=1}^{n} 3r^2 + \dfrac{3n}{2}(n+1) + n = \sum_{r=1}^{n} [(r+1)^3 - r^3]$

By substituting successive values of r into the right-hand side, most values will cancel, as usual.

$$\therefore \sum_{r=1}^{n} 3r^2 + \dfrac{3n}{2}(n+1) + n = [2^3 \qquad\quad - 1^3$$
$$3^3 \qquad\quad - 2^3$$
$$4^3 \qquad\quad - 3^3$$
$$\cdots\cdots$$
$$n^3 \qquad\quad - (n-1)^3$$
$$(n+1)^3 \qquad\quad - n^3]$$

or, $\quad \sum_{r=1}^{n} 3r^2 + \dfrac{3n}{2}(n+1) + n = (n+1)^3 - 1^3$

$$\Rightarrow \quad 3 \sum_{r=1}^{n} r^2 = n^3 + 3n^2 + 3n - \dfrac{3n}{2}(n+1) - n$$

$$= n^3 + \dfrac{3}{2}n^2 + \dfrac{1}{2}n = \dfrac{1}{2}n(2n^2 + 3n + 1)$$

$$\Rightarrow \quad \boxed{\sum_{r=1}^{n} r^2 = \dfrac{1}{6}n(n+1)(2n+1)}$$

Investigation 4

Show that $(r + 1)^4 - r^4 = 4r^3 + 6r^2 + 4r + 1$. Use this to show that,

$$4 \sum_{r=1}^{n} r^3 = \sum_{r=1}^{n} [(r + 1)^4 - r^4] - 6 \sum_{r=1}^{n} r^2 - 4 \sum_{r=1}^{n} r - \sum_{r=1}^{n} 1$$

By using the previous results for $\sum r^2$ and $\sum r$ and substituting successive values of r into the first term on the right-hand side, show that

$$\sum_{r=1}^{n} r^3 = 1^3 + 2^3 + 3^3 + \cdots + n^3 = \frac{1}{4}n^2(n + 1)^2$$

Summary

The sum of the natural numbers $= \sum_{r=1}^{n} r = \frac{1}{2}n(n + 1)$

The sum of the squares of natural numbers $= \sum_{r=1}^{n} r^2 = \frac{1}{6}n(n + 1)(2n + 1)$

The sum of the cubes of natural numbers $= \sum_{r=1}^{n} r^3 = \frac{1}{4}n^2(n + 1)^2$

Note that $\sum_{r=1}^{n} r^3 = \frac{1}{4}n^2(n + 1)^2 = \left[\frac{1}{2}n(n + 1)\right]^2 = \left[\sum_{r=1}^{n} r\right]^2$

These standard results can be used to sum other series.

WORKED EXAMPLES

1 Find the sum of the first $(n + 3)$ terms of the series $2^2 + 3^2 + 4^2 + \cdots$.

This is part of the series of the squares of the natural numbers with the first term, 1^2, missing. Taking $(n + 3)$ terms of the given series, means taking $(n + 4)$ terms if we include 1^2.

Hence, $\quad 1^2 + 2^2 + 3^2 + 4^2 + \cdots + m^2 = \frac{1}{6}m(m + 1)(2m + 1)$

$\Rightarrow \quad 1^2 + 2^2 + 3^2 + 4^2 + \cdots + (n + 4)^2 = \frac{1}{6}(n + 4)(n + 5)(2n + 9)$

i.e. $\quad 2^2 + 3^2 + 4^2 + \cdots (n + 4)^2 = \frac{1}{6}(n + 4)(n + 5)(2n + 9) - 1$

2 Find the sum of the series $1^3 + 3^3 + 5^3 + \cdots + (2n-1)^3$.

This series is formed from alternate terms of the series,

$$1^3 + 2^3 + 3^3 + 4^3 + 5^3 + \cdots + (2n-1)^3 = \frac{1}{4}(2n-1)^2(2n)^2 = n^2(2n-1)^2$$

Now the extra terms, $2^3 + 4^3 + 6^3 + \cdots + (2n-2)^3$, form a series which can be arranged into the sum of the cubes of the natural numbers.

$$2^3 + 4^3 + 6^3 + \cdots (2n-2)^3 = 2^3[1^3 + 2^3 + 3^3 + \cdots (n-1)^3]$$

$$= 8\left[\frac{1}{4}(n-1)^2 n^2\right] \quad \text{(since there are } n-1 \text{ terms)}$$

$$= 2n^2(n-1)^2$$

Hence, $\quad 1^3 + 3^3 + 5^3 + \cdots (2n-1)^3 = n^2(2n-1)^2 - 2n^2(n-1)^2$

$$= n^2[(2n-1)^2 - 2(n-1)^2]$$

$$= n^2[4n^2 - 4n + 1 - 2n^2 + 4n - 2]$$

$$= n^2(2n^2 - 1)$$

3 Find the sum of the series $1.2.3 + 2.3.4 + 3.4.5 + \cdots$ to n terms.

The general term, $u_r = r(r+1)(r+2)$. $\quad$ Hence, we need $\displaystyle\sum_{r=1}^{n} r(r+1)(r+2)$.

Now, $\qquad\qquad \displaystyle\sum_{r=1}^{n} r(r+1)(r+2) = \sum_{r=1}^{n}(r^3 + 3r^2 + 2r) = \sum_{r=1}^{n} r^3 + 3\sum_{r=1}^{n} r^2 + 2\sum_{r=1}^{n} r$

Using the standard results,

$$\sum_{r=1}^{n} r(r+1)(r+2) = \frac{1}{4}n^2(n+1)^2 + 3\left[\frac{1}{6}n(n+1)(2n+1)\right] + 2\left[\frac{n}{2}(n+1)\right]$$

$$= \frac{1}{4}n^2(n+1)^2 + \frac{1}{2}n(n+1)(2n+1) + n(n+1)$$

$$= \frac{1}{4}n(n+1)[n(n+1) + 2(2n+1) + 4]$$

$$= \frac{1}{4}n(n+1)(n^2 + 5n + 6)$$

$$= \frac{1}{4}n(n+1)(n+2)(n+3)$$

Exercise 17.5

1 Find the sums of the following series.

(a) $1 + 2 + 3 + \cdots + 121$

(b) $1^2 + 2^2 + 3^2 + \cdots + 14^2$

(c) $1^3 + 2^3 + 3^3 + \cdots + 9^3$

(d) $1 + 2 + 3 + \cdots + (2n-1)$

(e) $1^2 + 2^2 + 3^2 + \cdots (n-2)^2$

(f) $1^3 + 2^3 + 3^3 + \cdots + (2n+3)^3$

(g) $4 + 5 + 6 + \cdots + 32$

(h) $3^2 + 4^2 + 5^2 + \cdots + 15^2$

(i) $n^3 + (n+1)^3 + \cdots (2n)^3$

(j) $1 + 3 + 5 + \cdots + (2n-1)$

(k) $4^2 + 6^2 + 8^2 + \cdots (2n)^2$

(l) $3^3 + 5^3 + 7^3 + \cdots (2n+1)^3$

2   Find the general terms of the following series and use the results of the last section in this chapter, to find the sums.

   (a) $(1 \times 2) + (2 \times 3) + \cdots$ to n terms

   (b) $(4 \times 6) + (6 \times 8) + (8 \times 10) + \cdots$ to $(2n)$ terms

   (c) $(3 \times 7) + (5 \times 9) + (7 \times 11) + \cdots$ to n terms

   (d) $(1 \times 4) + (2 \times 7) + (3 \times 10) + \cdots$ to n terms

3   Show that the general term of the series $5(1)^2 + 6(2)^2 + 7(3)^2 + \cdots$ is $(r+4)r^2$. Hence, find the sum of the series to n terms.

4   Find,   **(a)** $\sum_{r=1}^{n} r(r+2)$ **(b)** $\sum_{r=1}^{n} (r+1)(2r^2 - 1)$

5   Find the sum of the first $(2n)$ terms of the series $1^2 + 3(2^2) + 5(3^2) + \cdots$.

6   Find **(a)** $\sum_{r=1}^{n} (r+2)(r+3)(2r+1)$ **(b)** $\sum_{r=1}^{n} (r^3 + r^2 + r)$

7   Find the sum of the first n term of the series $2.3.6 + 3.4.7 + 4.5.8 + \cdots$.

8   Show that $r(r+1)(r+3) = r^3 + 4r^2 + 3r$. Hence find, using the results of the last section in this chapter, $\sum_{r=1}^{n} r(r+1)(r+3)$.

18 Loci and co-ordinate geometry

Introduction

The locus of a point is the path it follows according to a certain rule.

A landscape gardener creating a circular flower bed would probably use a string (or tape measure) fixed to a stick (centre) to trace the outline of the flower bed.

How would he create the shape of an ellipse?

If a child flying a kite keeps the string taut and of fixed length, what path would the kite follow?

What shape does a skipping rope adopt when being used? (curve and surface)

What shapes can you see by looking at a garden hose or sprinkler? What path does a droplet of water follow? What shape is made by the water on the ground?
    What shape is made by a sprinkler on a large garden wall?

Have you seen the 'cusp' produced on the surface of a cup of tea by the light reflected from the rim of the cup?

Can you generate this curve?

What shape are the ripples on a pond when a stone is thrown in?

Look at the shadows on a wall, produced by a table lamp.

What shape is the reflector in a car or bicycle headlight?

Why are wheels circular? Too obvious?

Then what happens to the cup of liquid on this board as the 'wheels' rotate?

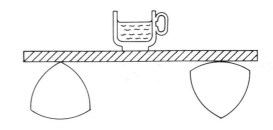

What shape is a 50 p coin?

What shape is a 20 p coin, and why?

Why are most coins round? Why is a room and furniture mostly square and rectangular?

In this chapter we shall be considering lines, curves and shapes and how we describe them. In A-level mathematics, this has always been done algebraically for exactness and precision, but with programmable calculators and computer graphics, we can appreciate more loci.

Exercise 18.1

In each of the following questions a locus is defined. **(a)** Sketch the path, **(b)** find its Cartesian equation in two dimensions, and **(c)** describe the locus in three dimensions.

1　The set of points P, which are equidistant from O $(0,0)$ and A $(4,0)$.

2　The set of points P, which are equidistant from A $(4,0)$ and B $(0,2)$.

3　The set of points P, which are equidistant from the lines $x = 2$ and $y = 1$.

4　The set of points P, equidistant from the x-axis and the y-axis.

5　The set of points which are two units from the point C $(3,4)$.

6　The bisector of the angle AOC, where A is $(4,0)$, O is $(0,0)$ and C is $(3,4)$.

7　The set of points P, where $A\hat{P}B = 90°$, A is $(4,0)$, B is $(0,2)$.

8　The set of points P, where $A\hat{P}B = 60°$.

9　The set of points P, where area $\triangle ABP =$ area $\triangle ABC$.

10　The set of points P, where $PO = 2 \times PD$ where D is $(3,0)$.

Loci

Investigation 1

The falling ladder　You will need graph paper in cm and a 15 cm ruler. Draw x and y axes from 0 to 15 cm. Place the cm scale along the y-axis, with 0 cm at $(0,0)$ and the 15 cm mark at $(0,15)$. The 15 cm ruler represents the ladder.

Move the base 2 cm along the x-axis to $(2,0)$, keeping the 15 cm mark on the y-axis. Record the position of **(a)** the 5 cm mark with $+$, **(b)** the 7.5 cm mark with $\odot$, **(c)** the 10 cm mark with $\times$, and **(d)** rule the line joining $(2,0)$ to $(0,14.9)$, the line of the ruler.

Move the base another 2 cm to $(4,0)$ keeping the 15 cm mark on the y-axis at $(0,14.5)$ and repeat **(a)(b)(c)** and **(d)**.

The ruler represents the falling ladder. What are the loci of **(a)(b)** and **(c)**? **(d)** consists of straight lines which 'envelop' a curve. What is the curve?

You have drawn four **loci** in the positive quadrant. Can you complete the loci by using the other three quadrants?

Investigation 2

Copy the figure, right, and mark on the graph any points P, which are the same distance from the point $F(0, 1)$ and the line D, whose equation is $y = -1$.

Three points have been given, $P_0(0, 0)$ $P_1(2, 1)$ and $P_2(4, 4)$.

What shape is the locus of P?

Can you find the x, y equation for P?

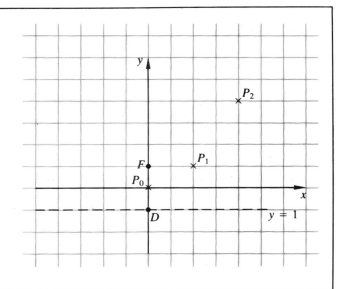

Investigation 3

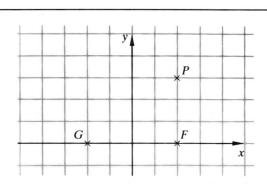

How are all the crosses in the figure, left, related to the origin O? Can you find more crosses?

What locus do all the crosses lie on?
(There is more than one answer)

Can you find an equation relating x and y for each locus?

Answer the same questions for the three points marked with a ringed dot, $(5, 0)$, $(4, 3)$, $(3, 4)$.

Investigation 4

In the figure, right, the point $P(2, 3)$ satisfies the relation $PF + PG = 8$

Can you find more points satisfying this relation?

Do they all lie on a curve? What shape is the curve?

Can you find its x, y relation?

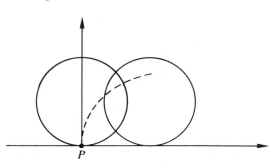

Investigation 5

Roll a circular coin along a straight edge, marking the locus traced by the initial lowest point *P*, on the coin. (The author followed the 'P' on a One Pound coin.) Where is *P* after one revolution?

What is the shape called?

Its equation is difficult, you need trigonometry and parametric form.

Circles

One of your answers in Investigation 3 should have been, a circle centre *O*.

All points on the circle are the same distance from the centre (figure, right).

$$OP = r, \text{ the radius of the circle}$$

Using Pythagoras' Theorem for any point $P(x, y)$ on the circle,

$$x^2 + y^2 = r^2$$

$(2, 1)$ lies on the circle $\Rightarrow 2^2 + 1^2 = r^2 = 5$

$$\Rightarrow r = \sqrt{5} \simeq 2.24$$

The equation of the circle, centre $(0,0)$ radius 1, is $x^2 + y^2 = 1$.
The equation of the circle, centre $(0,0)$ radius 2, is $x^2 + y^2 = 2^2 = 4$.

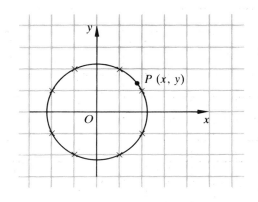

Consider the circle, figure right, having the same radius $(\sqrt{5})$ but with its centre at $C(3, 2)$.

Any point $P(x, y)$ on the circle is the same distance $(\sqrt{5})$ from *C*, so $CP = \sqrt{5} \Rightarrow CP^2 = 5$.

Using the formula for the distance between two points, $P(x, y)$ and $C(3, 2)$,

$$CP^2 = (x-3)^2 + (y-2)^2 = 5 \cdots \cdots (1)$$

Expanding the brackets gives,

$$x^2 - 6x + 9 + y^2 - 4y + 4 = 5 \quad \Rightarrow \quad x^2 + y^2 - 6x - 4y + 8 = 0 \cdots \cdots (2)$$

Both equations (1) and (2) are used to describe the circle, radius $\sqrt{5}$ centre $(3, 2)$.

285

Exercise 18.2

1 Find the equations, in both forms, of the following circles.

(a) Centre $(3, 4)$, radius 2   (d) Centre $(0, 2)$, radius 3   (g) Centre $(-1, -2)$, radius 3

(b) Centre $(3, 4)$, radius 5   (e) Centre $(3, -4)$, radius 2   (h) Centre $(5, 12)$, radius 13

(c) Centre $(3, 0)$, radius 4   (f) Centre $(-3, 2)$, radius 4

2 What are the centres and radii of the following circles?

(a) $(x + 3)^2 + (y - 2)^2 = 25$ (c) $(x - 4)^2 + (y + 5)^2 = 169$

(b) $(x - 3)^2 + (y - 4)^2 = 16$ (d) $x^2 + (y + 3)^2 = 196$

3 Find the centre and radius of each of these circles.

(a) $x^2 + y^2 - 2x - 4y + 4 = 0$ (d) $x^2 + y^2 + x + 3y - 3\frac{3}{4} = 0$ (g) $x^2 + y^2 + 6x - 7 = 0$

(b) $x^2 + y^2 - 6x - 8y + 21 = 0$ (e) $2x^2 + 2y^2 - 8x + 24y = 248$ (h) $3x^2 + 3y^2 - 27 = 0$

(c) $x^2 + y^2 + 4x + 2y - 11 = 0$ (f) $x^2 + y^2 - 10x - 12y = 0$

Intersection of a straight line with a circle

Consider the intersection of a straight line $L, (y = x + 1)$ with the circle C, centre $(4, 3)$ radius 2, whose equation is $(x - 4)^2 + (y - 3)^2 = 4$.

At the points where L and C meet, the x and y values of L and C coincide, so we solve the equations for L and C simultaneously.

Substituting $y = x + 1$ for y in the C equation gives,

$$(x - 4)^2 + (x + 1 - 3)^2 = 4$$

$$\Rightarrow \quad x^2 - 8x + 16 + x^2 - 4x + 4 = 4 \quad \Rightarrow \quad 2x^2 - 12x + 16 = 0$$

$$\Rightarrow \quad x^2 - 6x + 8 = 0 \quad \Rightarrow \quad (x - 2)(x - 4) = 0 \quad \Rightarrow \quad x = 2 \text{ or } x = 4$$

$$x = 2 \quad \Rightarrow \quad y = x + 1 = 3 \quad \text{and } x = 4 \quad \Rightarrow \quad y = 5$$

So L and C intersect at $(2, 3)$ and $(4, 5)$

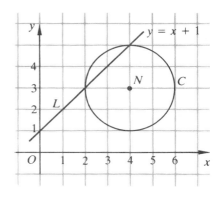

WORKED EXAMPLES

1 Find the intersection of the straight line L with gradient 2 passing through $(0, 1)$, with the circle C centre $(3, 2)$ which passes through $(5, 1)$, and explain your result.

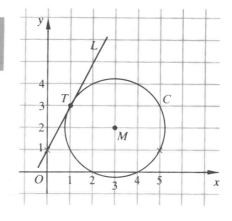

Using $y - y_1 = m(x - x_1)$ to find L gives,

$$y - 1 = 2(x - 0) \quad \Rightarrow \quad y = 2x + 1$$

The radius, r, of C is the distance from $(3, 2)$ to $(5, 1)$.

$$r^2 = (3 - 5)^2 + (2 - 1)^2 = 4 + 1 = 5 \quad \Rightarrow \quad r = \sqrt{5}$$

Equation of C is $(x - 3)^2 + (y - 2)^2 = 5$

Solving L and C simultaneously $\Rightarrow$ $(x-3)^2 + (2x+1-2)^2 = 5$

$\Rightarrow$ $x^2 - 6x + 9 + 4x^2 - 4x + 1 = 5$ $\Rightarrow$ $5x^2 - 10x + 5 = 0$

$\Rightarrow$ $x^2 - 2x + 1 = 0$

$\Rightarrow$ $(x-1)^2 = 0$ $\Rightarrow$ $x = 1$, a repeated root

$x = 1$, $y = 3$ at $T(1,3)$ (see figure)

This means L meets C twice at $(1,3)$, i.e. L is a tangent to the circle at T.

If M is the centre of C, M is $(3,2)$ and TM is the radius through T.

The gradient of TM is $\dfrac{2-3}{3-1} = -\dfrac{1}{2}$ and since the gradient of L is 2, TM is perpendicular to L.

(Straight lines are perpendicular if the product of their gradients is -1.)

Since, for a circle, the tangent is perpendicular to the radius through the point of contact, this confirms that L is a tangent to C.

From the diagram, the line $y = 2x + 2$ would not intersect C, so the ensuing quadratic equation would have no solution.

i.e. $y = 2x + 2$ in $(x-3)^2 + (y-2)^2 = 5$ $\Rightarrow$ $(x-3)^2 + (2x+2-2)^2 = 5$ $\Rightarrow$ $5x^2 - 6x + 4 = 0$

'$b^2 - 4ac$' $= 36 - 4 \times 5 \times 4 = 36 - 80 < 0$ $\Rightarrow$ no real roots.

2   Can we find the intersection of the two circles just considered?

C_1 is given by $(x-4)^2 + (y-3)^2 = 4$ C_2 is given by $(x-3)^2 + (y-2)^2 = 5$

$\Rightarrow$ $x^2 - 8x + 16 + y^2 - 6y + 9 = 4$ $\Rightarrow$ $x^2 - 6x + 9 + y^2 - 4y + 4 = 5$

$\Rightarrow$ $x^2 + y^2 - 8x - 6y + 21 = 0$ $\Rightarrow$ $x^2 + y^2 - 6x - 4y + 8 = 0$

Subtracting C_1 from C_2 $\Rightarrow$ $2x + 2y - 13 = 0$ $\Rightarrow$ $y = -x + 6\frac{1}{2}$

Substitute $y = -x + 6\frac{1}{2}$ in C_2

$\Rightarrow$ $x^2 + (6\frac{1}{2} - x)^2 - 6x - 4(6\frac{1}{2} - x) + 8 = 0$

$\Rightarrow$ $x^2 + 42\frac{1}{4} - 13x + x^2 - 6x - 26 + 4x + 8 = 0$

$\Rightarrow$ $2x^2 - 15x - 24\frac{1}{4} = 0$

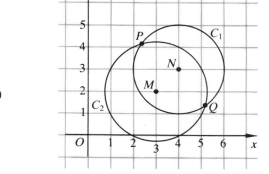

Using the quadratic formula gives,

$$x = \frac{15 \pm \sqrt{225 - 4 \times 2(-244)}}{4} = \frac{15 \pm \sqrt{225 - 194}}{4} = \frac{15 \pm \sqrt{31}}{4} = 5.14 \text{ or } 2.36$$

$x = 5.14$ $\Rightarrow$ $y = 6.5 - 5.14 = 1.36$ $x = 2.36$ $\Rightarrow$ $y = 4.14$

Q is $(5.14, 1.36)$ P is $(2.36, 4.14)$

This Example shows how a simple problem can lead to a complicated solution.

The equation $y = -x + 6\frac{1}{2}$ is a straight line on which both P and Q lie, so it is the straight line PQ. This is the common chord of both circles and is perpendicular to MN the line of centres.

Exercise 18.3

In Questions **1** to **4**, find the points of intersection of the given lines with the circle, centre $(3, 2)$ and radius $\sqrt{5}$.

1 $y = x$ **2** $y = 1$ **3** $x + y = 4$ **4** $y = \frac{1}{2}x + 3$

5 Find the points of intersection of the line $y = x - 3$ with the circle $x^2 + y^2 - 8x - 6y + 21 = 0$.

6 Find the points of intersection of the lines $y = 2x$ and $y = \frac{1}{2}x$ with the circle $(x - 5)^2 + (y - 5)^2 = 25$

7 Sketch the circle $x^2 + y^2 - 10x - 10y + 45 = 0$. Find the points of contact, T and S, where the lines $y = 2x$ and $y = \frac{1}{2}x$ touch this circle. With the origin O and centre of the circle C, find OC and check that $OT = OS$. Check also that $OT^2 + TC^2 = OC^2$.

8 Using the circle in Question **7**, find the equation of SC and the point opposite S where this line meets the circle again. From your sketch, identity the points on the circumference with integer co-ordinates and, using these, confirm the result that angles in a semi-circle are right angles.
   What is the area of the largest rectangle that will fit inside the circle?

9 Find the points of intersection of the circles $x^2 + y^2 - 10x - 10y + 45 = 0$ and $(x - 4)^2 + (y - 2)^2 = 5$.

10 P and Q are the points of intersection of $y = x$ with the circle, centre $(3, 2)$ radius $\sqrt{5}$. R and S are the points of intersection of $y = \frac{1}{3}(x - 2)$ with the same circle.
   Show that $\triangle s\, XPS$ and XRS are similar, using co-ordinates. (X is the point $(-1, -1)$.) Show also that $\triangle s\, XRP$ and XQS are similar.

Tangents and normals

What is the locus of a 'hammer' thrown by a hammer thrower?
   Does it fly off at a tangent? Why?

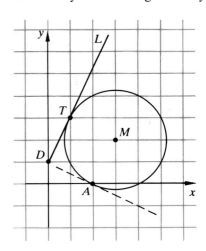

In Worked Example **1**, we found that the line TL was a tangent to the circle
$$(x - 3)^2 + (y - 2)^2 = 5$$

Can we find the equation of the tangent to the circle which passes through $A(2, 0)$?

Method 1 Knowing that the tangent is perpendicular to the radius,

gradient $AM = 2$

$\Rightarrow$ grad tangent $= -\dfrac{1}{2}$.

Equation of tangent through $A(2, 0)$ is
$$y - 0 = -\frac{1}{2}(x - 2) = -\frac{1}{2}x + 1$$

which also passes through $D(0, 1)$.

Method 2 finds the gradient of the tangent to the circle at A without using the radius. This is the method used when the curve is not circular.

Differentiate the equation of the curve to find the gradient $\dfrac{dy}{dx}$ at the point A.

$$(x-3)^2 + (y-2)^2 = 5 \quad \Rightarrow \quad 2(x-3) + 2(y-2)\dfrac{dy}{dx} = 0 \quad \Rightarrow \quad \dfrac{dy}{dx} = -\dfrac{(x-3)}{y-2}$$

At $A(2,0)$, $\dfrac{dy}{dx} = \dfrac{-(2-3)}{-2} = -\dfrac{1}{2}$ leading to the tangent equation being $y - 0 = -\dfrac{1}{2}(x-2)$, and $y = -\dfrac{1}{2}x + 1$.

You may prefer to expand $(x-3)^2 + (y-2)^2 = 5$ before differentiating,

to get, $x^2 - 6x + 9 + y^2 - 4y + 4 = 5$

Differentiating $\quad \Rightarrow \quad 2x - 6 + 2y\dfrac{dy}{dx} - 4\dfrac{dy}{dx} = 0 \quad \Rightarrow \quad \dfrac{dy}{dx}(2y-4) = -(2x-6)$

$\quad \Rightarrow \quad \dfrac{dy}{dx} = \dfrac{-(x-3)}{y-2}$, the same result.

The line AM is the normal (at right angles to tangent) to the curve at the point A.

The line TM is the normal to the circle at T, where DTL is the tangent to the circle at T.

$DT = DA$ illustrates the fact that tangents to a circle from on external point are equal in length.

WORKED EXAMPLES

1 Find the equations of the normal and the tangent to $y = x^2$ at the point $P(2,4)$.

To find the gradient of the tangent at $P(2,4)$, find $\dfrac{dy}{dx}$.

$$y = x^2 \quad \Rightarrow \quad \dfrac{dy}{dx} = 2x$$

At P, $x = 2 \quad \Rightarrow \quad \dfrac{dy}{dx} = 4 \quad \Rightarrow \quad$ Equation of PT has gradient 4.

Equation of PT is $y - 4 = 4(x-2) \quad \Rightarrow \quad y = 4x - 4$

The normal to the curve at P i.e. PN, has gradient $-\dfrac{1}{4}$, so the equation of PN is

$$y - 4 = -\dfrac{1}{4}(x-2)$$

$$\Rightarrow \quad y = 4 - \dfrac{1}{4}x + \dfrac{1}{2}$$

$$\Rightarrow \quad y = 4\tfrac{1}{2} - \dfrac{x}{4}$$

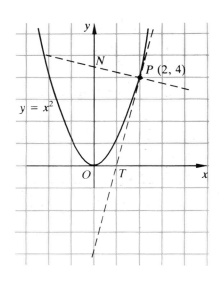

289

2 Find the tangent to the curve, described parametrically by $x = t - 2$, $y = \frac{t}{3}(t-2)(t-4)$ at the point given by $t = 1$. Find where this tangent meets the curve again. Find the equation of the normal at $t = 2$ and where this line meets the curve.

$$t = 1 \quad \Rightarrow \quad x = -1, \quad y = \frac{1}{3}(-1)(-3) = 1$$

To find the gradient $\quad \dfrac{dy}{dx} = \dfrac{dy}{dt} \times \dfrac{dt}{dx}$ and $\dfrac{dx}{dt} = 1 \quad \Rightarrow \quad \dfrac{dy}{dx} = \dfrac{dy}{dt}$

$$y = \frac{t}{3}(t-2)(t-4) = \frac{1}{3}(t^3 - 6t^2 + 8t) \quad \Rightarrow \quad \frac{dy}{dt} = \frac{1}{3}(3t^2 - 12t + 8) = \frac{1}{3}(3 - 12 + 8) = -\frac{1}{3} \quad \text{when } t = 1$$

Equation of the tangent at $(-1, 1)$ is $\quad y - 1 = -\frac{1}{3}(x+1) \quad \Rightarrow \quad y = 1 - \frac{1}{3}x - \frac{1}{3} = -\frac{1}{3}x + \frac{2}{3} = -\frac{1}{3}(x-2)$

x, y equation of the curve is found by eliminating t to give $y = \dfrac{x+2}{3}(x)(x-2) = \dfrac{x(x^2-4)}{3}$

The tangent meets the curve where

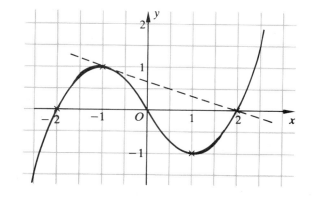

$$\frac{x(x^2-4)}{3} = \frac{-(x-2)}{3}$$

$\Rightarrow \quad x(x^2 - 4) + x - 2 = 0$

$\Rightarrow \quad (x-2)[x(x+2) + 1] = 0$

$\Rightarrow \quad (x-2)(x^2 + 2x + 1) = 0 \quad \Rightarrow \quad (x-2)(x+1)^2 = 0$

$\Rightarrow \quad x = 2$ at $(2, 0)$

or $x = -1$ (repeated) tangent at $(-1, 1)$

$t = 2$ gives $x = 0$, $y = 0$ the origin where $\dfrac{dy}{dx} = -\dfrac{4}{3}$

Gradient of normal at the origin is $\dfrac{3}{4} \quad \Rightarrow \quad$ the equation of the normal at the origin is $y = \dfrac{3}{4}x$.

This meets the curve where $\quad \dfrac{x(x^2-4)}{3} = \dfrac{3}{4}x$

$\Rightarrow \quad 4x(x^2 - 4) = 9x \quad \Rightarrow \quad x(4x^2 - 16 - 9) = 0$

$\Rightarrow \quad x = 0$ (origin)

or $\quad 4x^2 = 25 \quad \Rightarrow \quad x^2 = \dfrac{25}{4}$

$\Rightarrow \quad x = \pm\dfrac{5}{2} = \pm 2\tfrac{1}{2}$

$\Rightarrow \quad$ Normal meets the curve again at $(2\tfrac{1}{2}, 1\tfrac{7}{8})$ or $(-2\tfrac{1}{2}, -1\tfrac{7}{8})$.

Exercise 18.4

1 Find the equations of the tangent and normal to the given curve at the stated point.

(a) $y = x^2 - 3x + 2$; (3, 2)   **(d)** $y = x^2 + \dfrac{1}{x}$; (1, 2)   **(f)** $y = \dfrac{2x - 1}{x - 2}$; (1, -1)

(b) $y = x^3 - x$; (1, 0)   **(e)** $x^2 - xy + y = 0$; (2, 4)

(c) $xy = 4$; (4, 1)

2 Find the values of m for which $y = mx$ is a tangent to the circle $(x - 3)^2 + (y - 2)^2 = 5$.

3 Find the values of m for which $y = mx$ is a tangent to the curve $y = (x - 1)(x - 2)$.

4 Find the equation of the tangent and normal to the given curve at the stated point.

(a) $x = 2t, y = t^2$; (2, 1)   **(d)** $x = t^2, y = t^3$; $t = 1$

(b) $x = t + 1, y = t^2 + 1$; (2, 2)   **(e)** $x = 5\cos\theta, y = 5\sin\theta$; $\theta = \tan^{-1}\frac{3}{4}$

(c) $x = \dfrac{2}{t}, y = 2t$; (2, 2)

5 Find the equations of the tangents to the curve $y = x^2 - 3x + 2$ which are parallel and perpendicular to the line $y = 2x$.

6 Find the values of c, for which $y = x + c$ is a tangent to the circle $(x - 1)^2 + (y - 2)^2 = 3$.

The parabola

Investigation 1 led to the shape of a **parabola**. These occur frequently in mathematics.

The first parabola you met was probably $y = x^2$ and the curve in Investigation 1 (see diagram) is similar.

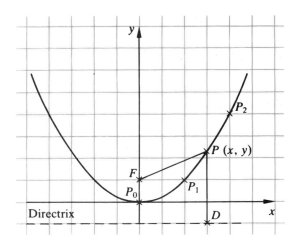

Points on the parabola are equidistant from F and the line $y = -1$ (the **directrix**), so

$$FP = PD$$

If P has co-ordinates (x, y), $FP^2 = (x - 0)^2 + (y - 1)^2$ and $PD = y + 1$.

$$FP^2 = PD^2 \;\Rightarrow\; x^2 + (y - 1)^2 = (y + 1)^2$$
$$\Rightarrow\; x^2 + y^2 - 2y + 1 = y^2 + 2y + 1$$
$$\Rightarrow\; x^2 = 4y \text{ or } y = \frac{1}{4}x^2$$

F is called the **focus** and P_0 the **vertex** of the parabola.

$x = 0$ (FP_0) is the **axis**, being the line of symmetry.

WORKED EXAMPLE

Find the equation of the parabola with focus $(1, 0)$ and directrix $x = -1$.

See figure, right. Any point $P(x, y)$ on the parabola is equidistant from $F(1, 0)$ and D (nearest point to P on the directrix).

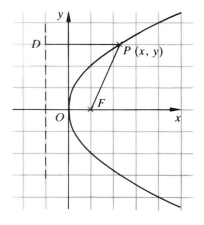

$$FP = PD \quad \Rightarrow \quad \sqrt{(x-1)^2 + y^2} = x + 1$$
$$\Rightarrow \quad (x-1)^2 + y^2 = (x+1)^2$$
$$\Rightarrow \quad x^2 - 2x + 1 + y^2 = x^2 + 2x + 1$$
$$\Rightarrow \quad y^2 = 4x$$

This parabola is similar to that at the beginning of this section, except that the axes x and y are interchanged; so changing y for x gives the equation.

For the first parabola, $x^2 = 4y$ and interchanging $\Rightarrow \quad y^2 = 4x$.

More properties of the parabola will be discussed in Book 2.

Exercise 18.5

1 Why is the focus so called? Draw a parabola accurately ($y^2 = 4x$ will do) and reflect lines drawn parallel to the axis ($y = 0$) as though the parabola were a mirror. What happens?

2 With the notation in the Worked Example, draw an accurate diagram of $y^2 = 4x$. Take a point P on the parabola and draw the tangent to the parabola at P. Produce PF to meet the parabola at Q and draw the tangent at Q.

What do you notice about these two tangents at P and Q? (You may need to take another point P.) Measure the angles between the tangent at P and PD and the tangent at P and PF. What do you notice? You can check this with the tangents at $(1, 2)$ and $(1, -2)$.

3 Find the focus and directrix of the parabola $y = x^2$.

4 Find the focus, axis and directrix of the parabola $y = (x - 1)(x - 2)$.

5 Find the equations of the tangent and normal to the parabola $y^2 = 4x$ at the point $(1, 2)$.

6 Find the equation of the parabola with

    (a) focus $(2, 0)$ and directrix $y = -2$

    (b) focus $(2, 0)$ and directrix $x = -2$

    (c) focus $(2, 1)$ and vertex $(1, 1)$

    (d) focus $(2, 1)$ and vertex $(2, 2)$.

7 Find the equation of the parabola passing through $(1, 2)$, $(3, 4)$ and $(0, 3)$.

$y^2 = 4ax$ and parametric form

In general three points determine a parabola, but we can specify two special points, the focus and vertex, to fix its shape and position. For convenience choose the vertex at the origin $(0,0)$ and focus $(a,0)$, which means that the directrix is $x = -a$, to make $FO = OD_1$.

For a general point on the parabola $P\,(x, y)$, $PF = PD_2$,

$$PF = \sqrt{(x-a)^2 + (y-0)^2} \text{ and } PD_2 = x + a.$$

$$PF^2 = PD_2{}^2 \quad \Rightarrow \quad (x-a)^2 + y^2 = (x+a)^2$$

$$\Rightarrow \quad x^2 - 2ax + a^2 + y^2 = x^2 + 2ax + a^2$$

$$\Rightarrow \quad y^2 = 4ax$$

$$x = a \quad \Rightarrow \quad y^2 = 4a^2 \quad \Rightarrow \quad y = \pm 2a \text{ and } x = 4a \quad \Rightarrow \quad y = \pm 4a$$

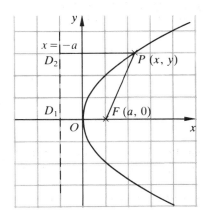

A table of values for x and y gives,

x	a	$4a$	$9a$	$16a$	$t^2 a$
y	$\pm 2a$	$\pm 4a$	$\pm 6a$	$\pm 8a$	$\pm 2ta$

which leads to the parametric form $x = at^2$, $y = 2at$ allowing t to take positive and negative values.

$$t = 0 \text{ gives } (0,0); \quad t = 1 \text{ gives } (a, 2a); \quad t = 2 \text{ gives } (4a, 4a)$$

while $t = -1$ gives $(a, -2a)$; $\quad t = -2$ gives $(4a, -4a)$

So the **parabola** $y^2 = 4ax$ has **focus** $(a, 0)$, **directrix** $x = -a$, **vertex** $(0,0)$ and its **parametric form** is $x = at^2$, $y = 2at$.

Tangent at the general point $(at^2, 2at)$

The gradient $\dfrac{dy}{dx} = \dfrac{dy/dt}{dx/dt} = \dfrac{2a}{2at} = \dfrac{1}{t}$

Equation of the tangent at the point $P(at^2, 2at)$ is $\quad y - 2at = \dfrac{1}{t}(x - at^2) \quad \Rightarrow \quad ty - 2at^2 = x - at^2$

Equation of tangent is $\quad \boxed{ty = x + at^2}\quad$ and $t = 1 \quad \Rightarrow \quad$ tangent at $(1, 2)$ is $y = x + a$.

Normal at the general point $(at^2, 2at)$

Gradient of normal at P is $-t$, so the equation of the normal is

$$y - 2at = -t(x - at^2) = -tx + at^3$$

Equation of normal is $\quad \boxed{y + tx = 2at + at^3}$

WORKED EXAMPLE

Prove that a line parallel to the axis through P is reflected in the tangent at P to pass through the focus.

With the notation in the figure, $\tan T\hat{P}R = \tan \alpha = \text{gradient of tangent} = \dfrac{1}{t}$.

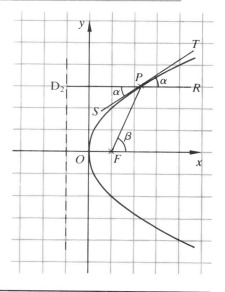

$T\hat{P}R = S\hat{P}D_2 = \alpha$ and $\tan \beta = \text{grad } FP = \dfrac{2at}{at^2 - a} = \dfrac{2t}{t^2 - 1}$

$\tan 2\alpha = \dfrac{2\tan\alpha}{1 - \tan^2\alpha} = \dfrac{2/t}{1 - (1/t^2)} = \dfrac{2t}{t^2 - 1} = \tan\beta \quad \Rightarrow \quad \beta = 2\alpha$

$\therefore$ Angle $F\hat{P}D_2 = \beta = 2\alpha \quad \Rightarrow \quad S\hat{P}F = \alpha$

The line RP makes the same angle with the tangent as the line PF, so RP is reflected through the focus.

Exercise 18.6

1 Find the Cartesian equation of the parabola $x = t^2$, $y = 2t$ and plot the curve. Find the equation of the tangent and normal at the points $P\,(t^2, 2t)$ and $(1, 2)$.

2 Find the gradient of the chord joining the points $P(ap^2, 2ap)$ and $Q(aq^2, 2aq)$ which both lie on the parabola $y^2 = 4ax$, and the equation of the line PQ. If this line passes through the focus $F\,(a, 0)$, show that $pq = -1$. Find the equations of the tangent at P and Q and show that if PQ passes through F, the tangents intersect at right angles on the directrix.

3 Find the intersection of the line $y = x + 1$ with the parabola $y^2 = 4x$.

4 Find the intersection of the circle $x^2 + y^2 = 5$ with the parabola $y^2 = 4x$.

5 Find the value of c for which $y = \frac{1}{2}x + c$ is a tangent to the parabola $y^2 = 4x$ and the point of contact.

6 Find the intersection of $y^2 = 4x$ and the circle $(x - 3)^2 + y^2 = 8$. What is particular about this intersection?

7 Find the intersection of $y^2 = 4x$ with the circle $(x - 3)^2 + y^2 = 9$.

8 A line AB of length 8 units moves, keeping A on the y-axis and B on the x-axis. Find the locus of **(a)** M the mid-point of AB, **(b)** the point Q where $AQ:QB = 3:1$, **(c)** the point R, where $AR:RB = 1:2$. Find the Cartesian equations in each case.

9 Find the locus of P, where $AP = OP$ and A is $(-1, 3)$. O is the origin.

10 Find the locus of P, where $OP:BP = 3:1$ and where B is $(4, 0)$ and O the origin.

11 Show that the mediators of the sides of $\triangle ABC$, where $A = (2, 4)$, $B = (3, -1)$, $C = (7, 3)$, are concurrent.

12 Repeat Question **11** for the altitudes of $\triangle ABC$.

13 Find the locus of P, where $AP + BP = 8$ and A is $(3, 0)$, B is $(-3, 0)$.

14   A circle of radius 1 cm rolls without slipping along the circumference of a circle, radius 4 cm. Trace the locus of a fixed point on the smaller circle as it rolls around the inside and outside of the larger circle.

15   Draw a circle, radius 4 cm, and mark a point A on the circumference. Take another point, Q, on the circle, and with radius QA draw a circle centre Q. Repeat this for points Q spaced evenly around the circle. What is the resulting envelope?

Curve sketching

Investigation 6

Can you identify these curves, most of which you have met already? Give an equation for each curve and note whether it is odd or even.

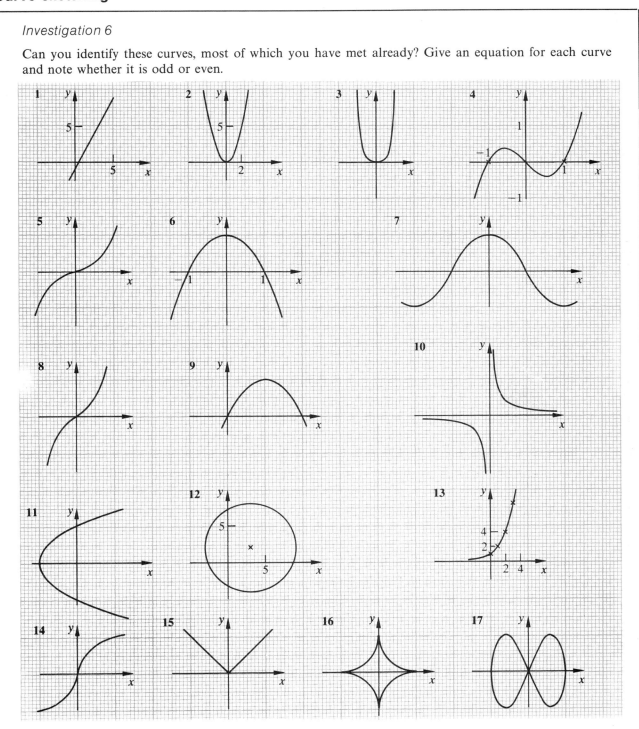

Here are the answers to the graphs in **Investigation 6**, with comments.

1 A straight line with equation $y = 2x - 1$ (gradient 2, intercept -1).

2 $y = x^2$, which is symmetrical about the y-axis and is an even function $f(-x) = f(+x)$.

3 $y = \frac{1}{2}x^4$ or $y = x^4$ according to scale; an even function.

4 A cubic curve, in fact $y = x^3 - x$; from the values where it cuts the x-axis, $y = k(x-1)x(x+1) = k(x^3 - x)$. The gradient at the origin is -1; $\frac{dy}{dx} = 3kx^2 - k = -k$ when $x = 0$, so $k = 1$. $y = x^3 - x$ is an odd function; it has rotational symmetry order 2 (180°) about $(0, 0)$; $f(-x) = -f(x)$.

Note that the maximum and minimum points do **not** occur at $x = \pm\frac{1}{2}$.

5 $y = x^3$, an odd function, whose gradient at the origin is zero.

6 Negative quadratic cutting the x-axis at ± 1; $y = -x^2 + 1 = 1 - x^2 = (1 + x)(1 - x)$.

7 $y = \cos x$, or possibly part of a quartic (x^4). Try $y = (x^2 - 1)(x^2 - 9)$ and discuss finer points.

8 $y = \tan x$, or a cubic. Try $y = x^3 + x$; what are the differences? Both are odd.

9 $y = \sin x$ or a negative quadratic like $y = -(x-1)^2 + 1 = 2x - x^2 = x(2 - x)$; $\sin x$ is odd.

10 $y = \dfrac{1}{x}$ or $xy = 1$; $xy = k$, depending on the scale; an odd function. Discontinuous at $x = 0$, $y \to 0$ as $x \to \pm \infty$, $y \to \pm \infty$ as $x \to 0$, from below or above; discuss.

11 Strictly not a function, $x = y^2 - 1 \Rightarrow y^2 = x + 1$, a parabola. Focus? Directrix?

12 Circle centre $(3, 2)$ radius 6 $\Rightarrow (x - 3)^2 + (y - 2)^2 = 36 \Rightarrow x^2 + y^2 - 6x - 4y - 23 = 0$.

13 $y = 2^x$; y always positive for all values of x; $y \to 0$ as $x \to -\infty$.

14 $y = \sqrt[3]{x} = x^{\frac{1}{3}}$, an odd function.

15 $y = |x|$ the value of x irrespective of sign; $|-2| = 2$; an odd function.

16 $x = \cos^3 t$, $y = \sin^3 t \Rightarrow x^{\frac{2}{3}} + y^{\frac{2}{3}} = 1$. Astroid (hypocycloid with 4 cusps). Look it up!

17 $x = \cos t$, $y = \sin 2t \Rightarrow y = 2x\sqrt{1 - x^2} \Rightarrow y^2 = 4x^2(1 - x^2)$.

Notes

The last two examples are both odd and even but are many-valued and best described parametrically.

Zeros In **4**, $y = x^3 - x = x(x^2 - 1) = x(x - 1)(x + 1)$, and $y = 0$ when $x = 0$, 1 or -1.
These values of x are called the **zeros** of the function $f(x) = x^3 - x$. The zeros are the values of x where the graph cuts the x-axis.
In **6**, the zeros are at $x = \pm 1$.

Asymptotes are lines to which the graph approaches.

In **10**, as x approaches 0 from above (positive side), $y(=1/x)$ approaches ∞, and as x approaches 0 from below (negative side), y approaches $-\infty$. At $x = 0$, y is infinite.

In **13**, as $x \to -\infty$, y approaches, or tends to, 0 from above (the positive side). $y = 0$ (x-axis) is a horizontal asymptote.

Rational functions

Investigations 7–11

7 Sketch the graphs of **(a)** $y = -\dfrac{1}{x}$ **(b)** $y = \dfrac{1}{x-1}$ **(c)** $y = \dfrac{1}{x+1}$ and **(d)** $y = \dfrac{1}{2-x}$.

Name the asymptotes both horizontal and vertical.

8 Sketch the graphs of $y = \dfrac{1}{x-1} + 1 = \dfrac{x}{x-1}$; $\quad y = \dfrac{1}{x+1} + 1 = \dfrac{x+2}{x+1}$; $\quad y = 1 - \dfrac{1}{x-1} = \dfrac{x-2}{x-1}$.

Specify the zeros and the asymptotes.

9 Sketch the graphs of $y = \dfrac{2}{x}$; $\quad y = \dfrac{2}{x+1}$; $\quad y = \dfrac{-2}{x+1}$; $\quad y = 1 - \dfrac{2}{x+1} = \dfrac{x-1}{x+1}$.

Specify the zeros and the asymptotes.

10 Sketch $y = \dfrac{1}{x-1}$ and $y = \dfrac{1}{x+1}$ on the same graph and use these to draw

$$y = \dfrac{1}{x-1} - \dfrac{1}{x+1} = \dfrac{2}{(x-1)(x+1)} = \dfrac{2}{x^2-1}.$$

11 Specify the zeros and asymptotes of $y = \dfrac{(x-1)(x+2)}{(x+1)(x-2)}$. How does this help to draw the graph?

Results

The graphs in **Investigation 7** are all modifications of $y = \dfrac{1}{x}$ involving translations and, in those, involving $-x$ reflections.

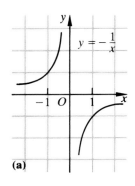

(a)

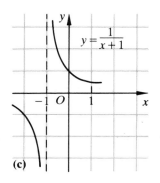

(b)

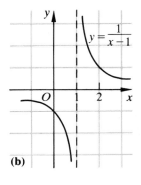

(c)

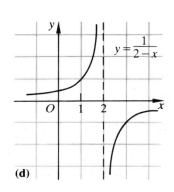

(d)

Similarly, the graphs in **Investigation 8** are transformations of $y = \dfrac{1}{x}$, the first one, $y = \dfrac{1}{x-1} + 1$, being the same as

$y = \dfrac{1}{x-1}$ but lifted up by 1 unit to give a horizontal asymptote, $y = 1$.

$y = \dfrac{x+2}{x+1} = \dfrac{1}{x+1} + 1$ is a translation 1 unit up of $y = \dfrac{1}{x+1}$, or $y = \dfrac{1}{x}$ moved 1 back and 1 up.

See figure, right.

$y = \dfrac{x-2}{x-1} = 1 - \dfrac{1}{x-1}$ can be achieved by reflecting $y = \dfrac{1}{x-1}$ in the x-axis and lifting up 1 unit, so its horizontal asymptote is $y = 1$, its vertical asymptote is $x = 1$ and its zero is at $x = 2$.

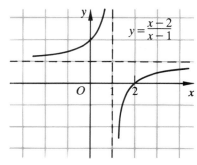

In **Investigation 9**, $y = \dfrac{2}{x} \Rightarrow xy = 2$ and is further from the origin than $\dfrac{1}{x}$ (figure **(a)** below).

The others, **(b)**, **(c)** and **(d)**, are all similar to $y = \dfrac{2}{x}$ but with asymptote $x = -1$.

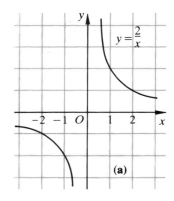

(a)

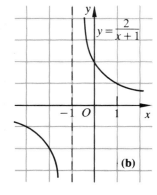

(b)

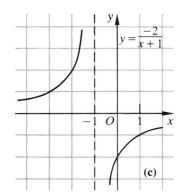

(c)

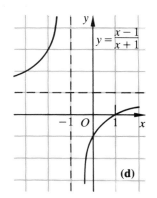

(d)

In **Investigation 10**, the approach suggested is rather cumbersome (but instructive) and the following approach may be easier.

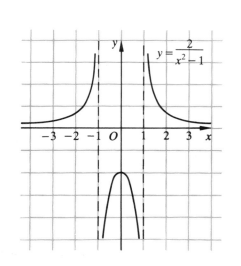

$y = \dfrac{2}{x^2 - 1} = \dfrac{2}{(x-1)(x+1)}$ has asymptotes at $x = 1$ and $x = -1$.

$y = \dfrac{2}{x^2 - 1}$ is an even function, so is symmetrical about the y-axis.

When x is large, $y \simeq \dfrac{2}{x^2}$ which $\to 0$ as $x \to \infty$.

When $x > 1$, y is positive.

When $x < -1$, y is positive.

When $-1 < x < 1$, y is negative. $|x^2 - 1| < 1 \Rightarrow y < -2$

298

$$\frac{dy}{dx} = \frac{-4x}{(x^2 - 1)^2} \quad \Rightarrow \quad \text{stationary point at } x = 0$$

$$x > 0 \quad \Rightarrow \quad \frac{dy}{dx} < 0 \text{ and } x < 0 \quad \Rightarrow \quad \frac{dy}{dx} > 0$$

i.e. **maximum** at $x = 0$.

The general shape is now determined and a few values for $x = \frac{1}{2}$, $x = 1\frac{1}{2}$, 2 and $2\frac{1}{2}$, specifies the graph (remember symmetry).

For **Investigation 11**, $y = \dfrac{(x-1)(x+2)}{(x+1)(x-2)}$ has xeros at $x = 1$ and $x = -2$ and asymptotes at $x = -1$ and $x = 2$.

For large x, $y = \dfrac{x^2 + x - 2}{x^2 - x - 2} \simeq \dfrac{x^2}{x^2} = 1$ so we have a horizontal asymptote at $y = 1$.

i.e. for large $+x$, y approaches 1 from above, and for large $-x$, y approaches 1 from below.

Consider the signs of the brackets in $y = \dfrac{(x-1)(x+2)}{(x+1)(x-2)}$

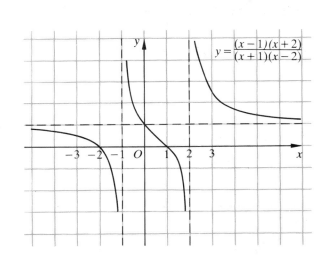

For $\quad x > 2$; $\quad y = \dfrac{(+)(+)}{(+)(+)}$ i.e. $y > 0$

$\quad 1 < x < 2$; $\quad y = \dfrac{(+)(+)}{(+)(-)} \quad \Rightarrow \quad y < 0$

$-1 < x < +1$; $\quad y = \dfrac{(-)(+)}{(+)(-)} \quad \Rightarrow \quad y > 0 \quad \text{and} \quad x = 0, y = 1$

$-2 < x < -1$; $\quad y = \dfrac{(-)(+)}{(-)(-)} \quad \Rightarrow \quad y < 0$

$\quad x < -2$; $\quad y = \dfrac{(-)(-)}{(-)(-)} \quad \Rightarrow \quad y > 0$

Reciprocal functions

To draw $y = \dfrac{1}{x}$, first draw $y = x$ (figure, right),

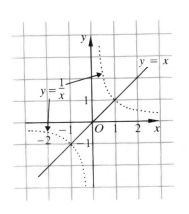

$x > 0 \quad \Rightarrow \quad \dfrac{1}{x} > 0 \text{ and } x < 0 \quad \Rightarrow \quad \dfrac{1}{x} < 0$

$x = 1 \quad \Rightarrow \quad \dfrac{1}{x} = 1 \text{ and } x = -1 \quad \Rightarrow \quad \dfrac{1}{x} = -1$

For each x value, plot its reciprocal value; remember $\dfrac{1}{0} = \infty$.

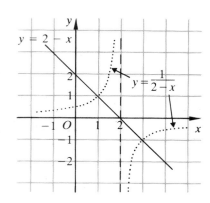

See figure, left.

For $y = \dfrac{1}{2-x}$, draw $y = 2 - x$ (straight line through $2, 0$).

$x = 2 \;\Rightarrow\; 2 - x = 0 \;\Rightarrow\; y = \dfrac{1}{2-x}$ is infinite i.e. asymptote.

$x > 2 \;\Rightarrow\; 2 - x < 0 \;\Rightarrow\; y = \dfrac{1}{2-x} < 0$ curve below x-axis.

$x < 2 \;\Rightarrow\; 2 - x > 0 \;\Rightarrow\; y = \dfrac{1}{2-x} > 0$ curve above x-axis.

WORKED EXAMPLE

Use the reciprocal graph plotting idea to draw **(a)** $y = \dfrac{1}{x^2}$ and **(b)** $y = \dfrac{2}{x^2 - 1}$.

(a) $y = \dfrac{1}{x^2}$. Draw $y = x^2$ (full line in figure, right)

Plot reciprocal values, $x^2 = 1 \;\Rightarrow\; x = \pm 1 \;\Rightarrow\; \dfrac{1}{x^2} = 1$

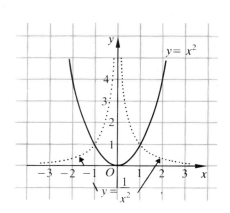

$x^2 > 0$ for all $x \;\Rightarrow\; \dfrac{1}{x^2} > 0$ for all x.

Where $x^2 = \tfrac{1}{2}$, plot $\dfrac{1}{x^2} = 2$ (no need to find x).

Where $x^2 = 2$, plot $\dfrac{1}{x^2} = \tfrac{1}{2}$.

$x^2 = 0$ at the origin $\;\Rightarrow\; \dfrac{1}{x^2} = \infty$, asymptote at $x = 0$.

$y = \dfrac{1}{x^2}$ is shown dotted in figure.

(b) $y = \dfrac{2}{x^2 - 1}$. Plot $y = \dfrac{x^2 - 1}{2} = \dfrac{1}{2}(x^2 - 1)$

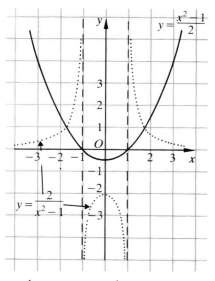

i.e. think of $y = x^2$, subtract 1 (down) then halve values.

$\dfrac{x^2 - 1}{2}$ has zeros at $x = \pm 1 \;\Rightarrow\; \dfrac{2}{x^2 - 1}$ has asymptotes at $x = \pm 1$.

$\dfrac{x^2 - 1}{2}$ has minimum value $(0, -\tfrac{1}{2}) \;\Rightarrow\; \dfrac{2}{x^2 - 1}$ has maximum $(0, -2)$.

$\dfrac{x^2 - 1}{2} > 0$ for $x > 1$ and $x < -1 \;\Rightarrow\; \dfrac{2}{x^2 - 1}$ also positive for $x > 1$, $x < 1$.

$y = \dfrac{2}{x - 1}$ is shown dotted in figure, right.

Look back at the reciprocal trigonometrical functions, $\sec x = \dfrac{1}{\cos x}$; $\operatorname{cosec} x = \dfrac{1}{\sin x}$; $\cot x = \dfrac{1}{\tan x}$.

The modulus function $y = |x|$

$|x|$ stands for the value of x irrespective of sign

$|2| = 2$ and $|-2| = 2$; if $x < 0$ $|x| > 0$.

To draw $y = |x|$, draw $y = x$ and reflect negative (dashed) parts of the graph in the x-axis (see figure, right).

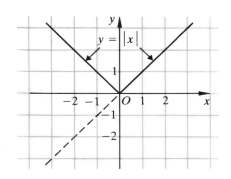

WORKED EXAMPLES

---

1   Draw the graphs of **(a)** $y = |x - 2|$ and **(b)** $y = |2x - 1|$.

(a) $y = |x - 2|$

Draw $y = x - 2$ and, where $y = x - 2$ is negative (dashed), change negative values to positive i.e. reflect dashed line in x-axis.

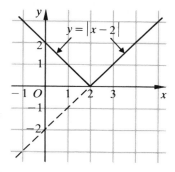

(b) $y = |2x - 1|$

Draw $y = 2x - 1$ and reflect negative part (dashed) in the x-axis.

The result is a V-shaped graph which is positive for all x (except $x = \frac{1}{2}$).

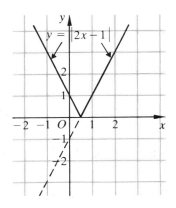

---

2   Use the graphs of $y = |x - 2|$ and $y = |x + 1|$ to **(a)**   solve $|x - 2| > |x - 1|$ and **(b)**   draw $y = |x - 2| + |x + 1|$.

(a)   We require the values of x for which $y = |x - 2|$ lies above $y = |x + 1|$ (dashed in figure, right).

This is true for $x < \frac{1}{2}$.

(b)   For $x > 2$, $|x - 2| = x - 2$ and $|x + 1| = x + 1$
so, $|x - 2| + |x + 1| = x - 2 + x + 1 = 2x - 1$

For $x < -1$, $|x - 2| = 2 - x$ and $|x + 1| = -x - 1$
so, $|x - 2| + |x + 1| = 2 - x - x - 1 = -2x + 1$

For $-1 < x < 2$, $|x - 2| = 2 - x$ and $|x + 1| = x + 1$
so, $|x - 2| + |x + 1| = 2 - x + x + 1 = 3$

The graph of $y = |x - 2| + |x + 1|$ is shown dotted.
    It is easier to plot the graph from the values in the diagram.

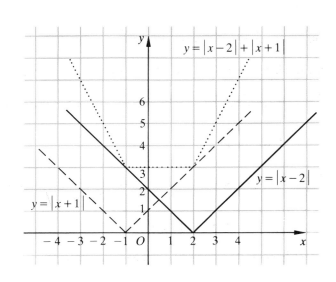

19 Motion in a circle

Introduction

How do we measure record speeds? Singles, L.P.s and the old 78s.

Do the tyres of a small car wear out before those of a large car?

Do racing cars have all four wheels replaced?

Which rotates faster the front (pedal) cog of a bicycle or the back cog?

How fast does a video-tape rotate?

How fast does an audio-tape rotate?

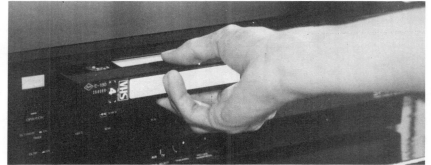

Interpret the washing machine spin speed.

How fast does the second hand of a watch rotate?

How fast does a kitchen blender rotate?

How fast does a car wheel rotate? Is this faster than wheels on a roller-skate or skate board? How is the (linear) speed of a car travelling at $50\,\mathrm{km\,h^{-1}}$ related to the rotational speed of the wheels?

Investigations

Estimate rotational speeds of everyday rotating objects. Express them in r.p.m., degrees per second and radians per second. (See Table, below.)

Note   1 revolution per minute = 2π radians per minute = $2\pi/60$ radians per second.
1 radian per second = 60 radians per minute = $60/2\pi$ r.p.m. $\simeq 9.55$ r.p.m.

Examples	r.p.m.	Degrees per second	Radians per second
A 'single' rotating at	45 r.p.m.	$270° \, \mathrm{s}^{-1}$	$\dfrac{3\pi}{2} \simeq 4.7^c \, \mathrm{s}^{-1}$
Second hand on watch	1 r.p.m.	$6° \, \mathrm{s}^{-1}$	$0.1^c \, \mathrm{s}^{-1}$
Car wheel at 30 m.p.h.			
Bicycle wheel at 10 m.p.h.			
Rollerskate at 5 m.p.h.			
Casette-tape on play			
Casette-tape on playback			
Electric drill			

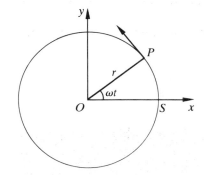

Circular motion with constant speed

Consider a point $P(x, y)$, rotating in a circle of radius r with constant angular speed ω radians per second.

If at time $t = 0$ the particle is at $S(r, 0)$, then at time t seconds, P has turned through an angle ωt and its co-ordinates are,

$$x = r \cos \omega t, \quad y = r \sin \omega t$$

$$OP = \mathbf{r} = x\mathbf{i} + y\mathbf{j} = r \cos \omega t \, \mathbf{i} + r \sin \omega t \, \mathbf{j}$$

$$\mathbf{v} = \frac{d\mathbf{r}}{dt} = -r\omega \sin \omega t \, \mathbf{i} + r\omega \cos \omega t \, \mathbf{j}$$

if ωt measured in radians (for differentiating).

The magnitude of $\mathbf{v}$ denoted by $v = \sqrt{(r^2\omega^2 \sin^2 \omega t + r^2\omega^2 \cos^2 \omega t)} = \sqrt{r^2\omega^2} = r\omega$

Since the vectors $a\mathbf{i} + b\mathbf{j}$ and $-b\mathbf{i} + a\mathbf{j}$ are perpendicular, $\mathbf{v}$ and $\mathbf{r}$ are perpendicular, which confirms that the velocity of P is along the tangent to the circle.

If r is measured in metres and ω in radians per second, then v is in metres per second.

To find the acceleration of P, we differentiate our expression for $\mathbf{v}$.

$$\mathbf{a} = \frac{d\mathbf{v}}{dt} = -r\omega^2 \cos \omega t \, \mathbf{i} - r\omega^2 \sin \omega t \, \mathbf{j} = -\omega^2(r\cos \omega t \, \mathbf{i} + r \sin \omega t \, \mathbf{j}) = -\omega^2 \mathbf{r}$$

This shows that the acceleration of a particle moving in a circle with constant angular speed ω, is directed towards the centre of the circle and has magnitude $a = r\omega^2$.

Since $v = r\omega$, $a = r\omega^2 = r\dfrac{v^2}{r^2} = \dfrac{v^2}{r}$, where v is the speed along the tangent.

The velocity is **not** constant as its direction (along tangent) is always changing.

WORKED EXAMPLES

---

1 Find the speed and acceleration of a point of the rim of a turn-table of radius 13 cm, rotating at $33\frac{1}{3}$ r.p.m.

$$r = 13\,\text{cm} = 0.13\,\text{m}, \quad \omega = 33\frac{1}{3}\,\text{r.p.m.} = \frac{100}{3} \times \frac{2\pi}{60}\,\text{rad s}^{-1} = 3.49\,\text{rad s}^{-1}$$

$$v = r\omega = 0.13 \times 3.49\,\text{m s}^{-1} = 0.45\,\text{m s}^{-1} \text{ along the tangent}$$

$$a = r\omega^2 = 0.13 \times (3.49)^2 = 1.58\,\text{m s}^{-2} \text{ towards the centre}$$

---

2 Compare the angular speeds of a rollerskate wheel moving at 5 m.p.h. and a car wheel moving at 30 m.p.h., the radii being 2 cm and 28 cm, respectively.

The angular speeds are proportional to the linear speeds and inversely proportional to the radii.

$$\frac{\omega\,\text{skate}}{\omega\,\text{car}} = \frac{v_s/r_s}{v_c/r_c} = \frac{v_s}{v_c} \times \frac{r_c}{r_s} = \frac{1}{6} \times \frac{28\,\text{cm}}{2\,\text{cm}} = \frac{7}{3} = 2\frac{1}{3}$$

So the rollerskate rotates $2\frac{1}{3}$ times faster than the car wheel.

---

3 Find the angular velocity of the vector $\mathbf{OP} = t\mathbf{i} + t^2\mathbf{j}$ about the origin at time t, in terms of t.

$$\mathbf{OP} = x\mathbf{i} + y\mathbf{j} \text{ and } OP \text{ makes an angle } \theta \text{ with } Ox \text{ where } \tan\theta = \frac{y}{x} = \frac{t^2}{t} = t$$

$$\text{Differentiating} \quad \Rightarrow \quad \sec^2\theta\,\frac{d\theta}{dt} = 1 \quad \Rightarrow \quad \omega = \frac{d\theta}{dt} = \frac{1}{\sec^2\theta} = \frac{1}{1 + \tan^2\theta} = \frac{1}{1 + t^2}$$

Exercise 19.1

1 Find the angular velocity of a 26″ (66 cm) bicycle wheel when the bicycle travels at 10 m.p.h.

2 Find the angular velocity of an old 78 record, and the speed of a point on the rim (diameter = 10 inches (25.4 cm))

3 The plate in my microwave cooker has diameter 30 cm and rotates three times in 6 seconds. What is the speed of a point on the rim of the plate?

4 At what time after noon do the hands of a clock next coincide?

5 On my exercise bicycle I did 32 revolutions of the pedals in 20 seconds and the speed dial indicated that I was theoretically moving at 20 m.p.h. Is this realistic?
   Does it help if I tell you that 32 pedal revolutions cover one tenth of a mile (according to the dial)?

6 The radius of the pedal on my exercise bicycle is 15 cm. While I was pedalling at 96 r.p.m. (32 revs in 20 seconds) what was the speed of my feet and what was their acceleration?

7 In my microwave oven I placed a pea 10 cm from the centre of the plate. What was its linear speed as it cooked? (the plate revolves at 1 rev per 2 seconds). What was its acceleration? Why didn't the pea move towards the centre of the plate? Why does food further from the centre, cook faster than food at the centre?

8 A particle moves in a circle whose equation is $x^2 + y^2 = 4$, with constant speed. At time 3 seconds, the particle has position vector $2\mathbf{j}$ and at time $t = 9$ its position vector is $-2\mathbf{j}$. Find its velocity, acceleration and angular velocity.

9   Find the angular velocity of a particle about the origin at time t, given that its position vector is given by

(a) $\mathbf{r} = \cos 3t\,\mathbf{i} + \sin 3t\,\mathbf{j}$ **(b)** $\mathbf{r} = t(\mathbf{i} - \mathbf{j})$ **(c)** $\mathbf{r} = t^2\mathbf{i} - 2t\mathbf{j}$ **(d)** $\mathbf{r} = \mathbf{i} + t\mathbf{j}$

10   Find the velocity of a point on the equator of the Earth (radius 6400 km) as the Earth spins about its axis, and also the velocity of a person in London (Lat. $51\frac{1}{2}°$ N).
   What happens to a person standing at the North Pole?

11   If the average car travels 10 000 miles per year, find the number of revolutions of the car's wheels in covering that distance. Compare **(a)**   The average car with **(b)**   a Rolls Royce and **(c)**   a Mini. You will have to find out the wheels' radii.

Angular acceleration

So far we have considered only situations where the angular (and tangential) speed is constant.

If my car decelerates uniformly from $10\,\text{m s}^{-1}$ to a speed of $30\,\text{m s}^{-1}$ in 10 seconds, its linear acceleration is given by

$$v = u + at \quad \Rightarrow \quad 30 = 10 + 10a \quad \Rightarrow \quad a = 2\,\text{m s}^{-2}$$

What is the angular acceleration of the wheels? (radius 20 cm)

You may find it easier to imagine the car stationary and the road moving backwards with the speed of the car, so that the hub of the wheel is stationary, but the wheels still rotate at the same rate.

$$v = r\omega \quad \Rightarrow \quad \text{when } u = 10\,\text{m s}^{-1}, \quad \omega_0 = \frac{u}{r} = \frac{10}{0.2} = 50\,\text{rad s}^{-1}$$

$$\text{when } v = 30\,\text{m s}^{-1}, \quad \omega = \frac{v}{r} = \frac{30}{0.2} = 150\,\text{rad s}^{-1}$$

The angular velocity increases uniformly from $50\,\text{rad s}^{-1}$ to $150\,\text{rad s}^{-1}$ in 10 seconds.

The angular acceleration, $\alpha = \dfrac{\text{increase in angular velocity}}{\text{change in time}} = \dfrac{150 - 50}{10} = 10\,\text{rad s}^{-2}$.

In general, $\alpha = \dfrac{\omega - \omega_0}{t} \quad \Rightarrow \quad \omega - \omega_0 = \alpha t \quad \Rightarrow \quad \boldsymbol{\omega = \omega_0 + \alpha t}$ (compare with $v = u + at$)

During the 10 seconds, my car covers a distance of $s = \dfrac{(u + v)t}{2} = \dfrac{(10 + 30)10}{2} = 200\,\text{m}$.

   The formula $s = ut + \frac{1}{2}at^2$ gives $s = 10 \times 10 + \frac{1}{2} \times 2 \times 10^2 = 200\,\text{m}$, as you would expect.

In covering 200 m, the car wheel turns through $\dfrac{200}{2\pi r}$ revolutions $= \dfrac{200}{2\pi r} \times 2\pi^c = 1000$ radians.

The average angular velocity is $\dfrac{50 + 150}{2} = 100\,\text{rad s}^{-1}$, and this multiplied by the time, 10 seconds, gives the total angular displacement (distance) as $100 \times 10 = 1000$ radians.

   Thus, the angle turned through, $\boldsymbol{\theta = \dfrac{\omega + \omega_0}{2}\,t.}$

Similarly, the formula $\theta = \omega_0 t + \frac{1}{2}\alpha t^2$ gives $\theta = 50 \times 10 + \frac{1}{2} \times 10 \times 10^2 = 1000$ radians.

The fourth formula (to compare with $v^2 = \omega^2 + 2as$) can be deduced from $\omega = \omega_0 + \alpha t$ and $\theta = \omega_0 t + \frac{1}{2}\alpha t^2$.

$$\omega^2 = (\omega_0 + \alpha t)^2 = \omega_0^2 + 2\omega_0 \alpha t + \alpha^2 t^2 = \omega_0^2 + 2\alpha(\omega_0 t + \frac{1}{2}\alpha t^2) = \boldsymbol{\omega_0^2 + 2\alpha\theta}$$

We now have formulae in rotational motion to compare with linear motion, as long as the acceleration is uniform (constant).

Linear motion	(Constant acceleration)	Rotational motion
$v = u + at$		$\omega = \omega_0 + \alpha t$
$s = \dfrac{(u+v)t}{2}$		$\theta = \dfrac{(\omega_0 + \omega)t}{2}$
$s = ut + \frac{1}{2}at^2$		$\theta = \omega_0 t + \frac{1}{2}\alpha t^2$
$v^2 = u^2 + 2as$		$\omega^2 = \omega_0^2 + 2\alpha\theta$
$s = vt - \frac{1}{2}at^2$		$\theta = \omega t - \frac{1}{2}\alpha t^2$

A particle moving in a circle with uniform angular acceleration, will have an acceleration towards the centre (equal to $r\omega^2$) and also a tangential acceleration.

WORKED EXAMPLE

A turn-table takes 3 seconds to reach its running speed of $33\frac{1}{3}$ r.p.m., starting from rest. Find its acceleration (assumed constant) and the angle turned through during this time.

$$\omega_0 = 0; \quad \omega = \frac{100}{3} \times \frac{2\pi}{60} \text{ rad s}^{-1}; \quad t = 3 \text{ s}$$

$$\alpha = \frac{\omega - \omega_0}{t} = \frac{10\pi}{27} = 1.16 \text{ rad s}^{-2}$$

$$\theta = \frac{(\omega_0 + \omega)t}{2} = \frac{200\pi}{180} \times \frac{3}{2} = \frac{10\pi}{6} \text{ radians} = \frac{10\pi}{6 \times 2\pi} = \frac{5}{6} \text{ revolution}$$

Exercise 19.2

1   The plate in my revolving microwave oven takes 2 seconds to reach its operating speed of 2 revs per second from the moment I switch it on. If it accelerates uniformly, find **(a)** the acceleration and **(b)** the angle turned through during this time; **(c)** if the plate takes double the time to stop, find the deceleration (assumed constant) and **(d)** the angle turned through while it is slowing down.

2   While mending a bicycle puncture with my bicycle upside-down, I turned the pedals for 5 s, accelerating uniformly until the wheel speed was 60 rad s^{-1}, maintained this speed for 10 s and then let the pedals go. The back wheel took 50 s to come to rest, decelerating uniformly. Find the number of revolutions of the wheel.

3   On my exercise bicycle I can reach my top speed of 120 revolutions per minute in 20 seconds from rest. What is my acceleration (assumed constant) in revs min^{-2} and in rad $\sec^{-2}$?

4   A wheel completes 6 complete revolutions in 3 seconds while rotating with constant angular acceleration starting from rest. Find the acceleration and the angular velocity at the end of 3 seconds.

5   A wheel decelerates, changing its speed from $5 \,\mathrm{rad\,s^{-1}}$ to $3 \,\mathrm{rad\,s^{-1}}$ in 4 seconds. Find **(a)** the deceleration (assumed constant), **(b)** the time it takes in coming to rest from $5 \,\mathrm{rad\,s^{-1}}$, **(c)** the number of revolutions it makes in total.

6   A particle starting from rest moves in a circle with constant angular acceleration of $\pi \,\mathrm{rad\,s^{-2}}$. Find the angle it turns through during the second and third seconds.

Forces causing motion in a horizontal circle

We have seen that a particle moving in a circle radius r, with constant angular speed ω, has a tangential velocity of magnitude $v = r\omega$, and an acceleration towards the centre of the circle equal to $r\omega^2$ or $\dfrac{v^2}{r}$.

It follows from Newton's second law ($\mathbf{F} = m\mathbf{a}$), that there must be a force producing this acceleration towards the centre of the circle.

WORKED EXAMPLE

A particle of mass m is attached to the centre of a smooth table by a light inextensible string. If it is moving in a horizontal circle with speed $2 \,\mathrm{m\,s^{-1}}$ and the string is 1 m long, find the tension in the string.

In the vertical direction, there is no motion so the forces balance.

The reaction $R = mg$ (the weight).

Horizontally, the particle has an acceleration towards the centre of

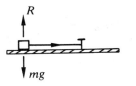
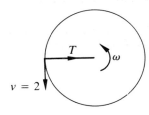

$$a = \frac{v^2}{r} = \frac{2^2}{1} = 4 \,\mathrm{m\,s^{-2}}$$

The force producing the acceleration is the tension in the string, so $T = ma = \dfrac{mv^2}{r} = 4m \,\mathrm{N}$

Exercise 19.3

1   A particle of mass m, moves in a horizontal circle of radius r, with angular velocity ω and tangential velocity v.

   (a)   If $m = 10 \,\mathrm{kg}$, $r = 2 \,\mathrm{m}$ and $v = 3 \,\mathrm{m\,s^{-1}}$, find ω and the force F towards the centre.

   (b)   If $m = 10 \,\mathrm{kg}$, $r = 4 \,\mathrm{m}$ and $\omega = 3 \,\mathrm{rad\,s^{-1}}$, find v and F.

   (c)   If $m = 10 \,\mathrm{kg}$, $r = 5 \,\mathrm{m}$ and $F = 8 \,\mathrm{N}$, find v and ω.

2   A car of mass $800 \,\mathrm{kg}$ is negotiating a roundabout of radius $20 \,\mathrm{m}$. If its speed is $30 \,\mathrm{m\,s^{-1}}$, find the friction force between the wheels and the road. If the coefficient of friction between the wheels and the road is 0.5, find the maximum speed the car can travel around the roundabout with no tendency to slip or skid sideways.

3   My microwave oven rotates at 2 revs per second. A pea placed $10 \,\mathrm{cm}$ from the centre of the rotating plate is on the point of slipping outwards when the plate rotates. Calculate the coefficient of friction between the pea and the plate.

4 When I drive my car around the right-angled bend at the end of my road, my packet of sweets on the dashboard slips unless I keep my speed down to $10\,\mathrm{m\,s^{-1}}$. Find the coefficient of friction, given that the bend is an arc (90°) of a circle of length $30\,\mathrm{m}$.

5 A hammer-thrower twirls his hammer (7 kg shot on a wire of length 1 m) in a circular path before letting go. Assuming the hammer is projected at 45° (for maximum range) and travels 70 m, find the speed of projection and the tension in the wire before release.

6 Repeat Question **5** for a discus-thrower achieving 60 m with a discus weighing 1 kg. What is the force (tension) in this thrower's arm (of length 1 m)?

Conical pendulum

You may have seen a tennis trainer consisting of a tennis ball attached by a string to a post where the player can hit the ball alternately backhand and forehand, causing it to rotate around the post, following the path of a horizontal circle.

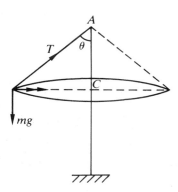

The acceleration of the ball is towards the centre of the circle C (post) and the force producing this acceleration will be a component of the tension in the string connecting the ball to the post at A (figure, right).

The ball moves in a horizontal circle, so in the vertical direction there is no movement and the forces balance.

$$\uparrow \quad T\cos\theta = mg \quad \Rightarrow \quad T = mg\sec\theta$$

In the horizontal direction, the force towards the centre of the circle produces the acceleration towards the centre, so using $F = ma$,

$$\longrightarrow \quad T\sin\theta = ma = mr\omega^2 \text{ or } \frac{mv^2}{r}$$

As the ball rotates around the post the string traces out the surface of a cone, and this constitutes an example of a **conical pendulum**.

> WORKED EXAMPLES

1 For the tennis trainer above, the string has length 1 m and the angle it makes with the vertical is 60°. Find the speed of the ball and the tension in the string if the ball has mass 0.15 kg.

$$\uparrow \quad T\cos\theta = mg \quad \text{and} \quad \leftrightarrow \quad T\sin\theta = \frac{mv^2}{r} \quad \Rightarrow \quad \frac{T\sin\theta}{T\cos\theta} = \frac{mv^2}{mgr} = \frac{v^2}{gr}$$

$$r = 1\times\sin 60° = \frac{\sqrt{3}}{2}m \quad \Rightarrow \quad \tan 60° = \frac{v^2}{\frac{\sqrt{3}}{2}g} \quad \Rightarrow \quad v^2 = \frac{\sqrt{3}}{2}g\tan 60° = \frac{\sqrt{3}}{2}\times 10\times\sqrt{3} = 15$$

$$\Rightarrow \quad v = 3.87\,\mathrm{m\,s^{-1}}$$

$$T = \frac{mg}{\cos\theta} = \frac{0.15\times 10}{\frac{1}{2}} = 3\,\mathrm{N}$$

2 A particle of mass 5 kg on the end of an inextensible string of length 1.3 m describes a horizontal circle of radius 1.2 m, the centre of the circle being directly below the point of suspension. Find its speed and the tension in the string.

With the notation in the figure,

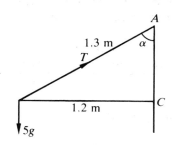

$$\uparrow \quad T\cos\alpha = 5g \quad \Rightarrow \quad T = \frac{5g}{\cos\alpha} = \frac{5g}{\frac{5}{13}}$$

$$\text{(since } AC = 0.5 \quad (5, 12, 13 \,\triangle\,))$$

$$\Rightarrow \quad T = 130\,\text{N}$$

$$\longrightarrow \quad T\sin\alpha = \frac{mv^2}{r} = \frac{5v^2}{1.2} \quad \Rightarrow \quad 5v^2 = 1.2 \times 130 \times \frac{12}{13}$$

$$\Rightarrow \quad 5v^2 = 1.2 \times 120 = 144 \quad \Rightarrow \quad v^2 = \frac{144}{5} \quad \Rightarrow \quad v = 5.4\,\text{m s}^{-1}$$

3 A new tennis trainer has been designed to keep the ball at the same height from the ground, 1 m. The ball (mass 0.15 kg) is attached to the post with two strings, *BA* and *BD* (both of length 1 m), and *DG* = 0.5 m. Find the least velocity required to keep the ball rotating at this height and the tension in the strings at double this speed.

When both strings are taut, $DG = 0.5 = DC = CA$ (by symmetry) and

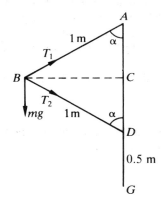

$$\cos\alpha = \frac{0.5}{1} \quad \Rightarrow \quad \alpha = 60°$$

If $\alpha < 60°$, the lower string, *BD*, is slack $\Rightarrow T_2 = 0$
The critical speed occurs when T_2 just becomes 0.

$$\uparrow \quad T_1\cos 60° = mg + T_2\cos 60° \quad \text{and} \quad T_2 = 0 \quad \Rightarrow \quad T_1 = 2mg$$

$$\longleftarrow \quad T_1\sin 60° + T_2\sin 60° = \frac{mv^2}{r} \quad \text{and} \quad T_2 = 0 \quad \Rightarrow \quad 2mg\sin 60° = \frac{mv^2}{\sin 60°}$$

$$\Rightarrow \quad v^2 = 20\sin^2 60° = 15 \quad \Rightarrow \quad v = 3.87\,\text{m s}^{-1}$$

For double speed, $v = 7.74\,\text{m s}^{-1} \quad \Rightarrow \quad v^2 = 60$

$$\uparrow \quad T_1\cos 60° - T_2\cos 60° = mg \qquad \Rightarrow \quad T_1 - T_2 = 2mg = 3 \left.\begin{array}{l}\\\\\end{array}\right\} \quad \begin{array}{l} T_1 = 7.5\,\text{N} \\\\ T_2 = 4.5\,\text{N}\end{array}$$

$$\leftrightarrow \quad (T_1 + T_2)\sin 60° = \frac{mv^2}{r} = \frac{m \times 60}{\sin 60°} \quad \Rightarrow \quad T_1 + T_2 = \frac{60 \times 0.15}{\frac{3}{4}} = 12 \quad \Rightarrow$$

Exercise 19.4

---

1 A particle of mass 4 kg is attached by a light inextensible string of length 80 cm to a fixed point. The particle moves in a horizontal circle with constant speed, with the string making an angle of $\theta = \sin^{-1}\frac{4}{5}$ with the vertical. **(a)** Find the speed of the particle and the tension in the string, **(b)** if $v = 4\,\text{m s}^{-1}$, find the angle θ and the tension in the string.

2 An inextensible string will support a mass of 2 kg without breaking. With half of this mass on the string (of length 0.5 m), how fast can the mass move in a conical pendulum before the string breaks?

3 A sling of length 1 m (remember David and Goliath!) projects a stone of mass 0.1 kg from a leather pouch at the end of a string (assume radius 2 m, arm plus sling). To throw a stone 50 m (at 45° for maximum range), how fast is the sling travelling in its circular path before projection? (Compare with hammer-throwing problem.) What is the tension in the sling just before the stone is released?

4   Another new tennis trainer (see figure in Worked Example **3**) has strings of length 0.8 m and 0.6 m (lower) tied to the post at A and D, where $AD = 1$ m. Find the least velocity for both strings to be taut and the tensions when this least velocity is trebled.

5   A particle moves in a horizontal circle with constant angular velocity 3 rad s^{-1} on the smooth inner surface of a sphere. Find the depth of the circle below the centre of the sphere.

6   Interpret the motion of a particle whose position vector at time t is $\mathbf{r} = 3 \cos 2t\,\mathbf{i} + 3 \sin 2t\,\mathbf{j} + 4\mathbf{k}$.

7   Interpret the motion of a particle whose position vector at time t is $\mathbf{r} = 2 \cos t\,\mathbf{i} + 2 \sin t\,\mathbf{j} + 4t\,\mathbf{k}$.

Banked tracks

As a car travels around a roundabout it moves in a circle, so its acceleration must be towards the centre of the circle (see figure, right). The friction of the tyres on the ground produces the necessary inwards force to produce the inwards acceleration. If there is no inwards force, there is a tendency for the car to move outwards, as there is for a parcel lying on the back seat.

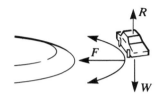

To help the car negotiate the corner (especially on racing circuits) the road (or track) is banked to produce a force towards the centre from the reaction (second figure, right).

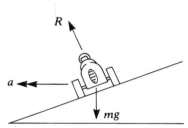

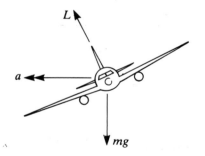

Similarly an aircraft 'banks' to turn so that a component of the **lift** force on its wings acts inwards towards the centre of its turning circle (see figure, left).

Railway tracks, also, are canted towards the centre of a curve (the outer rail is higher) to reduce the pressure on the rails.

Cyclists, motorcyclists, skiers and water skiers lean inwards to negotiate bends, so that the reaction force has a component inwards towards the centre of the bend.

WORKED EXAMPLES

---

1 The corner of a racetrack in the shape of a semi-circle of radius 100 m, is banked at 30° to the horizontal. What is the speed at which a car rounds the bend with no tendency to side-slip?

If the car travels too fast it will tend to slip up the bank, while if it travels too slowly it tends to slide down the bank.

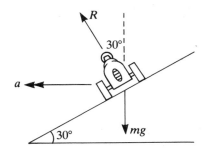

$\uparrow$ no acceleration $\Rightarrow$ $R \cos 30° = mg$ $\cdots\cdots$ **(1)**

$\leftarrow$ $F = ma$ $\Rightarrow$ $R \sin 30° = \dfrac{mv^2}{r} = \dfrac{mv^2}{100}$ $\cdots\cdots$ **(2)**

Dividing **(2)** by **(1)** $\Rightarrow$ $\tan 30° = \dfrac{v^2}{gr}$

$\Rightarrow$ $v^2 = gr \tan 30°$

$\Rightarrow$ $v^2 = 10 \times 100 \times \tan 30° = 577.4$

$\Rightarrow$ $v = 24.0 \,\text{m s}^{-1} \simeq 54.1 \,\text{m.p.h.}$

---

2 On the level, a car rounds a bend of radius 100 m at a speed of $20 \,\text{m s}^{-1}$ without slipping. At what angle must the road be banked so that a speed of $30 \,\text{m s}^{-1}$ is achieved without slipping, if the coefficient of friction is the same in both cases?

On the level, the friction force provides the inwards force (see upper figure, right).

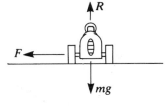

$\uparrow$ $R = mg$ $\leftarrow$ $F = \dfrac{mv^2}{r}$ and $F = \mu R = \mu mg$

$\therefore \dfrac{mv^2}{r} = \mu mg$ $\Rightarrow$ $\mu = \dfrac{v^2}{gr} = \dfrac{20^2}{1000} = 0.4$

With banking (see lower figure, right) the friction force F acts down the slope.

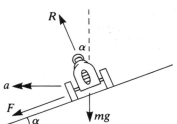

$\uparrow$ $R \cos \alpha = mg + F \sin \alpha$ $\cdots\cdots$ **(1)**

$\leftarrow$ $R \sin \alpha + F \cos \alpha = \dfrac{mv^2}{r}$ $\cdots\cdots$ **(2)**

$F = \mu R = 0.4R$ in **(1)** $\Rightarrow$ $R \cos \alpha - 0.4R \sin \alpha = mg$

$F = 0.4R$ in **(2)** $\Rightarrow$ $R \sin \alpha + 0.4R \cos \alpha = \dfrac{mv^2}{r}$

$\left. \phantom{\dfrac{mv^2}{r}} \right\}$ $\Rightarrow$ $\dfrac{\sin \alpha + 0.4 \cos \alpha}{\cos \alpha - 0.4 \sin \alpha} = \dfrac{v^2}{gr} = \dfrac{900}{1000}$

Cross-multiplying $\Rightarrow$ $10 \sin \alpha + 4 \cos \alpha = 9 \cos \alpha - 3.6 \sin \alpha$ $\Rightarrow$ $13.6 \sin \alpha = 5 \cos \alpha$

$\Rightarrow$ $\tan \alpha = \dfrac{5}{13.6}$ $\Rightarrow$ $\alpha = 20.2°$

Exercise 19.5

1 A road banked at 15° is curved in the arc of a circle of radius 80 m. At what speed can a car travel round the bend with no tendency to side-slip?

2 At what angle should an aircraft bank when flying at $120 \, \text{m s}^{-1}$ in a horizontal circle of radius 3000 m?

3 A car negotiates a curve of radius 100 m sloped at $\tan^{-1} \frac{1}{4}$ to the horizontal. At what speed can the car travel with no slipping?

4 A car rounds a bend of radius 80 m at $20 \, \text{m s}^{-1}$ with no slipping. Find the coefficient of friction. How fast could the car travel round a bend of 100 m banked at 20°, with the same coefficient of friction, with no slipping?

5 A train of mass 70 000 kg travels at $50 \, \text{km h}^{-1}$ round a bend of radius 1000 m. If the track is level, find the lateral thrust on the outer rail. At what height above the inner rail should the outer rail be set to eliminate sideways thrust at this speed, if the rails are 1.4 m apart?

6 A circular racetrack bend of radius 200 m is banked at 40°. At what speed does a car have no tendency to side-slip? If the coefficient of friction between the wheels and track is $\frac{1}{3}$, find the maximum speed for the car to negotiate the bend without slipping.

7 A railway track is banked on a circular curve of radius 600 m so that there is no sideways force on the rails when a train of mass 10^5 kg travels at $40 \, \text{km h}^{-1}$ around the curve. Find the lateral force when the train travels at **(a)** $30 \, \text{km h}^{-1}$, **(b)** $50 \, \text{km h}^{-1}$ and **(c)** is at rest on the curve.

Answers

Chapter 1

Exercise 1.1

1 $x = \dfrac{y-3}{2}$ **2** $r = \dfrac{C}{2\pi}$ **3** $r = \sqrt{\dfrac{A}{\pi}}$ **4** $h = \dfrac{S}{2\pi r}$ **5** $h = \dfrac{V}{\pi r^2}$ **6** $r = \sqrt{\dfrac{V}{\pi h}}$ **7** $h = \dfrac{A - 2\pi r^2}{2\pi r}$

8 $a = \dfrac{v-u}{t}$ **9** $u = \dfrac{s - \frac{1}{2}at^2}{t}$ **10** $s = \dfrac{v^2 - u^2}{2a}$ **11** $f = \dfrac{uv}{u+v}$ **12** $u = \dfrac{fv}{v-f}$ **13** $x = \dfrac{h}{3P-2}$

14 $x = \dfrac{y+1}{y-1}$ **15** $r = \sqrt{\dfrac{A + \pi h^2}{\pi}} - h$ **16** $t = \dfrac{\sqrt{2as - u^2}}{a} - \dfrac{u}{a}$

Exercise 1.2

1 $x = 1\frac{1}{2}$ **2** Many: $x = 1, y = 4$ or $(2,2)$ or $(3,0)$ **3** -1 **4** 1.22 **5** $x = 1, y = 4$

6 $x = 0$ or -2 **7** 1 or -5 **8** 1 or 4 **9** -1 or -2 **10** 0 or -3 **11** 1

12 -0.57 or -1.77 **13** all x

Exercise 1.3

1 $x > 3.5$ **2** $x < -1$ **3** $x \geqslant -18$ **4** $x \geqslant 1$ **5** $0 < x < \frac{1}{3}$ **6** $0 < x < 2$ **7** $x > 4$ or $x < 0$

8 $0 < x < 1$ **9** $x > 2$ or $x < -2$ **10** $x > 2$ or $x < -2$

Exercise 1.4

1 $x = \frac{3}{2}$ **2** (A) $x - y = -3$ (B) $2x + 4y = 18$ (C) correct   (D) $-3x = -3$

(E) $x + 2y = 9$ drawn wrongly   **3** $x + 2 = \pm 3$ (2 solutions)   **4** first $\Rightarrow$ wrong

5 $x = -5$ (not -1)   **6** $(x+2)^2 \neq x^2 + 4$ **7** wrong working

Exercise 1.5

1 16 **2** 16 **3** 64 **4** 81 **5** x^5 **6** a^{14} **7** b^6 **8** b^6 **9** c^8 **10** $2^4 = 16$

11 2^{x+1} **12** d^4 **13** e^2 **14** $f^1 = f$ **15** $g^0 = 1$ **16** h^{-1} **17** j^5 **18** k^5 **19** m

20 n^7 **21** $4^{(3^2)} = 4^9 = (2^2)^9 = 2^{18}$ or $(4^3)^2 = 64^2 = (2^6)^2 = 2^{12}$

Exercise 1.6

1 p^{-1} **2** q^2 **3** q^2 **4** r^4 **5** s^{-4} **6** t^{-6} **7** t^{-6} **8** x^5 **9** $x^0 = 1$ **10** $x^2 y^2$

11 $x^2 y^3$ **12** y **13** $\frac{1}{9}$ **14** $\frac{1}{4}$ **15** $5^0 = 1$ **16** $\frac{1}{4}$ **17** $2\frac{1}{4}$ **18** $6\frac{1}{4}$ **19** $\frac{4}{9}$ **20** $\dfrac{1}{x}$

21 $\dfrac{1}{x^2}$ **22** $\dfrac{1}{x^m}$ **23** $a^1 = a$ **24** $a^1 = a$ **25** 1 **26** 0 **27** $\to 1$

Exercise 1.7

1 2^6 **2** 5^6 **3** a^6 **4** a^6 **5** x^8 **6** x^8 **7** b^{21} **8** 2^{-2} **9** 2^{-6}

Exercise 1.8

1 ± 2 **2** ± 6 **3** 3 **4** $\frac{2}{3}$ **5** ± 3 **6** x^2 **7** 10 **8** 2 **9** $\frac{1}{7}$ **10** $\pm\frac{1}{2}$

11 ± 2 **12** ± 1.5 **13** ± 3.5 **14** 1.25 **15** $\pm\frac{2}{3}$ **16** ± 48 **17** ± 48 **18** $\pm\frac{2}{3}$

19 ± 14 or ± 2 **20** ± 10 **21** ± 5 **22** $x^{\frac{5}{6}}$ **23** 9×3^x **24** 3^{2x+1} **25** 4×3^x

26 $2^x(2^x - 2)$ **27** 2^{3x+1} **28** 2^{x-1} **29** x **30** 8

Exercise 1.9

1 ± 27 **2** ± 8 **3** 4 **4** 16 **5** 9 **6** ± 27 **7** $\pm\frac{1}{3}$ **8** $\pm\frac{1}{27}$ **9** ± 32

10 $\frac{1}{25}$ **11** $\frac{4}{9}$ **12** $2\frac{1}{4}$ **13** $\pm\frac{64}{27}$ **14** ± 0.064 **15** ± 0.001 **16** ± 0.5 **17** 0.5

18 $\frac{1}{4}$ **19** x^3 **20** x^2 **21** a^6 **22** $a^3 b^3$ **23** $a^4 b^4 c^4$ **24** $\pm(a + b)$ **25** x^2 **26** x

27 x **28** x^2 **29** $x + 1$ **30** $x(x + 1)^{\frac{1}{2}}$ **31** $x + 1$ **32** $x^{\frac{1}{2}} + x^{-\frac{1}{2}}$ **33** $1 + \dfrac{1}{x} = \dfrac{x + 1}{x}$

34 2.5 **35** $x^{\frac{1}{2}} - x^{-\frac{1}{2}}$ **36** $1 + \dfrac{1}{x}$

Exercise 1.10

1 **(a)** $4 = \log_2 16$ **(b)** $2 = \log_4 16$ **(c)** $\log_4 64 = 3$ **(d)** $\log_2 32 = 5$ **(e)** $\log_2 1 = 0$

   (f) $7^2 = 49$ **(g)** $3^3 = 27$ **(h)** $10^2 = 100$ **(i)** $2^x = 8$ **(j)** $a^c = b$

   (k) $3^0 = 1$ **(l)** $\log_2 \frac{1}{2} = -1$ **(m)** $\log_3 \frac{1}{9} = -2$ **(n)** $x = \log_e y$ **(o)** $x = e^y$

2 **(a)** 2 **(b)** 3 **(c)** 4 **(d)** 5

Exercise 1.11

1 **(a)** $\log 24$ **(b)** $\log\frac{8}{3}$ **(c)** $\log\frac{9}{8}$ **(d)** $\log 2$ **(e)** $\log 2$ **(f)** $\log 5$ **(g)** $\log 3$

   (h) $\log 27$ **(i)** $\log 2$

2 **(a)** $\log a$ **(b)** $\log 6a^2$ **(c)** $\log(1 - x^2)$ **(d)** 0 **(e)** $\log y$ **(f)** $\log(a - b)$ **(g)** 0

   (h) $\log\dfrac{a^2}{b}$ **(i)** $\log b^3$

3 **(a)** 1.585 **(b)** 2.465 **(c)** 2.16 **(d)** 2 **(e)** 0.898 **(f)** 2.322

4 **(a)** 0 **(b)** 1 **(c)** 0 **(d)** ξ **(e)** 0 or 1 **(f)** $+0.694$

Exercise 1.12

1 **(a)** $2\sqrt{2}$ **(b)** $3\sqrt{3}$ **(c)** $5\sqrt{2}$ **(d)** $2\sqrt{5}$ **(e)** $4\sqrt{3}$ **(f)** $6\sqrt{2}$ **(g)** $7\sqrt{2}$

   (h) $5\sqrt{5}$ **(i)** $8\sqrt{2}$ **(j)** $a\sqrt{2}$ **(k)** $a\sqrt{a}$ **(l)** $r\sqrt{\pi}$

2 **(a)** $2 + \sqrt{2}$ **(b)** $3 + 2\sqrt{2}$ **(c)** 1 **(d)** $6 + 3\sqrt{3} + 2\sqrt{2} + \sqrt{6}$ **(e)** 1 **(f)** 141

   (g) 1 **(h)** $4 - 6\sqrt{5}$ **(i)** $4 + 3\sqrt{2}$ **(j)** $4 + 3\sqrt{2}$ **(k)** 2 **(l)** 6

Answers

3 (a) $\dfrac{\sqrt{2}}{2}$ (b) $\dfrac{\sqrt{3}}{3}$ (c) $\sqrt{2}+1$ (d) $\sqrt{2}-1$ (e) $2-\sqrt{3}$ (f) $\frac{1}{7}(3-\sqrt{2})$ (g) $\sqrt{5}+2$

(h) $\sqrt{3}+1$ (i) $3-\sqrt{3}$ (j) $\sqrt{3}$ (k) $3+2\sqrt{2}$ (l) $\sqrt{2}$ (m) 1 (n) 1

(o) 1 (p) 1

Exercise 1.13

1 $\dfrac{1}{3(x+1)}-\dfrac{1}{3(x+4)}$ 2 $\dfrac{1}{3(x-2)}-\dfrac{1}{3(x+1)}$ 3 $\dfrac{8}{x-3}-\dfrac{7}{x-2}$

4 $\dfrac{5}{4(x-2)}+\dfrac{3}{4(x+2)}$ 5 $1+\dfrac{2}{3(x-1)}-\dfrac{5}{3(x+2)}$ 6 $\dfrac{2}{x+2}-\dfrac{1}{2x+1}$

Exercise 1.14

1 $\dfrac{2}{x-1}+\dfrac{1}{x+1}$ 2 $\dfrac{3}{x+2}+\dfrac{1}{x-2}$ 3 $\dfrac{2}{3x-1}+\dfrac{3}{x-3}$

4 $2+\dfrac{1}{x-2}-\dfrac{2}{x+1}$ 5 $\dfrac{5}{6(x+1)}+\dfrac{4}{15(x-2)}+\dfrac{11}{10(x+3)}$ 6 $\dfrac{2}{x-2}-\dfrac{1}{x+1}-\dfrac{2}{2x-1}$

7 $1-\dfrac{4}{x-1}+\dfrac{9}{x-2}$ 8 $x+6-\dfrac{8}{x-1}+\dfrac{27}{x-2}$

Chapter 2

Exercise 2.1

1 (a) $0.\dot{1}, 0.\dot{2}, 0.\dot{3}, 0.\dot{4}, 0.\dot{5}, 0.\dot{6}, 0.\dot{7}, 0.\dot{8}, 0.\dot{9}$ (!!)

(b) $0.\dot{1}4285\dot{7}, 0.\dot{2}8571\dot{4}, 0.\dot{4}2857\dot{1}, 0.\dot{5}7142\dot{8}, 0.\dot{7}1428\dot{5}, 0.\dot{8}5714\dot{2}$

(c) $0.0\dot{9}, 0.1\dot{8}, 0.2\dot{7}, 0.3\dot{6}, 0.4\dot{5}, 0.5\dot{4}, 0.6\dot{3}, 0.7\dot{2}, 0.8\dot{1}, 0.9\dot{0}$

2 (a) 3.1415927 (b) $3.\dot{1}4285\dot{7}$ (c) 3.1415929

Exercise 2.2

1 $\mathbb{Z}$ 2 $\mathbb{F}$ 3 $\mathbb{Z}$ 4 $\mathbb{Z}$ 5 $\mathbb{C}$ 6 $\mathbb{Q}$ 7 $\mathbb{Q}$ 8 $\mathbb{Q}$ 9 $\mathbb{Z}$ 10 $\mathbb{C}$ 11 $\mathbb{C}$

12 $\mathbb{C}$ 13 $\mathbb{Z}$ 14 $\mathbb{Z}$ 15 $\mathbb{Z}$ 16 $\dfrac{1}{3}, \dfrac{3}{25}, \dfrac{4}{50}, \dfrac{5}{16}, \dfrac{4}{125}$ 17 $\dfrac{2}{3}, \dfrac{4}{11}, \dfrac{1}{27}, \dfrac{6}{7}$

18 Assume $\sqrt{5}=\dfrac{p}{q}$ $\Rightarrow$ $5q^2=p^2$ $\Rightarrow$ $p=5n$ $\Rightarrow$ $5q^2=25n^2$ $\Rightarrow$ $q^2=5n^2$ $\Rightarrow$ $q-5m$ $\Rightarrow$ $\dfrac{p}{q}=\dfrac{5n}{5m}$; contradiction

19 $\mathbb{N}\subset\mathbb{Z}\subset\mathbb{Q}\subset\mathbb{R}$ 20 (a) $\mathbb{N}$ (b) $\mathbb{Q}$ (c) $\mathbb{Z}$

Exercise 2.3

1 (a) Many to one (depending on how you measure age)
  (b) Many to one or one to one (depending on domain) (c) Many to one
  (d) Many to one (e) Many to many (depending on domain)

2 (a) $1\to1, 2\to3, 3\to5, 4\to7, 5\to9$
  (b) $1\to1, 2\to2, 3\to3, 4\to5, 5\to7$, (counting 1 as prime)

3 $1\to1, 2\to2, 3\to3, 4\to2, 5\to5, 6\to2$ and $3, 7\to7, 8\to2, 9\to3$

4 (a) One↔one (b) One↔one (c) One↔many

5 (a) Many↔one (b) One↔one

315

Answers

Exercise 2.4

1 **(a)** enlargement by 3　**(b)** multiply by 3　**(c)** enlargement by $\frac{1}{3}$　**(d)** divide by **3**　**(e)** $\frac{x}{3}$

2 **(a)** increase by 2　**(b)** add 2　**(c)** decrease by 2　**(d)** subtract 2　**(e)** $x - 2$

3 **(a)** decrease by 3　**(b)** subtract 3　**(c)** increase by 3　**(d)** add 3　**(e)** $x + 3$

4 **(a)** enlargement by $\frac{1}{4}$　**(b)** divide by 4　**(c)** enlargement by 4　**(d)** multiply by 4　**(e)** $4x$

5 **(a)** invert through O　**(b)** change sign　**(c)** invert through O　**(d)** change sign　**(e)** $-x$

6 **(a)** invert through 3　**(b)** subtract from 6　**(c)** invert through 3　**(d)** subtract from 6　**(e)** $6 - x$

7(a) and (c)

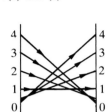

(b) and (d)
Divide into 1

(e) $\frac{1}{x}$

8(a) and (c)

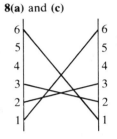

(b) and (d)
Divide into 6

(e) $\frac{6}{x}$

9(a) and (c)

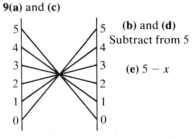

(b) and (d)
Subtract from 5

(e) $5 - x$

Exercise 2.5

1 ggf (or fg)　**2** $fggf$　**3** fg　**4** g^{-1}　**5** fg^{-1}　**6** f^{-1}　**7** $f^{-1}g$　**8** $g^{-1}f$　**9** g

10 $4(x + 1)$　**11** $4x + 1$　**12** $2x + 2$　**13** $2x + 3$　**14** $4x + 2$　**15** $8x$　**16** x　**17** x

18 $\frac{1}{2}x - 1$　**19** $\frac{1}{2}x - 1$　**20** $\frac{1}{2}(x - 1)$　**21** $\frac{1}{2}(x - 1)$　**22** $6x + 4$　**23** $6x + 2$　**24** $3x + 3$

25 $3x + 5$　**26** $(x + 1)^2$　**27** $x^2 + 1$　**28** $2x^2$　**29** $4x^2$　**30** $3x^2 + 2$　**31** $(3x + 2)^2$

32 $\frac{1}{3}(x - 2)$　**33** $\frac{1}{3}(x + 1)$　**34** $\frac{1}{3}(x - 3)$　**35** $\frac{1}{6}(x - 4)$　**36** $\frac{1}{6}(x - 2)$　**37** $\frac{1}{6}(x - 4)$

38 **(a)** 1　**(b)** 2　**(c)** 3　**(d)** x　**(e)** $S(x)$ is self-inverse

39 **(a)** 1　**(b)** 2　**(c)** 3　**(d)** x

Exercise 2.6

1 $x - 7$　**2** $\frac{1}{2}(x + 1)$　**3** $\frac{1}{2}(x - 3)$　**4** $\frac{1}{3}(x - 2)$　**5** $\frac{1}{3}x - 2$　**6** $\frac{1}{2}x + 3$　**7** $-x$　**8** $\frac{1}{x}$

9 $9 - x$　**10** $\frac{12}{x}$　**11** $9 - \frac{12}{x}$　**12** $\frac{12}{9 - x}$　**13** **(a)** $\frac{x - 2}{3}$　**(b)** x　**(c)** x

　(d) $f(x) = y \;\Rightarrow\; x = f^{-1}(y);\; ff^{-1}(y) = f(x) = y$

Exercise 2.7

1 $\dfrac{4x - 2}{3 - x}$　**2** $\dfrac{4x - 3}{x - 2}$　**3** $\dfrac{x}{x - 2}$　**4** $\dfrac{3 + x}{x}$　**5** $\dfrac{3}{x - 2}$　**6** $\dfrac{4x - 3}{2}$　**7** $\dfrac{2}{x - 5} + 1$　**8** $\dfrac{x - 4}{x - 3}$

9 $2 - \dfrac{1}{x}$　**10** $\dfrac{x - 10}{x - 7}$　**11** $\dfrac{x - 10}{x - 7}$　**12** $\dfrac{x - 4}{x - 3}$

13 **(a)** $\dfrac{7x - 4}{-5x + 3}$　**(b)** $\begin{pmatrix} a & b \\ c & d \end{pmatrix} \rightarrow \begin{pmatrix} d & -b \\ -c & a \end{pmatrix}$　**(d)** $\dfrac{dx - b}{-cx + a}$　**(e)** $\dfrac{x + 3}{x}$

　(f) $2 - \dfrac{1}{x};\ \dfrac{x - 2}{3};\ 6 - x;\ \dfrac{12}{x}$　**(g)** $\dfrac{ax + b}{cx + d} \equiv \begin{pmatrix} a & b \\ c & d \end{pmatrix} \rightarrow \begin{pmatrix} d & -b \\ -c & a \end{pmatrix} \equiv \dfrac{dx - b}{-cx + a}$

Exercise 2.8

1 $x^2 + 4x + 6$ **2** $x^2 + 2x + 4$ **3** $x^2 + 3x + 4$ **4** $x^2 + x + 2$ **5** $x^3 + 3x^2 + 5x + 3$

6 $x + 1 + \dfrac{2}{x + 1}$ **7** $(x + 1)^2$ **8** $(x^2 + 2x + 3)^2 = x^4 + 4x^3 + 10x^2 + 12x + 9$

9 $x^4 + 4x^3 + 12x^2 + 16x + 18$ **10** $x^3 + 3x^2 + 3x + 1$ **11** $x^4 + 4x^3 + 6x^2 + 4x + 1$

12 $x - 1$ **13** $\sqrt{x - 2} - 1$ **14** $x^2 + 2x + 2$ **15** $x^2 + 2$ **16** $x + 1 - \dfrac{2x + 2}{x^2 + 2x + 3}$

17 $1 + \dfrac{2}{(x + 1)^2}$ **18** $1 + \dfrac{2x + 2}{x^2 + 1}$ **19** $x + \dfrac{3}{x + 2}$ **20** $x + 4 + \dfrac{11}{x - 2}$

21 $a^2x^2 + 2abx + b^2$; $a^3x^3 + 3a^2bx^2 + 3ab^2x + b^3$; $a^4x^4 + 4a^3bx^3 + 6a^2b^2x^2 + 4ab^3x + b^4$

22 $a^2 + 2ab + b^2$; $a^2 + b^2 + c^2 + 2ab + 2bc + 2ac$

23 $a^3 + 3a^2b + 3ab^2 + b^3$; $a^3 + b^3 + c^3 + 3ab^2 + 3ac^2 + 3a^2b + 3a^2c + 3bc^2 + 3b^2c + 6abc$

Exercise 2.9

1 **(a)** -2 **(b)** -14 **(c)** 10 **(d)** 26 **(e)** 2; $x + a$ where $a \simeq 0.7$

2 **(a)** 2 **(b)** 4 **(c)** 4 **(d)** 8 **(e)** 2;   no real factors

3 **(a)** -24 **(b)** 0 **(c)** -60 **(d)** 0 **(e)** -120 **(f)** 0; $(x - 1)(x - 2)(x - 3)$

4 **(a)** $x(x + 1)(x - 1)$ **(b)** $(x - 2)(x^2 + 2x + 4)$ **(c)** $(x - 1)^3$ **(d)** $(x + 2)(x^2 + 4)$
   (e) $(x - 2)(x - 3)$ **(f)** $(3x - 1)(2x - 1)$ **(g)** $(a + b)(2a - b)$ **(h)** $(a + b)(a - b)$

5 **(a)** Yes, $(x + 6)$ **(b)** Yes, $(x + 1)$ **(c)** Yes, $(x^3 - x^2 + x + 2)$ **(d)** Yes, $(x + 1)(x^2 - 2)$
   (e) Yes, $(x - 1)(x^2 - 2)$ **(f)** Yes, $x(x + 1)$ **(g)** No   **(h)** No   **(i)** No   **(j)** No
   (k) No   **(l)** No

6 **(a)** $x + 1$, rem 0 **(b)** $x^2 + 4x + 7$, rem 8 **(c)** x, rem 4 **(d)** $x^2 + 3$, rem -2
   (e) $x^2 - x + 2$, rem -1 **(f)** $x - 1$, rem 6 **(g)** $x^2 + x - 1$, rem 0 **(h)** $x^2 + 2x + 4$, rem 0
   (i) $x^2 - 3x + 9$, rem 0

Exercise 2.10

	$q(x)$	Roots of $q(x) = 0$	Sum of roots	Product of roots
1	$x^2 - 5x + 6$	$x = 2$ or $x = 3$	5	6
2	$x^2 - 3x + 2$	$x = 1$ or $x = 2$	3	3
3	$x^2 + x - 12$	$x = 3$ or $x = -4$	-1	-12
4	$x^2 - 7x + 12$	$x = 3$ or $x = 4$	7	12
5	$12x^2 - 7x + 1$	$x = \frac{1}{3}$ or $x = \frac{1}{4}$	$\frac{7}{12}$	$\frac{1}{12}$
6	$x^2 + x - 3$	$x = 1.303$ or $x = -2.303$	-1	3
7	$x^2 - 12x + 20$	$x = 2$ or $x = 10$	12	20
8	$x^2 - 8x + 15$	$x = 3$ or $x = 5$	8	15
9	$x^2 + 5x + 6$	$x = -2$ or $x = -3$	-5	6
10	$x^2 + 4x + 5$	$x = -2 + i$ or $x = -2 - i$	-4	5
11	$2x^2 - 7x + 3$	$x = \frac{1}{2}$ or $x = 3$	$3\frac{1}{2} = \frac{7}{2}$	$\frac{3}{2}$
12	$3x^2 - 7x + 2$	$x = 2$ or $x = \frac{1}{3}$	$2\frac{1}{3} = \frac{7}{3}$	$\frac{2}{3}$
13	$x^2 - 4x + 5$	$x = 2 + i$ or $x = 2 - i$	4	5

Exercises 2.11 and 2.12

1 2 or -4 **2** 1 or -5 **3** 1 or -7 **4** -1 or $1\frac{1}{2}$ **5** 2 or -4 **6** $\frac{1}{3}$ or -3

7 2 or -2 **8** -1 or $\frac{1}{5}$ **9** $\frac{1}{2}$ or $\frac{1}{3}$ **10** $\frac{3}{2}$ or $-\frac{2}{3}$ **11** $\frac{3}{2}$ or $-\frac{5}{3}$ **12** $\dfrac{-b \pm \sqrt{b^2 - 4ac}}{2a}$

Exercise 2.13

1 **(a)** **(i)** $x = -2$ **(b)** **(i)** $x = 0$ **(c)** **(i)** $x = 3$ **(d)** **(i)** $x = 1$

         (ii) $(-2, -9)$ **(ii)** $(0, 12)$ **(ii)** $(3, -1)$ **(ii)** $(1, -3)$

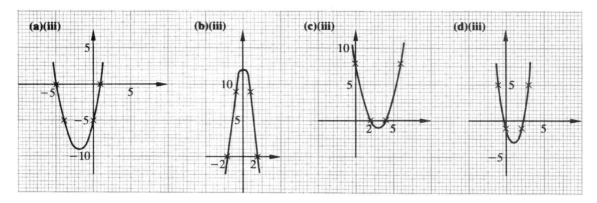

2 **(a)** complex   **(b)** equal and real   **(c)** real   **(d)** equal and real

3 **(a)** $0, 4$ **(b)** $\frac{1}{4}$ **(c)** -1 **(d)** $2, 18$

4 **(a)** $(x + 1)^2 + 1 > 0$ **(b)** $(x - 1)^2 + 1 > 0$ **(c)** $(x - 2)^2 + 1 > 0$ **(d)** $(x - 3)^2 + 1 > 0$

5 **(a)** $k \leqslant \frac{1}{4}$ **(b)** $k \leqslant -4$ or $k \geqslant 4$ **(c)** $k \geqslant -1$ **(d)** $k \leqslant 4$ or $k \geqslant 16$

Exercise 2.14

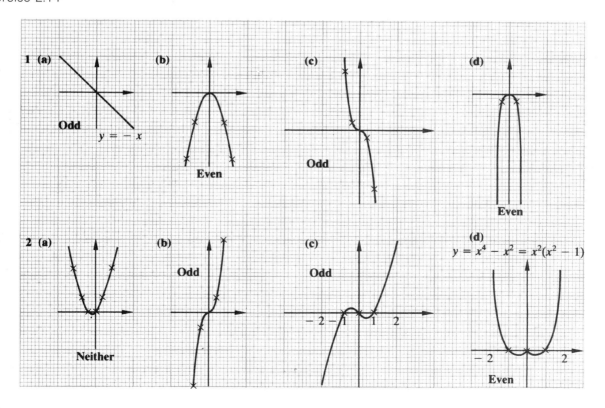

3 **(a)** $f(2) = 6$ **(b)** $f(2) = 10$ **(c)** $f(2) = 6$ **(d)** $f(2) = 2$
      $f(-2) = 2$ $f(-2) = -10$ $f(-2) = -6$ $f(-2) = 12$

4 Even if $f(-x) = f(x)$;   odd if $f(-x) = -f(x)$

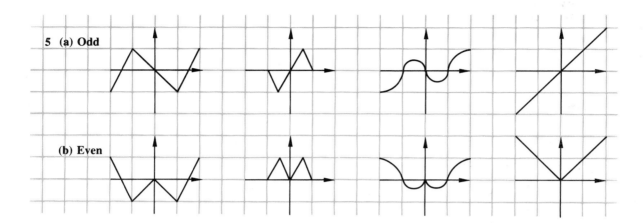

6 $y = 0$ (and $x = 0$?)   **7** $y = x^\circ = 1$ (even); $y = x^{-1} = \dfrac{1}{x}$ (odd); $y = x^{-2} = \dfrac{1}{x^2}$ (even); $y = x^{-3} = \dfrac{1}{x^3}$ (odd)

8 **(a)** even   **(b)** odd   **9** **(a)** odd   **(b)** even

Chapter 3

Exercise 3.1

1 **(i)** 2; $(0, 4)$; $(-2, 0)$ **(ii)** $\frac{3}{2}$; $(0, -3)$; $(2, 0)$ **(iii)** -1; $(0, 5)$; $(5, 0)$ **(iv)** -1; $(0, 3)$; $(3, 0)$
    (v) $-\frac{2}{3}$; $(0, 2)$; $(3, 0)$ **(vi)** $\frac{1}{3}$; $(0, -\frac{2}{3})$; $(2, 0)$

2 **(a)** $y = \frac{1}{2}x + 3$ **(b)** $2y + x = 9$ **(c)** $3y + x + 1 = 0$ **(d)** $2y = 5x + 16$

3 **(a)** $y = x - 1$ **(b)** $y = 2x - 3$ **(c)** $y + 3x = 7$ **(d)** $y = 1$

4 $y = 2x - 2$; $2y + x = 11$

5 **(a)** $AB = CD = \sqrt{10}$; $BC = AD = \sqrt{17}$; grad AB = grad $CD = \frac{1}{3}$; grad BC = grad $AD = 4$. $ABCD$ parallelogram.
   (b) $M = N = (3, 4\frac{1}{2})$ where diagonals bisect each other.   **(c)**   parallelogram   **(d)**   area $= 11$ units^2

6 **(a)** $AE = EF = FD = DA = \sqrt{17}$; grads AE and $DF = \frac{1}{4}$; grads EF and $AD = 4$; $AE \| DF$; $EF \| AD$.
   (b)   Mid-points of AF and DE are both $(3\frac{1}{2}, 4\frac{1}{2})$. Diagonals bisect each other.
   (c)   rhombus   **(d)**   area $AEFD = 15$; grad $AF = 1$; grad $DE = -1$. Diagonals are perpendicular.

7 **(a)** $AJ = HG = \sqrt{18}$; $AH = JG = \sqrt{8}$; grad AJ = grad $HG = 1$; grad AH = grad $JG = -1$. Opposite sides equal and parallel.
   (b)   Mid-points of diagonals coincide at $(3\frac{1}{2}, 2\frac{1}{2})$ **(c)**   rectangle; prove angles are $90°$; adjacent sides perpendicular.
   (d)   area $AHGJ = 2\sqrt{2} \times 3\sqrt{2} = 12$; equation of JG is $x + y = 9$ $\Rightarrow$ equation of AH is $x + y = k = 3$.

8 $y = \frac{1}{2}x + 3$; $y = \frac{1}{2}x - 2$; $2x + y = 3$; $2x + y = 13$; $3y + x = 9$; $y = 3x - 7$.

9 **(a)** $OA = \sqrt{40}$; $OB = \sqrt{50}$; $AB = \sqrt{50}$. $\triangle OAB$ is isoceles.
   (b) $\tan A\hat{O}X = \frac{1}{3}$ $\Rightarrow$ $A\hat{O}X = 18.4°$; $\tan B\hat{O}X = 7$ $\Rightarrow$ $B\hat{O}X = 81.9°$ $\Rightarrow$ $A\hat{O}B = 63.5° = O\hat{A}B$.
   (c) $O\hat{A}B = \arctan\frac{1}{7} + 45° = 8.1° + 45° = 53.1°$. Sum $= 126.9° + 53.1° = 180°$.
   (d)   No, but can you prove it? Rotate (a, b) through $60°$ and note co-ordinates of image (irrational, not integers).
   (e) M is $(3, 1)$; grad $BM = -3$; grad $OA = \frac{1}{3}$; grad $BM \times$ grad $OA = -1$ $\Rightarrow$ $BM \perp OA$.

10 **(a)** N is $(3\frac{1}{2}, 4\frac{1}{2})$; ON is $y = \frac{9}{7}x$ **(b)** BM is $3x + y = 10$; G is $(2\frac{1}{3}, 3)$.

    (c) AG is $3x + 11y = 40$; OB is $y = 7x$; K is $(\frac{1}{2}, 3\frac{1}{2})$.

    (d) Mid-point of OB is $(\frac{1}{2}, 3\frac{1}{2})$ i.e. K.

    (e) $ON = \frac{1}{2}\sqrt{130}$; OG is $\frac{1}{3}\sqrt{130}$; GN is $\frac{1}{6}\sqrt{130}$. $OG = \frac{2}{3}ON$; $OG:GN = 2:1$

    (f) $BG:GM = 4:2 = 2:1$ i.e. $BG = 2 \times GM$

    (g) G is the intersection of the medians; called the centroid of $\triangle OAB$.

11 $\operatorname{Grad} OD = \frac{1}{3}$; D_1 is $(2, -6)$; $\operatorname{grad} OD_1 = -3$; D_2 is $(-2, 6)$; $\operatorname{grad} OD_2 = -3$; $\operatorname{grad} OD_1 \times \operatorname{grad} OD = -1$.

12 $\operatorname{Grad} OP = \dfrac{b}{a}$; P_1 is $(b, -a)$; $\operatorname{grad} OP_1 = \dfrac{-a}{b}$; P_2 is $(-b, a)$; $\operatorname{grad} OP_2 = \dfrac{-a}{b}$;

    $\operatorname{grad} OP_1 \times \operatorname{grad} OP = \dfrac{b}{a} \times \dfrac{-a}{b} = -1$. Product of gradients of perpendicular lines is -1.

13 Parallel lines have the same gradient.

Exercise 3.2

1 S is $(2, 3\frac{1}{2})$ **2** J $(2.1, 3.6)$ **3** G $(5.8, 5.4)$ **4** H $(3.4, 4.2)$

5 **(a)** M $(\frac{1}{2}, 3)$ G $(4, 2)$ **(b)** L $(6, 3)$ H $(4, 2)$ **(c)** N $(4\frac{1}{2}, 0)$ I $(4, 2)$ **(d)** $G = H = I = $ centroid $\triangle OAB$

6 **(a)** $P(6, 0), Q(7, 2)$ **(b)** $\operatorname{grad} PQ = 2 = \operatorname{grad} OB$; $PQ \parallel OB$ **(c)** $PQ = \sqrt{5}, OB = 3\sqrt{5}$; $OB = 3 \times PQ$

    (d) $PQ \parallel OB$ and $OB = 4 \times PQ$

7 **(a)** G $(4, 2)$ **(b)** X $(5\frac{1}{4}, 3)$ **(c)** $D(7, 4)$ **(d)** Y $(5\frac{1}{4}, 3)$ **(e)** $X = Y$ **(f)** all same $X = (5\frac{1}{4}, 3)$;

    X is the centroid (centre of mass) of $OACB$ (tetrahedron).

    (g) Circumcentre, centroid, incentre, orthocentre.

    (h) X is the intersection of lines joining the mid-points of opposite sides $(5\frac{1}{4}, 3)$. X is the centroid (centre of mass). Intersection of diagonals is $(5.4, 3.6)$.

    (i) Centroid is $X = $ centre of mass (gravity).

8 **(b)** $G(5, 6); H(2, 3); I(1, 6)$ **(c)** $\left(\dfrac{25}{11}, \dfrac{54}{11}\right)$ **(d)** $D(6, 4); E(6, 1); F(2, 8); \dfrac{CD}{DB} \times \dfrac{BE}{EA} \times \dfrac{AF}{BC} = \dfrac{3}{2} \times \dfrac{1}{3} \times \dfrac{2}{1} = +1$;

    $(5, 4)$

Exercise 3.3

1 $K(13, 9)$ **2** $L(-5, 0)$ **3** $R(-3, 1)$ **4** $T(19, 12)$ **5** $R(3, 1\frac{1}{2})$; $S(2, 0)$; $T(6, 6)$. RST has equation $y = \frac{3}{2}x - 3$. **6** $QR_1:R_1P = -1:3 \Rightarrow R(0, 3)$; $PS_1:S_1O = 2:-1 \Rightarrow S_1(-6, 0)$; S_1R_1T is $y = \frac{1}{2}x + 3$.

Exercise 3.4

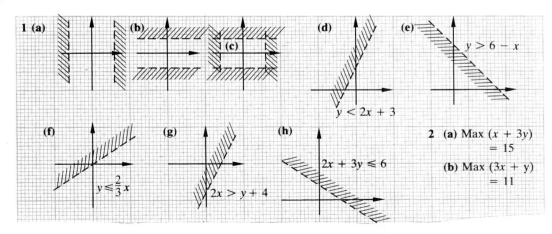

1 (a) **(b)** **(c)**

(d) $y < 2x + 3$ **(e)** $y > 6 - x$

(f) $y \leqslant \frac{2}{3}x$ **(g)** $2x > y + 4$ **(h)** $2x + 3y \leqslant 6$

2 **(a)** Max $(x + 3y) = 15$

    (b) Max $(3x + y) = 11$

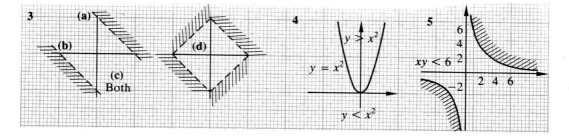

Exercise 3.5

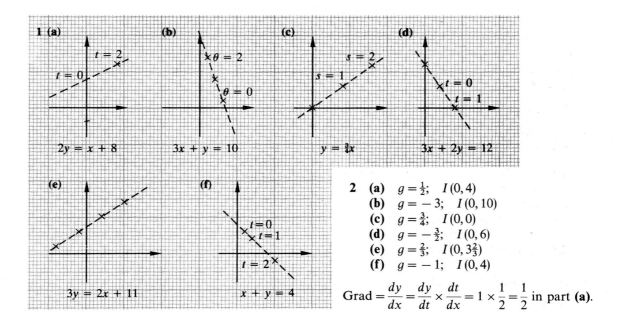

2 (a) $g = \frac{1}{2}$; $I(0, 4)$
   (b) $g = -3$; $I(0, 10)$
   (c) $g = \frac{3}{4}$; $I(0, 0)$
   (d) $g = -\frac{3}{2}$; $I(0, 6)$
   (e) $g = \frac{2}{3}$; $I(0, 3\frac{2}{3})$
   (f) $g = -1$; $I(0, 4)$

Grad $= \dfrac{dy}{dx} = \dfrac{dy}{dt} \times \dfrac{dt}{dx} = 1 \times \dfrac{1}{2} = \dfrac{1}{2}$ in part (a).

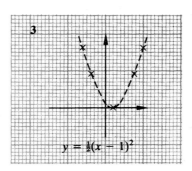

Chapter 4

Exercise 4.1

1 (a) 7   (b) 1   (c) $\frac{1}{4}$ (d) 5   2 (a) -11 (b) 0   (c) 6

3 (a) 2   (b) 1   (c) $1\frac{1}{2}$ 4 (a) 0   (b) 0   (c) $\frac{2}{3}$ (d) ∞

5 **(a)** approaches 2 from above as $x \to \infty$ and from below as $x \to -\infty$.

    (b) approaches ∞ as $x \to 1$ from above and $-\infty$ as $x \to 1$ from below.   **(c)** $(0,0)$

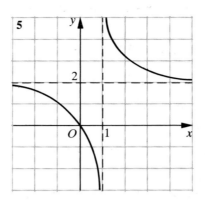

6 **(i)** **(a)** approaches $\frac{3}{2}$ from below as $x \to \infty$ and from above as $x \to -\infty$.

       (b) approaches 1 as $x \to 1$ from above or from below.   **(c)** $(0,0)$

   (ii) **(a)** approaches 1 from below as $x \to \infty$ and from above as $x \to -\infty$.

       (b) approaches 0 from above as $x \to 1$ from above and from below as $x \to 1$ from below.

       (c) $(1,0), (0,-1)$

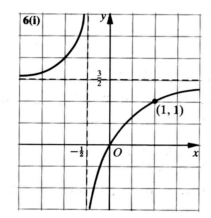 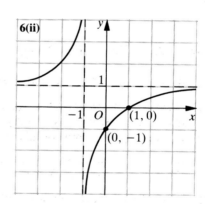

7 **(a)** approaches 3 from below as $x \to \infty$ and from above as $x \to -\infty$.

   (b) approaches $-\infty$ as $x \to 0$ from above and ∞ as $x \to 0$ from below.

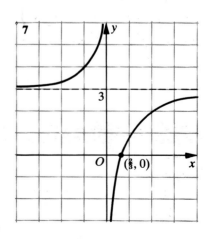

8 **(a)** 2   **(b)** -1 **(c)** 12   **(d)** $\frac{1}{2}$ **9** **(a)** ∞ **(b)** 3

10 **(a)** 4   **(b)** $\frac{1}{2}$ **(c)** 2   **11** 3; 2; $\dfrac{x(5x-1)}{x^2-1}$; 5

Exercise 4.2

1 **(a)** $3x^2$ **(b)** $20x^3$ **(c)** 0   **(d)** 2   **(e)** $-2x^{-3}$

   (f) $-\dfrac{2}{x^2}$ **(g)** $\dfrac{9}{x^4}$ **(h)** $2-2x$ **(i)** -1 **(j)** $6x^2-8x$

2 **(a)** 3 **(b)** $2x$ **(c)** $4x-3$ **(d)** $12x^2-12x$ **(e)** $1-\dfrac{1}{x^2}-\dfrac{2}{x^3}$ **(f)** $-\dfrac{2}{x^2}+\dfrac{12}{x^5}$

3 **(a)** $4x-4$ **(b)** x^3+x^2+x **(c)** $2x-2$ **(d)** $2x+\dfrac{2}{x^3}$ **(e)** 1

4 **(a)** $4x-1$ **(b)** $-\dfrac{1}{x^2}$ **(c)** $2x+3$ **(d)** $1-\dfrac{4}{x^2}$

5 **(a)** $6x-4,\ -10$ **(b)** $4x-5,\ -9$ **(c)** $4x+4,\ 0$

6 **(a)** $10, 24, -8$ **(b)** $5, \frac{1}{2}, -1$ **(c)** $0, 8, 0$

7 **(a)** 0 **(b)** -2 **(c)** 7 **(d)** 6

8 **(a)** $y=6x-9$ **(b)** $y=5-3x$ **(c)** $y+7x+2=0$ **(d)** $y=2x+3$
   (e) $y+2x=0$ **(f)** $4y=x+12$

9 **(a)** $\frac{1}{2}x^{-\frac{1}{2}}$ **(b)** $x^{-\frac{2}{3}}$ **(c)** $-\frac{1}{4}x^{-\frac{5}{4}}$ **(d)** $\frac{1}{2}x^{-\frac{3}{2}}$ **(e)** $\frac{1}{5}x^{-\frac{4}{5}}$
   (f) $\frac{1}{2}x^{-\frac{1}{2}}(3x-1)$ **(g)** $1-x^{-\frac{1}{2}}$ **(h)** $\frac{1}{2}x^{-\frac{3}{2}}(x-2)$ **(i)** $1+\frac{2}{3}x^{-\frac{1}{3}}+\frac{5}{3}x^{-\frac{1}{6}}$

10 $(-2, \frac{55}{3}), (6, -59)$ **11** **(a)** $(-1, 3)$ **(b)** $(\frac{1}{3}, \frac{5}{9}), (-\frac{1}{3}, \frac{13}{9})$

12 $(-1, 4);\ (5, 10);\ -5, 7;\ y+5x+1=0;\ y=7x-25$ **13** $6y+8x=\pm31$

14 $\frac{1}{3}, 3y=x+11;\ (-\frac{7}{3}, \frac{26}{9})$ **15** $y=1-x;\ y=\frac{1}{2}x;\ (\frac{2}{3}, \frac{1}{3})$

16 $\dfrac{ds}{dt}=2t-8,\ t=4,\ s=-9$ **17** $(0,0), (3,0);\ (1,4), (3,0)$

Exercise 4.3
1 **(a)** 2 **(b)** $12x^2-12x$ **(c)** $2-\dfrac{2}{x^3}$ **(d)** $-\frac{1}{4}x^{-\frac{3}{2}}-\frac{3}{4}x^{-\frac{1}{2}}$

2 **(a)** 18 **(b)** 42 **(c)** 1 **3** $24x+48$ **4** **(a)** $(2,-5)$ **(b)** $(-\frac{1}{2}, \frac{5}{4})$

5 **(a)** $(1,-1)$ Min   **(b)** $(1,1)$ Min, $(0,2)$ point of inflexion   **(c)** $(0,0)$ Max; $(\frac{2}{3}, -\frac{4}{27})$ Min
   (d) $(1,4)$ Max; $(2,3)$ Min   **(e)** $(0,0)$ Min   **(f)** $(3,0)$ Min
   (g) $(\frac{3}{4}, -\frac{27}{256})$ Min; $(0,0)$ point of inflexion

6 **(a)** $(1,2)$ Min; $(-1,-2)$ Max   **(b)** $(1,-4)$ Min; $(-1,0)$ Max
   (c) $(0,0)$ Min; $(1,0)$ Min; $(\frac{1}{2}, \frac{1}{16})$ Max   **(d)** $(2,0)$ Min; $(\frac{2}{3}, \frac{32}{27})$ Max

7 **(a)** $(2,12)$ Min   **(b)** $(0,0)$ point of inflexion; $(3,27)$ Max   **(c)** $(1,1)$ Min; $(-3,33)$ Max

8 **(a)** Yes $(0,0)$ **(b)** Yes $(\frac{2}{3}, \frac{2}{27})$ **(c)** Yes $(1,5), (-1,5)$

9 $7\frac{11}{27}\,\text{cm}^3$ **10** $-1\frac{1}{2}\,\text{m s}^{-1}$ **11** $40\,\text{m}\times40\,\text{m}$ **12** $50\,\text{m}^2$ **13** $6\sqrt{6}\pi\,\text{cm}^3$

14 $r=3\,\text{cm},\ h=6\,\text{cm}$ **15** $8\,\text{cm}\times8\,\text{cm}\times4\,\text{cm}$ **16** $18\,\text{cm}^3$ **17** $A=144x-36x^2;\ x=2$

Chapter 5

Exercise 5.1
1 **(a)** $3x+c$ **(b)** $\frac{5}{2}x^2+c$ **(c)** $\frac{1}{3}x^3+2x^2+c$ **(d)** $\frac{1}{3}x^3+x^2+x+c$ **(e)** $-\frac{1}{2}x^{-2}+c$

2 **(a)** $4t+c$ **(b)** $-t^2+c$ **(c)** t^3-7t+c **(d)** $-t^{-1}+c$ **(e)** $\frac{2}{5}t^{\frac{5}{2}}+c$

3 **(a)** $y=-x+c$ **(b)** $y=2x^2+c$ **(c)** $y=\dfrac{1}{x}+c$ **(d)** $y=x^4+c$

4 **(a)** $y = -\frac{5}{2}x + c$ **(b)** $y = \frac{2}{3}x^{\frac{3}{2}} + c$ **(c)** $y = \frac{1}{3}x^3 - x^2 + c$

5 **(a)** $\frac{1}{2}x^2 - 3x + c$ **(b)** $x^4 - \frac{2}{3}x^3 + \frac{3}{2}x^2 - 5x + c$ **(c)** $\frac{1}{3}x^3 - 2x^2 + 3x + c$

6 **(a)** $\frac{1}{3}x^3 - 4x^2 + c$ **(b)** $\frac{1}{8}x^8 - x^5 + c$ **(c)** $\frac{1}{3}x^3 + 3x^2 + 9x + c$

(d) $x - \dfrac{2}{x} + c$ **(e)** $\frac{1}{2}x^2 + \dfrac{1}{x} + c$ **(f)** $-\dfrac{1}{x} - \dfrac{3}{2x^2} + \dfrac{1}{3x^3} + c$

(g) $\frac{2}{3}x^{\frac{3}{2}} - 2x^{\frac{1}{2}} + c$ **(h)** $\frac{3}{4}x^{\frac{4}{3}} + 3x^{\frac{1}{3}} + c$ **(i)** $\frac{2}{5}x^{\frac{5}{2}} - \frac{2}{3}x^{\frac{3}{2}} + c$

(j) $x + \dfrac{1}{x} + c$ **(k)** $\frac{3}{4}x^{\frac{4}{3}} + \frac{2}{5}x^{\frac{5}{2}} + c$ **(l)** $\frac{4}{3}x^3 - \dfrac{4}{x} - 8x + c$

7 $y = x^2 + 3x - 11$ **8** $y = 1 + \dfrac{2}{x^2}$ **9** $s = \dfrac{3t^2}{2} + \dfrac{1}{t} - 2$

10 $v = \frac{1}{3}t^3 - t + 4$ **11** $y = x^3 - 3x^2 - x + 3;\ \ (1,0),(3,0)$

12 $(3,0),\ y = 5x - 15;\ \ (-2,0),\ y + 5x + 10 = 0$ **13** $s = t^3 - 2t^2 + 5t + 5$ **14** $v = 3 - 2t^2$

15 $\dfrac{dy}{dx} = 12x^2 - 6x;\ \ y = 4x^3 - 3x^2 + 7$ **16** $\dfrac{dx}{dt} = 5t^4 - 2t + 4;\ \ x = t^5 - t^2 + 4t + 12$

17 $y = x^3 - 2x^2 - x + 2;\ \ (-1,0),(1,0),(2,0)$

18 $y = x^4 + 2x^3 - 3x^2 - 4x + 4;\ \ (1,0)$ Min, $(-2,0)$ Min, $(-\frac{1}{2}, 5\frac{1}{16})$ Max

Exercise 5.2

1 **(a)** $3x + 2x^2 + 2x^3 + c$ **(b)** $\frac{1}{2}x^2 + \dfrac{1}{x} + c$ **(c)** $x + x^2 + \frac{1}{3}x^3 + c$ **(d)** $\frac{3}{7}x^{\frac{7}{3}} - \frac{3}{5}x^{\frac{5}{3}} + c$

2 **(a)** 28   **(b)** 4   **(c)** 0

3 **(a)** $46\frac{2}{3}$ **(b)** $7\frac{7}{10}$ **(c)** $2\frac{1}{2}$ **(d)** $1\frac{11}{15}$ **(e)** 9   **(f)** $2\frac{2}{3}$ **(g)** $3\frac{2}{3}$ **(h)** $16\frac{1}{4}$ **(i)** $34\frac{2}{15}$

4 **(a)** $\frac{7}{3}$ **(b)** 8   **(c)** $\frac{2}{15}$ **(d)** $21\frac{2}{3}$ **(e)** $\frac{1}{4}$ **(f)** $\frac{1}{4}$ **(g)** $10\frac{2}{3}$ **(h)** $1\frac{1}{6}$

5 **(a)** 4   **(b)** $6\frac{3}{4}$ **6** $4\frac{1}{2}$ **7** $1\frac{1}{3}$

8 **(a)** $10\frac{2}{3}$ **(b)** $\frac{1}{6}$ **(c)** $\frac{4}{3}$ **(d)** $4\frac{1}{2}$ **(e)** $4\frac{1}{2}$ **(f)** $\frac{1}{2}$

Exercise 5.3

1 **(a)** 15   **(b)** $\frac{10}{3}$ **(c)** $\frac{3}{4}$ **2** **(a)** 18   **(b)** $42\frac{2}{3}$ **(c)** $10\frac{2}{3}$ **(d)** 12

3 **(a)** $18\frac{2}{3}$ **(b)** 36   **(c)** $15\frac{1}{3}$ **4** $1\frac{1}{3}$ **5** **(a)** $\frac{5}{12}$ **(b)** $\frac{8}{3}$ **(c)** $2\frac{1}{4}$

6 $(0,3),(3,0);\ 4\frac{1}{2}$ **7** **(a)** $4\frac{1}{2}$ **(b)** $10\frac{2}{3}$ **(c)** $20\frac{5}{6}$ **(d)** $2\frac{2}{3}$

8 **(a)** $5\frac{1}{3}$ **(b)** $21\frac{1}{3}$ **(c)** $21\frac{1}{3}$ **(d)** $2\frac{2}{3}$ **9** $a^2 - \frac{3}{2}a^3;\ \frac{1}{2}a^3;\ \ a = \frac{1}{2}$

Exercise 5.4

1(a) 51π **(b)** 204.8π **(c)** 25.1π **(d)** 13.5π **(e)** $13\frac{1}{3}\pi$

2 **(a)** $\frac{1}{12}\pi$ **(b)** $\frac{1}{2}\pi$ **(c)** $\frac{64}{5}\pi$ **(d)** $34\frac{2}{15}\pi$ **(e)** $\frac{1}{2}\pi$

3 **(a)** $\frac{32}{3}\pi$ **(b)** $34\frac{2}{15}\pi$ **(c)** $\frac{2}{3}\pi$ **(d)** $1\frac{1}{15}\pi$ **(e)** $3\frac{11}{15}\pi$ **(f)** $4\frac{4}{15}\pi$

4 $(0,0),(4,16);\ 136.5\pi$ **5** $6\frac{3}{4}\pi$ **6** 34.1π **7** **(a)** 55.47π **(b)** 115.2π

8 $\frac{17}{4}\pi$ **9** $\frac{68}{21}\pi$ **12** 34.13π

Chapter 6

Exercise 6.1

1 **(a)** 0.5; 0.866; 0.577   **(b)** 0.866; 0.5; 1.732   **(c)** 0.707; 0.707; 1

   (d) 0.985; -0.174; -5.671 **(e)** -0.342; -0.940; 0.364   **(f)** -0.866; 0.5; -1.732

   (g) 0.866; -0.5; -1.732 **(h)** $1, 0, \infty$ **(i)** $0, 1, 0$

2 $\cos^2 A + \sin^2 A = 1$ for all A **3** **(a)** $1; 0.866$ **(b)** $\sqrt{2}; 1$ **(c)** $\sqrt{3}; \sqrt{3}/2$

4 **(a)** 1.366   **(b)** 1; $\sin(A+B) \neq \sin A + \sin B$

5 **(a)** 0.866   **(b)** 1   **(c)** 0.866   **6** **(a)** 0.5   **(b)** 0   **(c)** -0.5

7 **(a)** 0.5   **(b)** 0   **(c)** -0.5 **8** **(a)** 0.5   **(b)** 0   **(c)** -0.5

Exercise 6.2

1 $360n + 53.1°$ or $360n + 126.9°$ **2** $360n \pm 36.9°$ **3** $360n + 315.6°$ or $360n + 224.4°$

4 $180n + 38.7°$ **5** $360n + 36.9°$ or $360n + 143.1°$ **6** $180n + 50.2°$ **7** $180n + 90°$ **8** $180n + 135°$

9 No solution   **10** $\sin\theta$ **11** $-\cos\theta$ **12** $\sin\theta$ **13** $\cos\theta$ **14** $\cos\theta$ **15** $\tan\theta$

16 $\theta = 180n + 45°$ **17** $180n$ **18** $360n + 38.2°$ or $360n + 141.8°$ **19** $\theta = 0°, 90°, 270°$ or $360°$

20 $\theta = 45°, 135°, 225°$ or $315°$ **21** No solution   **22** $\theta = 70.5°$ or $289.5°$ **23** $\theta = 0°$ or $360°$

Exercise 6.3

1 **(a)** $C = 53°$, $c = 10.82\,\text{cm}$, $b = 13.29\,\text{cm}$ **(b)** $A = 63°$, $c = 3.18\,\text{cm}$, $a = 6.24\,\text{cm}$

   (c) $a = 15\,\text{cm}$, $A = 61.9°$, $c = 28.1°$ **(d)** $C = 69°$, $b = c = 13.95\,\text{cm}$

   (e) $a = 7.44\,\text{cm}$, $B = 67.2°$, $C = 53.8°$ **(f)** $A = 49.5°$, $B = 58.8°$, $C = 71.8°$

   (g) $B = 63.3°$, $C = 66.7°$, $c = 7.19\,\text{m}$; or $B = 116.7°$, $C = 13.3°$, $c = 1.80\,\text{m}$

2 **(a)** $43.3\,\text{cm}^2$ **(b)** $9.92\,\text{cm}^2$ **(c)** $60\,\text{cm}^2$ **(d)** $65.1\,\text{cm}^2$ **(e)** $24.0\,\text{cm}^2$ **(f)** $34.2\,\text{cm}^2$

   (g) $19.28\,\text{m}^2$; or $4.83\,\text{m}^2$

3 $11.3\,\text{km}$; $172.5°$ **4** $x = 4.22\,\text{km}$; $353.1°$

Exercise 6.4

1 $\sin 15° = \cos 75° = \dfrac{\sqrt{3}-1}{2\sqrt{2}}$; $\cos 15° = \sin 75° = \dfrac{\sqrt{3}+1}{2\sqrt{2}}$; $\tan 15° = 2 - \sqrt{3}$; $\tan 75° = \dfrac{1}{\tan 15°} = 2 + \sqrt{3}$

2 $\cos(A-B) = \pm\frac{63}{65}$ or $\pm\frac{33}{65}$; $\sin(A+B) = \frac{56}{65}$ or $-\frac{16}{65}$

3 **(a)** $30°, 90°, 150°, 270°$; $360n + 30°$ or $360n + 150°$ or $180n + 90°$

   (b) $30°, 270°, 150°$; $360n + 30°$ or $360n + 150°$ or $360n + 270°$

   (c) $67.5°, 157.5°, 247.5°, 337.5°$; $A = 90n + 67.5°$

   (d) $0°, 60°, 180°, 300°, 360°$; $180n$ or $360n \pm 60°$

   (e) $0°, 45°, 135°, 180°, 225°, 315°, 360°$; $180n$ or $90n + 45°$

   (f) $38.2°, 141.8°$; $360n + 38.2°$ or $360n + 141.8°$

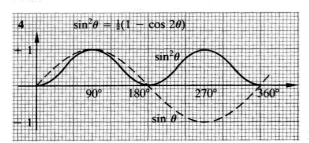

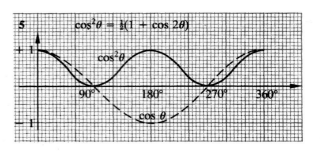

6 $\sin 3A = 3\sin A - 4\sin^3 A$ **7** $\cos 3A = 4\cos^3 A - 3\cos A$ **8** $2\sin A\cos B$

9 $2\cos A\cos B$ **10** $2\cos A\sin B$ **11** $-2\sin A\sin B$

Exercise 6.5
1 $s = \sqrt{5}$, $\beta = 26.6°$ **2** $r = 5$, $\alpha = 53.1°$

3 (i) $r = s = \sqrt{10}$; $\alpha = 18.4°$; $\beta = 71.6°$
 (ii) $r = s = 10$; $\alpha = 53.1°$; $\beta = 36.9°$
 (iii) $r = s = 13$; $\alpha = 67.4°$; $\beta = 22.6°$

Exercises 6.6 and 6.7
1 **(a)** $0°, 90°$ **(b)** $45°$ **(c)** No solution **(d)** $105°, 345°$

2 **(a)** $210°, 330°$ **(b)** $36.9°, 90°$ **(c)** $63.4°$ **(d)** $15.5°, 112.3°$

3 **(a)** $115.4°, 318.4°$ **(b)** $103.3°, 330.5°$ **(c)** $90°, 343.8°$ **(d)** $0°, 73.8°$

4 **(a)** $67.5°; 157.5°, 247.5°, 337.5°$ **(b)** $0°, 45°, 90°, 135°, 180°$ **(c)** No solution **(d)** No solution

Chapter 7

Exercise 7.1
1 (i) $-\mathbf{c}$ (ii) $\mathbf{a}+\mathbf{c}$ (iii) $\mathbf{b}+\mathbf{a}$ (iv) $\mathbf{a}-\mathbf{b}+\mathbf{c}$ (v) $-\mathbf{a}-\mathbf{c}$

2 (i) true (ii) false (iii) true (iv) true (v) true (vi) false

3 (i) $\mathbf{a}+\mathbf{b}$ (ii) $\mathbf{a}+\mathbf{c}$ (iii) $\mathbf{a}+\mathbf{b}+\mathbf{c}$ (iv) $\mathbf{c}-\mathbf{a}$
 (v) $\mathbf{c}-\mathbf{b}-\mathbf{a}$ (vi) $\mathbf{b}-\mathbf{c}-\mathbf{a}$ (vii) $\mathbf{c}+\mathbf{b}-\mathbf{a}$

4 $\overrightarrow{TS}$ **7** $\overrightarrow{OC} = \begin{pmatrix} 9 \\ -4 \end{pmatrix}$, $\overrightarrow{BO} = \begin{pmatrix} -8 \\ -7 \end{pmatrix}$, $\overrightarrow{AD} = \begin{pmatrix} 0 \\ -19 \end{pmatrix}$ where D is the point $(5, -17)$

8 $8.72\,\text{m}$, S $83°\,25'$ E **9** $9.45\,\text{cm}$, S $87°\,11'$ E

10 **(a)** $6.08\,\text{m s}^{-1}$, N $64°\,43'$ W **(b)** $6.65\,\text{m s}^{-2}$, S $77°\,56'$ W

11 $5.39\,\text{km}$, S $23°\,12'$ E **12** $14.03\,\text{m}$, S $87°\,57'$ E

Exercise 7.2
1 (i) $3a = \begin{pmatrix} 3 \\ 6 \end{pmatrix}$ (ii) $2\mathbf{a}+\mathbf{b} = \begin{pmatrix} -1 \\ 8 \end{pmatrix}$ (iii) $\mathbf{b}-\mathbf{c} = \begin{pmatrix} -8 \\ -1 \end{pmatrix}$

 (iv) $\mathbf{a}-2\mathbf{b}+\mathbf{c} = \begin{pmatrix} 12 \\ -1 \end{pmatrix}$ (v) $\mathbf{a}+\frac{1}{2}\mathbf{b} = \begin{pmatrix} -\frac{1}{2} \\ 4 \end{pmatrix}$

2 $\frac{1}{4}(\mathbf{c}+3\mathbf{a})$ **3** (i) $\mathbf{c}-\mathbf{a}$ (ii) $\frac{1}{2}(\mathbf{a}+\mathbf{b})$ (iii) $-\frac{1}{2}\mathbf{a}$ (iv) $\frac{1}{2}(\mathbf{c}-\mathbf{b}-\mathbf{a})$ (v) $\frac{1}{2}\mathbf{a}-\mathbf{c}$

4 $\frac{1}{2}(3\mathbf{b}-\mathbf{a})$

Exercise 7.3

1 (a) 13 (b) $\sqrt{17}$ (c) 25 (d) $\sqrt{6}$ (e) 5 (f) 3 (g) 7 (h) $\sqrt{6}$ (i) $\sqrt{29}$

2 (a) $53°\,8',\,36°\,52'$ (b) $73°\,24',\,115°\,23',\,31°$
   (c) $143°\,18',\,57°\,41',\,105°\,30'$ (d) $75°\,31',\,138°\,35',\,127°\,46'$

3 (i) $\sqrt{17}$ (ii) $\sqrt{26}$ (iii) $\sqrt{34}$ (iv) $\sqrt{45}$

4 (i) $\dfrac{1}{\sqrt{5}}(2\mathbf{i}+\mathbf{j})$ (ii) $-\mathbf{k}$ (iii) $\dfrac{1}{\sqrt{34}}(4\mathbf{i}-3\mathbf{j}+3\mathbf{k})$ (iv) $\dfrac{1}{\sqrt{329}}(3\mathbf{i}-16\mathbf{j}+8\mathbf{k})$

5 $\dfrac{2}{\sqrt{21}},\dfrac{-1}{\sqrt{21}},\dfrac{4}{\sqrt{21}},64°\,7',102°\,36',29°\,12'$

6 (a) $\mathbf{x}+\mathbf{y}$ (b) $-\mathbf{y}$ (c) $\mathbf{y}-\mathbf{x}$ (d) $\mathbf{x}+\mathbf{z}$ (e) $\mathbf{y}+\mathbf{z}$ (f) $\mathbf{x}+\mathbf{y}+\mathbf{z}$ (g) $-\mathbf{x}+\mathbf{y}+\mathbf{z}$

7 (a) $5\mathbf{i}+7\mathbf{k}$ (b) $-2\mathbf{i}+\mathbf{j}-5\mathbf{k}$ (c) $-\mathbf{i}-2\mathbf{j}+3\mathbf{k}$ (d) $6\mathbf{i}+2\mathbf{j}+11\mathbf{k}$ (e) $4\mathbf{i}+3\mathbf{j}+6\mathbf{k}$

8 $\mathbf{a}=10\cos 30°\,\mathbf{i}+10\sin 30°\,\mathbf{j}=5\sqrt{3}\mathbf{i}+5\mathbf{j}$

9 (a) $3\mathbf{i}+4\mathbf{j}$ (b) $-4\mathbf{i}+2\mathbf{j}$ (c) $-3\mathbf{i}-8\mathbf{j}$ (d) $5\mathbf{i}-2\mathbf{j}$

Exercise 7.4

1 $-\mathbf{i}+3\mathbf{j}-\tfrac{1}{2}\mathbf{k}$ **2** $\tfrac{3}{4}(\mathbf{a}+\mathbf{c})$

3 $\overrightarrow{AB}=2\mathbf{i}+4\mathbf{j}-5\mathbf{k},\,\overrightarrow{AC}=3\mathbf{i}+6\mathbf{j}-7\tfrac{1}{2}\mathbf{k}\;\Rightarrow\;\overrightarrow{AC}=\tfrac{3}{2}\overrightarrow{AB}$

5 (a) $\mathbf{r}_A=2\mathbf{i},\,\mathbf{r}_B=3\mathbf{i}+\sqrt{3}\mathbf{j},\,\mathbf{r}_C=2\mathbf{i}+2\sqrt{3}\mathbf{j}$ (b) $\mathbf{r}_X=\tfrac{1}{2}(5\mathbf{i}+3\sqrt{3}\mathbf{j})$ (c) $\mathbf{r}_Y=\tfrac{1}{4}(5\mathbf{i}+\sqrt{3}\mathbf{j})$

6 $\left(\dfrac{3+\lambda}{\lambda+1}\right)\mathbf{i}+\left(\dfrac{2-4\lambda}{\lambda+1}\right)\mathbf{j}+\left(\dfrac{\lambda-1}{\lambda+1}\right)\mathbf{k}$

7 $\mathbf{r}_A=\mathbf{i}+2\mathbf{j},\,\mathbf{r}_B=3\mathbf{i}+4\mathbf{j}+\mathbf{k},\,\mathbf{r}_D=2\mathbf{i}+\mathbf{j}-\mathbf{k},\,\mathbf{r}_X=\tfrac{5}{2}(\mathbf{i}+\mathbf{j}),\,C(4,3,0)$

8 $\mathbf{r}_A=5\mathbf{i}+12\mathbf{j},\,\mathbf{r}_B=3\mathbf{i}-4\mathbf{j},\,\tfrac{1}{13}(5\mathbf{i}+12\mathbf{j}),\,\tfrac{1}{5}(3\mathbf{i}-4\mathbf{j}),\,y=1$

9 $\overrightarrow{AB}=-4\mathbf{i}-3\mathbf{j}+\mathbf{k},\,\mathbf{r}=(1-4t)\mathbf{i}+(2-3t)\mathbf{j}+(1+t)\mathbf{k},\,(-3,-1,2),(\tfrac{29}{13},\tfrac{38}{13},\tfrac{9}{13})$

Chapter 8

Exercise 8.1

1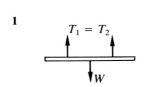

2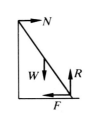

3

4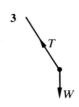

5 (a) and (b) 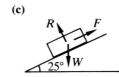 (c) 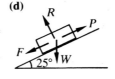 (d)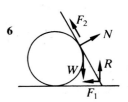

6

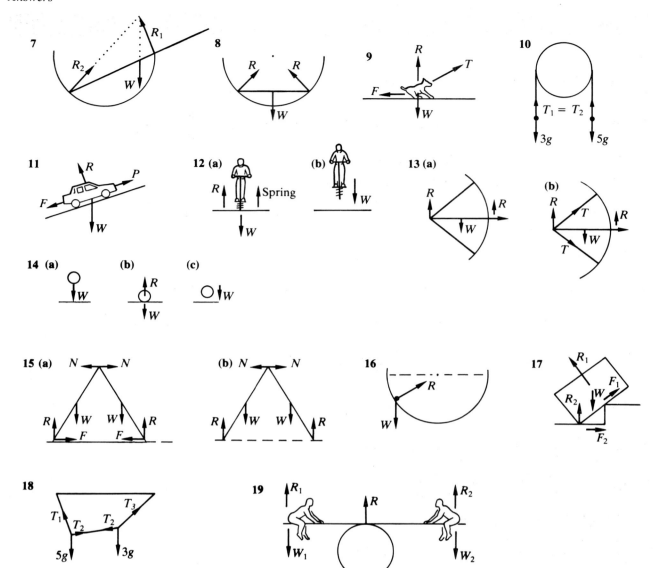

Exercise 8.2

1 **(a)** $9\mathbf{i} + 2\mathbf{j}$; $\sqrt{85} \simeq 9.22$ at $12.5°$ to Ox **(b)** $7.27\mathbf{i} + 16.31\mathbf{j}$; 17.86 at $66°$ to Ox

(c) $3.5\mathbf{i} + 8.62\mathbf{j}$; 9.3 at $67.9°$ to Ox **(d)** **(i)** $6\mathbf{i} + 5\mathbf{j}$; 7.8 at $39.8°$ to Ox

(ii) $8.5\mathbf{i} + 4.33\mathbf{j}$; 9.54 at $27°$ to Ox **(iii)** $11\mathbf{i}$; 11 along Ox

(iv) $6.87\mathbf{i} + 4.92\mathbf{j}$; 8.45 at $35.6°$ to Ox **(v)** $3.5\mathbf{i} + 4.33\mathbf{j}$; 5.57 at $51.1°$ to Ox

2 **(a)** $48.77\,\text{km}$ **(b)** $051.6°$ **(c)** $4\,\text{h}\,20\,\text{min}$ **(d)** $11.25\,\text{km}\,\text{h}^{-1}$

3 $P = 8.39\,\text{N}$; $T = 13.05\,\text{N}$ **4** $T = 20.04\,\text{N}$ **5** $P = 10$; $Q = 17.32$ **6** $P = 5.77$; $R = 17.12$

7 $P = 20$; $R = 25.98$ **8** $T_1 = 31.11$; $T_2 = 23.83$; $T_3 = 25.84$; $\alpha = 21.2°$

9 $T_1 = T_3 = 8.66\,\text{N}$; $T_2 = 5\,\text{N}$; $\alpha = 30°$

Chapter 9

Exercise 9.1

1 $gf(x) = (1 + x^2)^3$; $6x(1 + x^2)^2$ **2** $fg(x) = \dfrac{1}{1 + \sqrt{x}}$; $\dfrac{-1}{2\sqrt{x}(1 + \sqrt{x})^2}$

3 **(a)** $fg(x) = 3x - 6\sqrt{x} + 1$; $3 - \dfrac{3}{\sqrt{x}}$ **(b)** $gf(x) = \sqrt{3x^2 - 6x + 1}$; $\dfrac{3(x - 1)}{\sqrt{3x^2 - 6x + 1}}$

4 **(a)** $5(x + 2)^4$ **(b)** $8(2x + 3)^3$ **(c)** $8(5 - 4x)^{-3}$

   (d) $\frac{3}{2}(3x - 1)^{-\frac{1}{2}}$ **(e)** $\frac{4}{3}(3 - 4x)^{-\frac{4}{3}}$ **(f)** $40(5x + 7)^7$

5 **(a)** $5\left(2 - \dfrac{1}{x^2}\right)\left(2x + \dfrac{1}{x}\right)^4$ **(b)** $3\left(1 + \dfrac{1}{x^2}\right)\left(x - \dfrac{1}{x}\right)^2$ **(c)** $24x(3x^2 + 4)^3$

   (d) $\dfrac{-2x}{(1 + x^2)^2}$ **(e)** $\dfrac{-6x^2}{(x^3 + 1)^3}$ **(f)** $\dfrac{-6}{(4x - 7)^{\frac{3}{2}}}$

6 **(a)** $12x^3(1 + x^4)^2$ **(b)** $30x(2x - 1)(4x^3 - 3x^2 + 2)^4$ **(c)** $\frac{2}{3}(6x - 5)(3x^2 - 5x)^{-\frac{1}{3}}$

   (d) $12(1 - x)(2x^2 - 4x)^{-4}$ **(e)** $(7x - 4)(7x^2 - 8x + 4)^{-\frac{1}{2}}$ **(f)** $\dfrac{9}{2}\left(5x^2 + \dfrac{1}{x^4}\right)\left(5x^3 - \dfrac{1}{x^3}\right)^{\frac{1}{2}}$

7 **(a)** $6(4\sqrt{x} - 4x)\left(\dfrac{1}{\sqrt{x}} - 2\right)$ **(b)** $\dfrac{2(x - 1)}{\sqrt{x^3}}\left(\dfrac{1}{\sqrt{x}} + \sqrt{x}\right)^3$

   (c) $-\dfrac{1}{3}\left(\dfrac{1}{\sqrt[3]{x^2}} - \dfrac{3}{x^2}\right)\left(\sqrt[3]{x} + \dfrac{1}{x}\right)^{-2}$ **(d)** $-\dfrac{1}{2x^2}\left(1 + \dfrac{1}{x}\right)^{-\frac{1}{2}}$

8 **(a)** $\dfrac{6x}{(1 - x^2)^4}$ **(b)** $\dfrac{1}{4}\left(\dfrac{1}{\sqrt{x}} + 12x\right)(\sqrt{x} + 3x^2)^{-\frac{1}{2}}$

   (c) $\dfrac{-2x(x^2 + 1)}{[(x^2 + 1)^2 + 2]^{\frac{3}{2}}}$ **(d)** $\dfrac{-3\sqrt{x}}{(1 + x^{\frac{3}{2}})^2}$

9 $f'(x) = 12(3x + 1)^3$; $f''(x) = 108(3x + 1)^2$

10 $\dfrac{1}{2}\left(1 - \dfrac{1}{x^2}\right)\left(x + \dfrac{1}{x}\right)^{-\frac{1}{2}}$ **11** -24 **12** $3y = 2x + 5$

13 $(2, \frac{1}{4})$ min   **14** $(1, -\frac{1}{4})$ max

Exercise 9.2

1 $1.2\pi \, \text{cm}^2 \, \text{s}^{-1}$ **2** $\dfrac{3}{2\pi} \, \text{cm} \, \text{s}^{-1}$ **3** $6 \, \text{cm}^2 \, \text{s}^{-1}$ **4** $0.96 \, \text{cm}^3 \, \text{s}^{-1}$ **5** $7.2 \, \text{cm}^2 \, \text{s}^{-1}$

6 $50\pi \, \text{cm}^3 \, \text{s}^{-1}$ **7** **(a)** $\dfrac{1}{36\pi} \, \text{cm} \, \text{s}^{-1}$ **(b)** $\dfrac{2}{3} \, \text{cm}^2 \, \text{s}^{-1}$ **8** $\dfrac{2}{\pi} \, \text{cm} \, \text{s}^{-1}$ **9** $\dfrac{3}{\pi x^2} \, \text{cm} \, \text{s}^{-1}$

10 $\dfrac{1}{12} \, \text{m} \, \text{s}^{-1}$ **11** $243\pi \, \text{cm}^3 \, \text{s}^{-1}$ **12** $h = 10 - (100 - r^2)^{\frac{1}{2}}$; $0.87 \, \text{cm} \, \text{s}^{-1}$ **13** $0.8 \, \text{cm}^2 \, \text{s}^{-1}$

Exercise 9.3

1 **(a)** $x(3x - 8)$ **(b)** $3(4x^3 - 5x^2 + 4)$ **(c)** $2(3x + 1)(9x - 2)$

2 $2x(2x^2 + 15x + 24)$ **3** $2(x - 1)(2x^2 - x - 7)$ **4** $2x(x + 5)(2x + 5)$ **5** $(x + 2)^2(x - 1)(5x + 1)$

6 $(33x + 1)(x + 1)^3(3x - 5)^6$ **7** $-2x(1 - x^2)(1 + 3x^2)$ **8** $\frac{1}{2}x(4 - 3x - 7x^3)(1 - x^3)^{-\frac{1}{2}}$

9 $2x(1+7x)\sqrt{(1+4x)}$ **10** $\dfrac{1}{2\sqrt{x}}(1+\sqrt{x})^2(1-\sqrt{x})(1-5\sqrt{x})$ **11** $2(6-7x)(2x+3)^2(4-x)^3$

12 $\dfrac{-2x^3}{\sqrt{1-x^4}}$ **13** $\frac{1}{2}(16x-1)(2x+1)^{\frac{1}{2}}(x-1)^{\frac{3}{2}}$

14 **(a)** $\dfrac{2}{(3x+1)^2}$ **(b)** $\dfrac{-4}{(x+5)^2}$ **(c)** $\dfrac{6}{(x+5)^2}$ **(d)** $\dfrac{2x-x^4}{(x^3+1)^2}$

15 $\dfrac{1}{(x+1)^2}$ **16** $\dfrac{5}{(x+3)^2}$ **17** $\dfrac{x^2-2x-1}{(x-1)^2}$ **18** $\dfrac{-4x}{(x^2-1)^2}$

19 $\dfrac{5+6x-x^2}{(3-x)^2}$ **20** $\dfrac{-6x(x+1)}{(x-1)^5}$ **21** $\dfrac{1}{\sqrt{x}\sqrt{(x+2)^3}}$

22 $\dfrac{3x^2-x-3}{(2x+1)^{\frac{3}{2}}}$ **23** $\dfrac{2}{(x+1)^{\frac{1}{2}}(x+5)^{\frac{3}{2}}}$ **24** $\dfrac{8(x+1)(2x-1)^2}{(2x+1)^2}$

25 $\dfrac{2x(3-2x^2)(1+x^2)^{\frac{3}{2}}}{(1-x^2)^{\frac{3}{2}}}$ **26** $\dfrac{-6x(4x+21)(3x^2+1)^2}{(4x^3-7)^3}$

27 $3x(x+2); (0,-4)\min,(-2,0)\max$ **28** $(-2,25),(\frac{2}{3},\frac{11}{27})$

29 $\dfrac{-2x}{(1+x^2)^2};\ \dfrac{2-6x^2}{(1+x^2)^3}$ **30** $\dfrac{dy}{dx}=\dfrac{1}{(1+x)^2}\neq 0$

31 $y=60x+52$ **32** $(x+3)(3x+1);\ (x+3)(2x+1)^2(12x^2+17x-17)$

Exercise 9.4

1 **(i)** **(a)** $x\dfrac{dy}{dx}+y=0$ **(b)** $\dfrac{-2}{x^2}$ **(c)** $\frac{1}{2}y^2$

(ii) **(a)** $2y\dfrac{dy}{dx}=4$ **(b)** $x^{-\frac{1}{2}}$ **(c)** $\dfrac{2}{y}$

(iii) **(a)** $2x+2y\dfrac{dy}{dx}=0$ **(b)** $\dfrac{-x}{\sqrt{(1-x^2)}}$ **(c)** $-\dfrac{\sqrt{1-y^2}}{y}$

2 **(a)** $\dfrac{-3x}{4y}$ **(b)** $\dfrac{3-2x}{2y-4}$ **(c)** $\dfrac{3x^2+4x-2xy+1}{x^2+2y}$ **(d)** $\dfrac{y^2-3x^2+3}{3y^2-2xy}$

3 **(a)** $\dfrac{-1}{\sqrt{3}}$ **(b)** -1 **4** $(2,-1),(2,4)$ **5** $(-3,-\frac{3}{2})\max,(-3,3)\min$ **6** $3y+2x-11=0$

7 $y=x-2, y=6-x$ **8** $\dfrac{dy}{dx}=-1;\ x^2+2xy+y^2=5 \Rightarrow (x+y)^2=5 \Rightarrow x+y=\pm 5$ (a pair of lines)

9 $\dfrac{-(3y+2x)}{(3x+2y)}$

Exercise 9.5

1 **(a)** $\dfrac{2t}{3}$ **(b)** $\dfrac{2t^2}{t^2-1}$ **(c)** $3t^2$ **(d)** $2t^2$ **(e)** $\dfrac{1}{t+1}$ **(f)** $\dfrac{1}{2(t+1)}$

2 $\dfrac{1}{t},\ -\dfrac{1}{6t^3}$ **3** **(a)** $-\dfrac{3t}{2},\ -\dfrac{3}{4t}$ **(b)** $-2t,2$

4 (a) $x^3 = (1 - y^2)$, $\dfrac{dy}{dx} = \dfrac{3x^2}{2y - 2}$ **(b)** $y = 3 - 4x + x^2$, $\dfrac{dy}{dx} = 2x - 4$

5 (a) $-\dfrac{1}{t^2}$ **(b)** -1 **(c)** $2t^2$ **6** $\dfrac{dy}{dx} = -\dfrac{1}{t}$; $x^2 + y^2 = 1$; $\dfrac{dy}{dx} = -\dfrac{x}{y}$

7 $-\dfrac{1}{t^2}$; $t^2 y + x = 6t$ **8** $\dfrac{3t^2 - 1}{2t}$; $\left(\dfrac{2}{3}, \dfrac{2}{3\sqrt{3}}\right)$ max; $\left(\dfrac{2}{3}, -\dfrac{2}{3\sqrt{3}}\right)$ min

Exercise 9.6

1 0.01 **2** 1.28 **3** $\dfrac{1}{100}(9x^3 - 6x^2)$ **4** 2% decrease

5 (a) $\dfrac{24\pi r^2}{100}$ **(b)** 6%

6 1.131; 113.097; 114.232; change = 1.135; error = 0.35%

7 (a) 2.005 **(b)** 3.003 **(c)** 2.009 **(d)** 10.013

8 (a) $\dfrac{12\pi r^3}{100}$ **(b)** $\dfrac{24\pi r^2}{100}$

Chapter 10

Exercise 10.1

1 22.5°, 67.5°, 90°, 112.5°, 157.5°, 202.5°, 247.5°, 270°, 292.5°, 337.5°

2 (a) 0°, 40°, 80°, 120°, 160°, 200°, 240°, 280°, 320°, 360°

(b) 0°, 20°, 60°, 100°, 140°, 180°, 220°, 260°, 300°, 340°, 360°

(c) 0°, 60°, 90°, 120°, 180°, 240°, 270°, 300°, 360°

3 (a) $\dfrac{\sqrt{3}+1}{2\sqrt{2}}$ **(b)** $\dfrac{\sqrt{3}-1}{2\sqrt{2}}$ **(c)** $\dfrac{\sqrt{3}-1}{2\sqrt{2}}$ **(d)** $\dfrac{\sqrt{3}+1}{2\sqrt{2}}$ **(e)** $2+\sqrt{3}$ **(f)** $2-\sqrt{3}$

5 (a) 0°, 90°, 180°, 240°, 270°, 360° **(b)** 45°, 135°, 225°, 240°, 315°

7 (a) $\dfrac{\sec A \sec B}{1 - \sqrt{(\sec^2 A - 1)}\sqrt{(\sec^2 B - 1)}}$ **(b)** $\dfrac{\operatorname{cosec} A \operatorname{cosec} B}{\sqrt{(\operatorname{cosec}^2 A - 1)}\sqrt{(\operatorname{cosec}^2 B - 1)}}$

8 (a) 45°, 225°, **(b)** 45°, 135°, 225°, 315° **(c)** 38.2° or 141.8°

9 (a) $\cos x$ **(b)** $\sin \theta$ **(c)** $\sec \alpha$ **(d)** $\cot x$ **(e)** $\tan A$

10 (a) $\dfrac{8}{17}$ **(b)** $\dfrac{15}{17}$ **(c)** $\dfrac{15}{8}$ **(d)** $\dfrac{17}{15}$ **(e)** $\dfrac{17}{8}$ **(f)** $\dfrac{240}{161}$

Exercise 10.2

1 (a)

Radians	$\pi/180$	$\pi/18$	$\pi/12$	$\pi/6$	$\pi/4$	1^c	$\pi/3$	$5\pi/12$	$\pi/2$	2^c	$2\pi/3$	$3\pi/4$	$5\pi/6$	π	$3\pi/2$	2π	4π
Degrees	1°	10°	15°	30°	45°	57.3°	60°	75°	90°	114.6°	120°	135°	150°	180°	270°	360°	720°

(b)

Radians	$\pi/10$	$\pi/8$	$\pi/6$	$\pi/5$	$\pi/4$	$\pi/3$	$\pi/2$	π	$3\pi/4$	$5\pi/6$	1.5^c	1^c	1.22^c	0.88^c	0.357^c
Degrees	18°	22.5°	30°	36°	45°	60°	90°	180°	135°	150°	85.9°	57.3°	70°	50°30′	20°26′

2 87.3 cm² **3 (a)** 26.58 cm² **(b)** 287.6 cm² **(c)** 66.53 cm **4** 108.7 cm²

5 **(a)** 57.08 cm² **(b)** 21.46 cm² **6** **(a)** 748.6 cm **(b)** 124.9 cm²

7 **(a)** $2\pi n + 0.305$ or $2\pi n + 2.837$ **(b)** $2\pi n \pm 1.16$ **(c)** $n\pi + 0.464$ **(d)** $n\pi + 1.03$ **(e)** $2\pi n \pm 1.047$

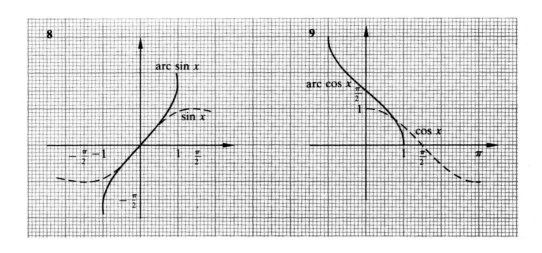

10 122.84 cm²

Exercise 10.3

1 **(a)** $\cos 10° = 0.984\,80$ **(b)** $\cos 5° = 0.996\,194\,7$ **(c)** $\cos 3° = 0.998\,629\,5$
$1 - \theta^2/2 = 0.984\,77$ $1 - \theta^2/2 = 0.996\,192\,2$ $1 - \theta^2/2 = 0.998\,629\,2$

(d) $\cos 1° = 0.999\,847\,7$
$1 - \theta^2/2 = 0.999\,847\,6$

2 $\cos(30 + 1) \simeq 0.857\,166\,7$; $\cos 31$ (directly) $\simeq 0.857\,167\,3$ **3** $\tan(60 + 1) = 1.804\,040\,1$

4 **(a)** $\dfrac{\theta}{1 - \theta^2} \simeq \theta$ **(b)** $\dfrac{1}{\theta}$ **5** **(a)** 6 **(b)** $\dfrac{\theta^2}{1 + \theta^2} \to 0$ **(c)** 1 **(d)** -3 **(e)** $2\theta \to 0$

(f) $\dfrac{2}{\theta} \to \infty$

Exercise 10.4

1 $-\operatorname{cosec}^2 x$ **2** $-\operatorname{cosec} x \cot x$ **3** $2 \cos 2x$ **4** $3 \cos 3x$ **5** $p \cos px$

6 $\cos(x + 3)$ **7** $-2 \sin(2x - 1)$ **8** $\cos(x + \pi/2)$ **9** $-\cos[(\pi/2) - x]$ **10** $\sec^2(x + \pi/4)$

11 $\sin 2x$ **12** $-\sin 2x$ **13** 0 **14** $3 \sin^2 x \cos x$ **15** $-3 \cos^2 x \sin x$

16 $2x \cos x^2$ **17** $3x^2 \sec^2(x^3)$ **18** $-4x^3 \sin x^4$ **19** $5x^4 \sec x^5 \tan x^5$

20 $-6x^5 \operatorname{cosec} x^5 \cot x^5$ **21** $8 \sin^7 x \cos x$ **22** $-13 \cos^{12} x \sin x$ **23** $4 \tan^3 x \sec^2 x$

24 $5 \sec^5 x \tan x$ **25** $-6 \cot^5 x \operatorname{cosec}^2 x$ **26** $3 \sin 6x$ **27** $6 \sin^2 2x \cos 2x$

28 $6 \tan^2 2x \sec^2 2x$ **29** $-16 \cos^3 4x \sin 4x$ **30** $2 \tan x \sec^2 x$ **31** $2 \sec^2 x \tan x$

32 $-2 \operatorname{cosec}^2 x \cot x$ **33** $-2 \cot x \operatorname{cosec}^2 x$ **34** $\cos x - \sin x$ **35** $\cos 2x$

36 $\sin x \sec^2 x + \sin x$ **37** $\cos x$ **38** $\dfrac{\cos \sqrt{x}}{2\sqrt{x}}$ **39** $\dfrac{\cos x}{2\sqrt{\sin x}}$ **40** $-\dfrac{1}{x^2} \cos \dfrac{1}{x}$

41 $-\operatorname{cosec} x \cot x$ **42** $\cos x - x \sin x$ **43** $x^2 \cos x + 2x \sin x$ **44** $(x^2 + 1) \sec^2 x + 2x \tan x$

45 $-3\operatorname{cosec}^2(3x+2)$ **46** $\dfrac{\sin x - x\cos x}{\sin^2 x}$ **47** $\dfrac{x\sec^2 x - 2\tan x}{x^3}$ **48** $\sec x \tan x$

49 $-\operatorname{cosec} x \cot x$ **50** $\cos x$

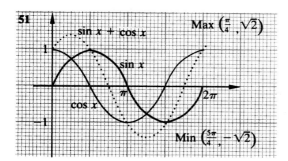

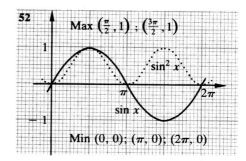

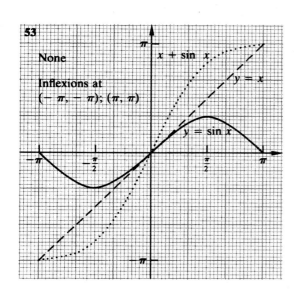

54 **(a)** $x\cos x + 2\sin x$
   (b) $x\sin x$

55 **(a)** $x^2 + y^2 = 25$

   (b) $\dfrac{y}{x} = \tan t$ **(c)** $-\cot t$

   (d) Gradient radius vector $\left(\dfrac{y}{x}\right) \times$ gradient of curve $\left(\dfrac{dy}{dx}\right) = -1 \;\Rightarrow\;$ curve is always $\perp$ to radius vector, i.e. curve is a circle.

   (e) $-\dfrac{1}{5\sin^3 t}$

56 $\pi/2$ **57** $3\cos 3x$; $\frac{1}{3}\sin 3x + c$ **58** $a\cos ax$; $\dfrac{1}{a}\sin ax$ **59** $-a\sin ax$; $-\dfrac{1}{a}\cos ax + c$

60 $a\sec^2 ax$;   **(a)** $\dfrac{1}{a}\tan ax$ **(b)** $\dfrac{1}{a}\tan ax - x$

61 $-a\operatorname{cosec}^2 ax$ **(a)** $\int \operatorname{cosec}^2 ax\,dx = -\dfrac{1}{a}\cot ax + c$ **(b)** $\int \cot^2 ax\,dx = \int(\operatorname{cosec}^2 ax - 1)dx$

$$= -\dfrac{1}{a}\cot ax - x + c$$

Exercise 10.5

1 **(a)** $\dfrac{-1}{\sqrt{1-x^2}}$ **(b)** $\dfrac{-2}{\sqrt{1-4x^2}}$ **(c)** $\dfrac{-1}{\sqrt{9-x^2}}$ **(d)** $\dfrac{1}{x\sqrt{x^2-1}}$ **(e)** $\dfrac{-1}{1+x^2}$

2 **(a)** $\dfrac{a}{1+a^2x^2}$; $\int \dfrac{1}{1+a^2x^2}\,dx = \dfrac{1}{a}\tan^{-1}(ax)$ **(b)** $\dfrac{b}{b^2+x^2}$; $\int \dfrac{1}{b^2+x^2}\,dx = \dfrac{1}{b}\tan^{-1}\dfrac{x}{b}$

3 **(a)** $\pi/2$ **(b)** $\pi/4$

4 $\sin^{-1}x + \dfrac{x}{\sqrt{1-x^2}}$; $\displaystyle\int_0^1 \sin^{-1}x\,dx = \left[x\sin^{-1}x\right]_0^1 - \int \dfrac{x}{\sqrt{1-x^2}}\,dx = \dfrac{\pi}{2} - 1$

Chapter 11

Exercise 11.1

1   **(a)** $v = 6 + 12t$; $a = 12$.   Motion with constant acceleration 12; starting velocity 6.

    (b) $v = 3t^2 - 12t + 11$; $a = 6t - 12$. Similar to Worked Example **1**.

    (c) $v = -3\sin t$; $a = -3\cos t$.   Oscillating motion (S.H.M.) between $-3 \leqslant s \leqslant +3$

2    $a = 2t - 3$; $s = \dfrac{t^3}{3} - \dfrac{3t^2}{2} + 2t + 1$.   Distance travelled $= 1\frac{1}{2} - 1 = \frac{1}{2}$

3    $v = t^2 + 3t + 2$; $s = \dfrac{1}{3}t^3 + \dfrac{3t^2}{2} + 2t + 3$; $3\frac{5}{6}$; $v_1 = 6$

4   **(a)** $v = 20$; $a = -10$ **(b)** $s = 20$; $t = 2$ **(c)** $s = 0$ when $t = 0$ or $t = 4$; $v_0 = 20$; $v_4 = -20$

    (d)   40; displacement 0 in first 4 seconds

5    $v = 20 - 10t$; $s = 20t - 5t^2$

6   **(a)** $v = 5\cos t$; $a = -5\sin t$

    (c)   Oscillation S.H.M. between
       $-5 \leqslant s \leqslant +5$

6(b)

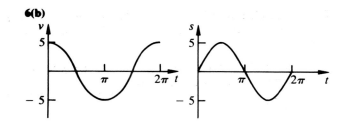

7    $v = 5 + 6t$; $s = 5t + 3t^2$; $t = \dfrac{v - 5}{6}$, $s = \dfrac{v^2 - 25}{12}$

8   **(a)** $v = u + at$ **(b)** $s = ut + \frac{1}{2}at^2$ **(c)** $a = \dfrac{v - u}{t}$ **(d)** $t = \dfrac{v - u}{a}$; $v^2 = u^2 + 2as$

Exercise 11.2

1   **(a)** $-0.8\,\mathrm{m\,s^{-2}}$ **(b)** $-2\,\mathrm{m\,s^{-1}}$ 2   **(a)** $-14\,\mathrm{m\,s^{-1}}$ **(b)** $s = -20\,\mathrm{m}$

3   **(a)** $9\,\mathrm{m\,s^{-1}}$ **(b)** $-0.75\,\mathrm{m\,s^{-2}}$ 4   **(a)** $65\,\mathrm{m}$ **(b)** $23\,\mathrm{m\,s^{-1}}$ 5   **(a)** $t = \frac{1}{2}\mathrm{s}$ **(b)** $2\,\mathrm{m}$

6   **(a)** $1\,\mathrm{s}$ **(b)** $5\,\mathrm{m}$ 7   **(a)** $-\frac{15}{8}\,\mathrm{m\,s^{-2}}$ **(b)** $\frac{8}{3} = 2\frac{2}{3}\mathrm{s}$ 8   **(a)**   **(i)** 1.82 s   **(ii)** 2 s   **(iii)** 2.16 s

    (b)   **(i)** $10.2\,\mathrm{m\,s^{-1}}$ **(ii)** $12\,\mathrm{m\,s^{-1}}$ **(iii)** $13.6\,\mathrm{m\,s^{-1}}$ 9   30.5 m; 37.5 m

10   $a = 1\,\mathrm{m\,s^{-2}}, u = \frac{1}{2}, v_1 = 1\frac{1}{2}; v_2 = 2\frac{1}{2}$ 11   2 s   12   **(a)**   150 m, 600 m, 150 m; 30 s, 30 s

    (b)   200 m, 600 m, 100 m; 50 s, 20 s

Exercise 11.3

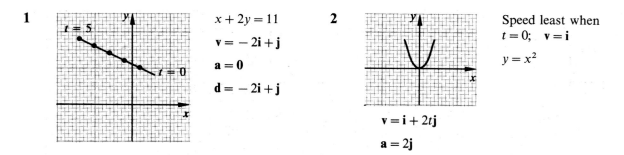

1
    $x + 2y = 11$

    $\mathbf{v} = -2\mathbf{i} + \mathbf{j}$

    $\mathbf{a} = 0$

    $\mathbf{d} = -2\mathbf{i} + \mathbf{j}$

2
    Speed least when
    $t = 0$; $\mathbf{v} = \mathbf{i}$

    $y = x^2$

    $\mathbf{v} = \mathbf{i} + 2t\mathbf{j}$

    $\mathbf{a} = 2\mathbf{j}$

3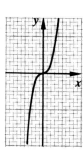

$$\mathbf{v} = 2t\mathbf{i} + \mathbf{j}$$
$$\mathbf{a} = 2\mathbf{i}$$
$$\mathbf{v}_{min} = \mathbf{j}$$
$$y^2 = x$$

4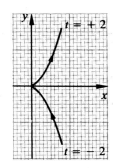

At origin, momentary at rest.

Graphs has a cusp (point).

5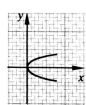

$$y = x^3$$
$$\mathbf{v} = \mathbf{i} + 3t^2\mathbf{j}$$
$$\mathbf{v}_{min} = \mathbf{i} \text{ at origin;}$$
$$t = 0$$

6 $x^2 + y^2 = 1;$ $\mathbf{v} \perp \mathbf{r}$

Circle starting at $(1, 0)$; when $t = 0$

$|\mathbf{v}| = 1$, uniform (constant). $\mathbf{a} = -\mathbf{r}$.

7 **(a)** Circle radius 2; same speed (angular).

(b) Circle radius 1; double angular speed. Speed along tangent same in **(a)** and **(b)**.

8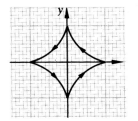

$x + y = 1$ in positive quadrant

$$\mathbf{v} = +\sin 2t(-\mathbf{i} + \mathbf{j})$$
$$\mathbf{a} = 2\cos 2t(-\mathbf{i} + \mathbf{j})$$

For a square, different equations for each section e.g. for $\pi/2 \leqslant t \leqslant \pi$

$$\mathbf{r} = -\sin^2 t\,\mathbf{i} + \cos^2 t\,\mathbf{j}$$

9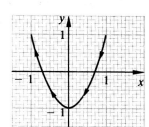

Astroid;

velocity and acceleration reverse at $\pi/2$.

10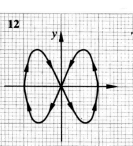

$|x| \leqslant 1$ $|y| \leqslant 1$

$$y = 2x^2 - 1$$

Only this part of parabola produced.

No t values will produce the whole curve.

For all $t = -1 \leqslant x \leqslant 1$ and $-1 \leqslant y \leqslant 1$.

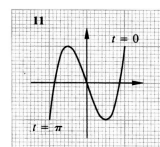

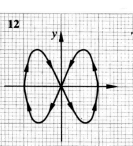

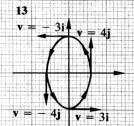

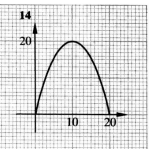

14 (cont) $y = 4x - \dfrac{x^2}{5};$ $\mathbf{v} = 5\mathbf{i} + (20 - 10t)\mathbf{j};$ $\mathbf{a} = -10\mathbf{j};$ $20.6\,\mathrm{m\,s}^{-1};$ $y\,\mathrm{max} = 20$ when $t = 2$.

Exercise 11.4

1 (a) $\mathbf{r}_1 = 10\mathbf{i} + 12.3\mathbf{j}, \mathbf{v}_1 = 10\mathbf{i} + 7.3\mathbf{j}$ (b) $\mathbf{r}_2 = 20\mathbf{i} + 14.6\mathbf{j}, \mathbf{v}_2 = 10\mathbf{i} - 2.7\mathbf{j}$

(c) $\mathbf{r}_3 = 30\mathbf{i} + 7\mathbf{j}, \mathbf{v}_3 = 10\mathbf{i} - 12.7\mathbf{j}$.   Hits ground when $t = 2\sqrt{3} \simeq 3.46\,\text{s}$.

2 (a) $1.25\,\text{m}$ (b) $5\sqrt{3} \simeq 8.66\,\text{m}$ (c) $1\,\text{s}$

3 $1.2\,\text{m}$ or $33.5\,\text{m}$ **4** $\sin^{-1}\frac{1}{3} = 19.5°$; $H = 1.25\,\text{m}$; $R = 14.2\,\text{m}$

5 $t = 2\,\text{s}$; $x = 12\,\text{m}$ **6** $20\sqrt{3} \simeq 34.64\,\text{m}$ **7** (a) $\mathbf{v}_1 = 6\mathbf{i} - 2\mathbf{j}$; $\mathbf{r}_1 = 6\mathbf{i} + 3\mathbf{j}$

(b) $\mathbf{v}_2 = 6\mathbf{i} - 12\mathbf{j}$; $\mathbf{r}_2 = 12\mathbf{i} - 4\mathbf{j}$ (c) $\mathbf{v}_3 = 6\mathbf{i} - 22\mathbf{j}$; $\mathbf{r}_3 = 18\mathbf{i} - 21\mathbf{j}$; $y = \dfrac{4x}{3} - \dfrac{5x^2}{36}$

8 $3.8°$ **9** $\mathbf{u} = 2.5\mathbf{i} + 11\mathbf{j}$; $H = 6.05\,\text{m}$; $R = 5.5$; $y = 4.4x - 0.8x^2$

10 $t = 2.59\,\text{s}$.   Only if batsman can run at $7.73\,\text{m}\,\text{s}^{-1}$ to cover $20\,\text{m}$ in $2.59\,\text{s}$.

11 $t = 1.83\,\text{s}$ **12** (a) $14\,\text{m}\,\text{s}^{-1}$ (b) $25\,\text{m}\,\text{s}^{-1}$ (c) $27\,\text{m}\,\text{s}^{-1}$ (d) $22\,\text{m}\,\text{s}^{-1}$

(e) $20\,\text{m}\,\text{s}^{-1}$ (f) $55\,\text{m}\,\text{s}^{-1}$

13 Ball passes through $9\mathbf{i} + 2\mathbf{j}$ and $30\mathbf{i} + 2.44\mathbf{j}$; $u = 28.4\,\text{m}\,\text{s}^{-1}$; $0.77\,\text{s}$

14 $67.8°$ or $-7.76°$ **15** $22.4\,\text{m}\,\text{s}^{-1} = 50\,\text{mph}!$

Chapter 12

Exercise 12.1

1 $40\,\text{km}\,\text{h}^{-1}$ **2** $280\,\text{km}\,\text{h}^{-1}$ in direction of A

3 (a) $3\mathbf{i} - \mathbf{j}$ (b) $-3\mathbf{i} - 5\mathbf{i}$ (c) $-6\mathbf{i} + 6\mathbf{j} + 4\mathbf{k}$

4 (a) $\sqrt{50}$, S $81°\,52'$ W   (b) 2 due west   (c) 5, S $53°\,8'$ E

5 $6.4\,\text{km}\,\text{h}^{-1}$ from N $51°\,20'$ E   **6** 25 knots on a bearing $171°\,52'$

7 9.17 knots on a bearing $120°\,54'$ **8** $75°\,4'$ to vertical

9 $5\,\text{m}\,\text{s}^{-1}$ at $53°\,8'$ to the vertical from south; $9.5\,\text{m}\,\text{s}^{-1}$ at $71°\,34'$ to the vertical from south.

10 $41.18\,\text{km}\,\text{h}^{-1}$ at $29°\,3'$ to direction of bus.

11 $3\mathbf{i} - 2\mathbf{j}, \mathbf{i} - \mathbf{j}$ **12** $3\mathbf{i} + 3\mathbf{j} + \mathbf{k}, 2\mathbf{j} + 2\mathbf{k}$ **13** 20 knots, N $30°$ E   **14** $\frac{1}{2}\mathbf{j}$

15 $1039\,\text{km}\,\text{h}^{-1}$ on bearing $120°$

Exercise 12.2

1 (a) $\mathbf{r}_A = 2t\mathbf{i} + t\mathbf{j}$, $\mathbf{r}_B = (4 + t)\mathbf{i} + (2t - 4)\mathbf{j}$; $t = 4$

(b) $\mathbf{r}_A = (1 - t)\mathbf{i} + (1 + 3t)\mathbf{j}$, $\mathbf{r}_B = (-1 + t)\mathbf{i} + 4t\mathbf{j}$; $t = 1$

2 2 seconds after P starts   **3** 10

4 (a) $10\,\text{km}\,\text{h}^{-1}$ at $53°\,8'$ to direction of man; 1.8 minutes   (b) $(100t^2 - 6t + \frac{1}{4})^{\frac{1}{2}}$; $0.4\,\text{km}$

5 $7.81\,\text{km}$; 11.7 minutes   **6** $167°\,48'$; 13.2 minutes   **7** 8.1 minutes   **8** $145°\,33'$; 7.7 seconds

9 5.4 minutes   **10** $53°\,8'$ **11** 6.32 seconds

Chapter 13

Exercise 13.1
1 (a) 5040 (b) 24 (c) 120 (d) 13!

2 (a) $\dfrac{7!}{3!} = 840$ (b) $\dfrac{15!}{5!}$ (c) $\dfrac{5!}{3!} = 20$ (d) $\dfrac{25!}{20!}$ (e) $\dfrac{120!}{118!} = 14\,280$

3 24 4 720 5 1320 6 120 7 120 8 1680

Exercise 13.2
1 (a) 60 (b) 990 (c) 2584 (d) 240 (e) 576 (f) 100 (g) $\frac{8}{5}$ (h) 105

2 (a) $\dfrac{8!}{5!}$ (b) $\dfrac{15!}{11!}$ (c) $\dfrac{8!}{6!3!}$ (d) $\dfrac{7!}{(3!)^2}$ (e) $\dfrac{14!}{10!4!}$ (f) $\dfrac{n!}{(n-3)!}$

3 (a) $5 \times 3!$ (b) $131 \times 10!$ (c) $9 \times 4!$ (d) $\dfrac{4 \times 8!}{3!}$ (e) $\dfrac{2 \times 10!}{3!}$

(f) $\dfrac{15!}{12!3!}$ (g) $(n-1)(n-1)!$ (h) $(n+4)(n+1)!$ (i) $\dfrac{11 \times 7!}{5!3!}$

Exercise 13.3
1 24 2 2520 3 (a) $2 \times 11!$ (b) $2 \times 10!$ 4 $6 \times 7!$

5 840 6 (a) 1680 (b) 5040 7 65 8 132 9 8640

10 (a) (i) 60 (ii) 840 (iii) 30

(b) (i) 60 (ii) 208 (iii) 30

11 480 12 1440 13 5040 14 480

15 168 16 (a) 14256 (b) 312 17 $\dfrac{10!}{4!(2!)^2} = 37\,800$

18 4320 19 $14!; 3!4!(5!)^2 = 2\,073\,600$ 20 $719 = 6! - 1$ 21 $2^{10} \times 10!$

Exercise 13.4
1 (a) 120 (b) 10 (c) 70 (d) 10 3 $\dfrac{10!}{6!4!} = {}^{10}C_4$ or ${}^{10}C_6$

4 (a) 10 (b) 56 (c) 462 (d) 20 5 (a) 126 (b) 35 (c) 3003 (d) 495

6 210 7 924 8 252 9 3150 10 11; 7 11 840 12 126 13 210

14 (a) 7700 (b) 26208 15 12348 16 (a) 352 (b) 344

Chapter 14

Exercise 14.1
1 (a) 14, 17 (b) $\frac{1}{5}, \frac{1}{6}$ (c) 9, 3 (d) 19, 25 (e) 1, −2 (f) $\frac{9}{2}, 6$

(g) 945, −10395 (h) 125, 216 (i) 42, 56 (j) 33, 65

2 (a) $3, 9, 27; 3^{20}$ (b) 0, 2, 0; 2 (c) 1, 0, −1; 0 (d) $0, -\frac{1}{2}, 0; \frac{1}{20}$ (e) $\frac{2}{3}, \frac{4}{3}, \frac{9}{5}; \frac{800}{231}$

3 (a) 2, 6, 18, 54 (b) $4, \frac{1}{2}, 4, \frac{1}{2}$ (c) 5, 10, 30, 120 4 $1, \frac{8}{5}, \frac{11}{5}, \frac{14}{5}, \frac{17}{5}, 4$

Answers

5 **(a)** $u_{r+1} = 3u_r - 1$, $u_1 = 1$ **(b)** $u_{r+1} = u_r + 4$, $u_1 = 3$

   (c) $u_{r+1} = \frac{1}{4}u_r$, $u_1 = 100$ **(d)** $u_{r+1} = \dfrac{2}{u_r}$, $u_1 = 8$

6 **(a)** $u_r = 2^r - 1$ for $r = 1, 2, 3, \ldots$.   **(b)** $u_r = \dfrac{1}{2^{r-1}}$ for $r = 1, 2, 3, \ldots$.

   (c) $u_r = (r-1)^2 - 1$ for $r = 1, 2, 3, \ldots$.   **(d)** $u_r = 2r^2 + 1$ for $r = 1, 2, 3, \ldots$.

Exercise 14.2

1 **(a)** $u_n \to 0$ **(b)** $u_n \to \infty$ **(c)** $u_n \to 4$ **(d)** $u_n \to 1$ **(e)** $u_n \to 1$

   (f) u_n oscillates finitely between 0 and 2   **(g)** $u_n \to \infty$ **(h)** $u_n \to 3$ **(i)** $u_n \to \infty$ **(j)** $u_n \to \frac{1}{2}$

2 **(a)** $u_n \to 0$ **(b)** $u_n \to \infty$ **(c)** u_n oscillates finitely between -2 and 2

3 **(a)** limit; 0   **(b)** no limit   **(c)** limit; 3   **(d)** limit; 1

   (e) no limit   **(f)** oscillates between -1 and 1   **(g)** no limit   **(h)** limit; 3

Exercise 14.3

1 **(a)** $2 + 5 + 10 + 17 + 26$ **(b)** $4 + \frac{25}{9} + \frac{9}{4} + \frac{49}{25} + \frac{16}{9}$

   (c) $-1 + 1 + 3 + 5 + 7$ **(d)** $\frac{1}{2} + \frac{1}{3} + \frac{1}{4} + \frac{1}{5} + \frac{1}{6} + \frac{1}{7}$

   (e) $-2 + 3 - 5 + 9 - 17 + 33$ **(f)** $n(n+1) + (n+1)(n+2) + (n+2)(n+3) + (n+3)(n+4)$

2 **(a)** $\displaystyle\sum_{r=1}^{99} r$ **(b)** $\displaystyle\sum_{r=1}^{20} r^3$ **(c)** $\displaystyle\sum_{r=1}^{20} (5r - 3)$ **(d)** $\displaystyle\sum_{r=1}^{12} (6r - 8)$

   (e) $\displaystyle\sum_{r=2}^{16} r(r+1)$ **(f)** $\displaystyle\sum_{r=1}^{20} (-1)^r r$ **(g)** $\displaystyle\sum_{r=1}^{n} (18 - 2r)$ **(h)** $\displaystyle\sum_{r=1}^{2n+1} (4r + 1)^2$

3 **(a)** Yes   **(b)** No   **(c)** No   **(d)** Yes   **(e)** Yes   **(f)** No

4 **(a)** $44; 89$ **(b)** $-14; 0$ **(c)** $64; -17$ **(d)** $7; 11\frac{1}{2}$

   (e) $49;\ 4n + 9$ **(f)** $3;\ 53 - 2n$ **(g)** $\frac{1}{2}(5n + 3);\ \frac{1}{2}(5n + 8)$ **(h)** $10 + 20n$

5 **(a)** 17   **(b)** 20   **(c)** 15   **(d)** 9   **(e)** 21   **(f)** 18   **(g)** 11

6 **(a)** -126 **(b)** 220   **(c)** 370   **(d)** 2550   **(e)** 5050   **(f)** 625   **(g)** n^2

7 10 100   **8** 32   **9** -40 **10** $2;\ n^2 + 7n$

11 $8; -9; 1340$ **12** $24; 24$ **13** 0   **14** $157\frac{1}{2}$

Exercise 14.4

1 **(a)** Yes   **(b)** Yes   **(c)** No   **(d)** Yes   **(e)** No   **(f)** Yes   **(g)** Yes   **(h)** Yes

2 **(a)** $26\,244;\ 4 \times 3^{19}$ **(b)** $\frac{1}{256};\ (\frac{1}{2})^{19}$ **(c)** $-\frac{1024}{625};\ -5(\frac{4}{5})^{13}$

   (d) $\frac{3}{7}(\frac{3}{2})^9;\ \frac{3}{7}(\frac{3}{2})^{n-1}$ **(e)** $0.000\,04;\ 4 \times (0.1)^{n-2}$

   (f) $\frac{1}{9};\ (-\frac{1}{3})^{n-5}$ **(g)** $(\frac{2}{3})^{2n-1}$

3 **(a)** 7   **(b)** 12   **(c)** 8

4 **(a)** 6138   **(b)** $200[1 - (\frac{1}{2})^{11}]$ **(c)** $\frac{8}{7}[1 - (\frac{3}{4})^8]$

   (d) $\frac{1}{3}(1 - 2^{50})$ **(e)** $\frac{4}{3}(4^{13} - 1)$ **(f)** $270[1 - (-\frac{1}{3})^n]$

5 $8;\ \frac{3}{128}$ **6** $-15\frac{1}{2}$ **7** $(\frac{3}{2})^8;\ \frac{4}{3}[(\frac{3}{2})^{10} - 1]$ **8** 7   **9** 1714.5   **10** $-\frac{1}{2}$ or $\frac{5}{3}$

11 216   **12** $3;\ 8, 24, 72, 216$

Exercise 14.5

1 **(a)** $26[1-(\frac{1}{2})^n]$; 26   **(b)** $\frac{2}{3}[1-(\frac{1}{10})^n]$; $\frac{2}{3}$

(c) $320[1-(-\frac{1}{4})^n]$; 320   **(d)** $\frac{1}{5}(2^n-1)$ **(e)** $\frac{16}{7}[1-(-\frac{3}{4})^n]$; $\frac{16}{7}$

2 **(a)** 8   **(b)** $\frac{4}{5}$ **(c)** 3   **(d)** $\frac{10}{11}$ **(e)** $\dfrac{1}{1-a}$ **(f)** $\dfrac{1}{1-\cos\theta}$ **3** $-\frac{1}{3}$ **4** $-2\frac{1}{4}$

5 **(a)** $|x|<\frac{2}{3}$ **(b)** $|x|<\frac{1}{4}$ **(c)** $|x|<\frac{1}{3}$ **(d)** $|x|<\frac{1}{2}$

6 **(a)** $\frac{32}{99}$ **(b)** $\frac{7}{9}$ **(c)** $\frac{7}{111}$ **7** $\frac{189}{16}$ **8** $1-\frac{1}{2}+\frac{1}{4}$

Chapter 15

Exercise 15.1

1 **(a)** $5g\,\text{N}=49\,\text{N}$ **(b)** $2.4g\,\text{N}=23.52\,\text{N}$ **(c)** $0.75g\,\text{N}=7.35\,\text{N}$

2 $10\mathbf{i}+8\mathbf{j}$ **3** $\mathbf{i}+2\mathbf{j}-\frac{2}{3}\mathbf{k}$ **4** $4\mathbf{i}-2\mathbf{j}-8\mathbf{k}$

5 **(a)** $\frac{1}{3}\,\text{m s}^{-2}$ **(b)** $20\,\text{m s}^{-2}$

6 $12\,\text{N}$ **7** $\mathbf{F}=4\mathbf{i}+2\mathbf{j}+2\mathbf{k}$; $\mathbf{a}=\mathbf{i}+\frac{1}{2}\mathbf{j}+\frac{1}{2}\mathbf{k}$ **8** $2.45\,\text{m s}^{-2}$ **9** $7\frac{2}{3}\,\text{m s}^{-2}$; $1470\,\text{N}$

10 $1\frac{1}{3}\,\text{m s}^{-2}$ **11** $294\,\text{N}$ **12** $0.4\,\text{m s}^{-2}$ **13** $0.9\,\text{m s}^{-2}$ **14** $3.3\,\text{m s}^{-2}$ **15** $1660\,\text{N}$

16 Reactions   **(a)** $30g=294\,\text{N}$ **(b)** $330\,\text{N}$ **(c)** $240\,\text{N}$
     Tensions   **(a)** $8134\,\text{N}$ **(b)** $9130\,\text{N}$ **(c)** $6640\,\text{N}$

17 $812\,\text{N}$; $9512\,\text{N}$ **18** **(a)** $32.88\,\text{N}$ **(b)** $7.34\,\text{N}$ **(c)** $34.80\,\text{N}$ **19** $2\,\text{m s}^{-2}$ downwards

Exercise 15.2

1 $7\,\text{m s}^{-2}$; $35\,\text{m s}^{-1}$ **2** $16\,\text{N}$ **3** $\frac{80}{3}$ seconds   **4** 10.8 seconds; $55.5\,\text{m s}^{-1}$

5 $75\,\text{m}$ **6** 2.12 seconds   **7** $0.8\,\text{m s}^{-2}$; $1.6\,\text{m s}^{-1}$; $2.18\,\text{m}$ **8** $600\,\text{N}$ **9** $3125\,\text{N}$; $30.6\,\text{m s}^{-1}$

10 $1.2\,\text{m s}^{-1}$; $72.5\,\text{m}$ **11** $2968.5\,\text{m}$

Exercise 15.3

1 $\frac{g}{5}\,\text{m s}^{-2}$; $\frac{24g}{5}\,\text{N}$; $\frac{48g}{5}\,\text{N}$ **2** $\frac{g}{7}\,\text{m s}^{-2}$; $\frac{24g}{7}\,\text{N}$; $\frac{48g}{7}\,\text{N}$ **3** $2000\,\text{N}$; $750\,\text{N}$

4 $1.55\,\text{m s}^{-2}$; $5488\,\text{N}$ **5** $0.66\,\text{m s}^{-2}$; $36.57\,\text{N}$; $70.66\,\text{N}$ **6** $\frac{2g}{5}\,\text{m s}^{-2}$; $\frac{6g}{5}\,\text{N}$; $\frac{6\sqrt{2g}}{5}=16.63\,\text{N}$

7 $\frac{g}{5}\,\text{m s}^{-2}$; $\frac{6g}{5}\,\text{N}$; $\frac{8g}{5}\,\text{N}$ **8** $2\,\text{m s}^{-2}$; $34.5\,\text{N}$ **9** $1.69\,\text{s}$; $2.37\,\text{m s}^{-1}$

10 $4.43\,\text{m s}^{-1}$; $0.45\,\text{s}$ **11** $0.032\,\text{m s}^{-2}$;   **(a)** $19\,286\,\text{N}$;   **(b)** $3214\,\text{N}$

12 $0.6\,\text{m s}^{-2}$; $25\,\text{s}$;   **(a)** $48\,\text{N}$ **(b)** $72\,\text{N}$ **13** $3.06\,\text{s}$

Chapter 16

Exercise 16.1

1 **(a)** $1.52\,\text{m s}^{-2}$ **(b)** $1.7\,\text{m s}^{-2}$ **(c)** $1.66\,\text{m s}^{-2}$ **(d)** $a=1\,\text{m s}^{-2}$

2 **(a)** $0.6\,\text{N}$ **(b)** $15.1\,\text{N}$ **3** **(a)** 0.7 **(b)** $26.6°$ **(c)** $44.7\,\text{N}$ **(d)** $89.4\,\text{N}$ **(e)** $200\,\text{N}$

4 (a) 0.36 **(b)** Paperweight accelerates **(c)** More friction on base than when up ended

5 (a) 5 N **(b)** $4\sqrt{2}\,\text{N} \simeq 5.66\,\text{N}$ **(c)** 9.43 N **6** $2.4\,\text{m s}^{-2}$ **7** 208.4 N; 24.9 N

8 $\mu = 0$; $\mu = \infty$; μ very large for a surfaces fitted with 'Velcro'; for small values, do we count linear motors (frictionless) or hovercraft?

9

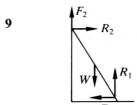

Friction and reactions at both ends, the weights of the ladder and climber.

With someone on the ladder, both reactions and therefore frictions, increase. As the climber climbs the ladder, F_1 has to increase to prevent slipping at the base, until F_1 reaches its limiting value, when the ladder is on the point of slipping.

10 (a) $3.54\,\text{m s}^{-2}$ **(b)** $3.54\,\text{m s}^{-2}$ **11** $F = 13\frac{1}{3}\,\text{N}$; $\mu = \frac{4}{15} \simeq 0.27$

12 $P = \dfrac{10}{\sqrt{17}} \simeq 2.42$, at an angle $\tan^{-1}(0.25) = 14°$ above horizontal. **13** $m = \frac{1}{3}\,\text{kg}$; $a = \frac{1}{3}\,\text{m s}^{-2}$

14

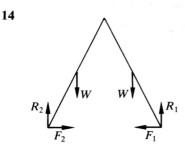

There are forces on each 'leg' at the top where each 'leg' pushes against the other, but for equilibrium they balance each other. By symmetry, $R_1 = R_2$ and $F_1 = F_2$.

Instead of the friction forces, or in addition to them, there may be a restraining bar or rope in tension holding the 'legs' together.

At the top, the force with which the heavier 'leg' pushes against the lighter one equals the force with which the lighter 'leg' pushes against the heavier 'leg'. But they may not be horizontal, so the lighter 'leg' can 'support' the heavier one.

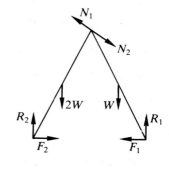

Resolving horizontally $\Rightarrow$ $F_1 = F_2$; $R_1 = 1\frac{1}{4}W$ and $R_2 = 1\frac{3}{4}W$

As someone climbs the ladder, R_2 increases but R_1 still helps,

i.e. $R_1 > W$

Chapter 17

Exercise 17.1

1 (a) $a^4 + 4a^3b + 6a^2b^2 + 4ab^3 + b^4$ **(b)** $x^6 + 6x^5y + 15x^4y^2 + 20x^3y^3 + 15x^2y^4 + 6xy^5 + y^6$

(c) $1 - 3x + 3x^2 - x^3$ **(d)** $32a^5 + 80a^4b + 80a^3b^2 + 40a^2b^3 + 10ab^4 + b^5$

(e) $x^4 - 8x^3y + 24x^2y^2 - 32xy^3 + 16y^4$ **(f)** $x^3 - 9x^2 + 27x - 27$ **(g)** $a^2 - 2ab + b^2$

(h) $128x^7 - 448x^6 + 672x^5 - 560x^4 + 280x^3 - 84x^2 + 14x - 1$ **(i)** $x^4 + 4x^2 + 6 + \dfrac{4}{x^2} + \dfrac{1}{x^4}$

(j) $x^{10} + 10x^8 + 40x^6 + 80x^4 + 80x^2 + 32$ **(k)** $x^{12} - 12x^9 + 60x^6 - 160x^3 + 240 - \dfrac{192}{x^3} + \dfrac{64}{x^6}$

(l) $8x^6 + 12x^2 + \dfrac{6}{x^2} + \dfrac{1}{x^6}$

2 $1 + 2x + \frac{3}{2}x^2 + \frac{1}{2}x^3 + \frac{1}{16}x^4$; 1.4641   **3** $32 + 80x + 80x^2 + 40x^3 + 10x^4 + x^5$; 32.808

4 $1 - 6x + 15x^2 - 20x^3 + 15x^4 - 6x^5 + x^6$; 0.970   **5** $1 + 2x - 26x^2 - 32x^3$

6 $32x^9 - 496x^8 + 2408x^7$ **7** $1 - 5x + 5x^2 + 10x^3$

Exercise 17.2

1 **(a)** $1 + 5x + 10x^2 + 10x^3 + 5x^4 + x^5$ **(b)** $64 - 192x + 240x^2 - 160x^3 + 60x^4 - 12x^5 + x^6$

  (c) $1 + 4x + 7x^2 + 7x^3 + \frac{35}{8}x^4 + \frac{7}{4}x^5 + \frac{7}{16}x^6 + \frac{1}{16}x^7 + \frac{1}{256}x^8$

  (d) $\frac{1}{128x^7} + \frac{7}{64x^5} + \frac{21}{32x^3} + \frac{35}{16x} + \frac{35}{8}x + \frac{21}{4}x^3 + \frac{7}{2}x^5 + x^7$ **(e)** $64 - 144x + 108x^2 - 27x^3$

  (f) $1 - 10x^2 + 45x^4 - 120x^6 + 210x^8 - 252x^{10} + 210x^{12} - 120x^{14} + 45x^{16} - 10x^{18} + x^{20}$

2 **(a)** $x^6 + 3x^3 + 3 + \frac{1}{x^3}$ **(b)** $1 - \frac{8}{x^2} + \frac{24}{x^4} - \frac{32}{x^6} + \frac{16}{x^8}$ **(c)** $x^{15} + 5x^{12} + 10x^9 + 10x^6 + 5x^3 + 1$

3 **(a)** $1 + 36x + 594x^2 + 5940x^3 + 40\,095x^4$ **(b)** $1024 - 1280x + 720x^2 - 240x^3 + \frac{105}{2}x^4 - \frac{63}{8}x^5$

  (c) $1 + 20x^2 + 190x^4 + 1140x^6 + 4845x^8$ **(d)** $1 - 9x + 36x^2 - 84x^3 + 126x^4 - 126x^5$

  (e) $3125 - 625x + 50x^2 - 2x^3$ **(f)** $\frac{1}{x^6} + \frac{12}{x^4} + \frac{60}{x^2} + 160$

4 **(a)** $-336x^5$ **(b)** $-\frac{15}{8}x^3$ **(c)** $15x^2$ **(d)** $\frac{15}{16}$

5 **(a)** $x^4 - 4x^3 + 6x^2 - 4x + 1$

  (b) $128x^7 - 448x^6 + 672x^5 - 560x^4 + 280x^3 - 84x^2 + 14x - 1$

  (c) $-3125x^5 + 6250x^4 - 5000x^3 + 2000x^2 - 400x + 32$

6 **(a)** $1 + 6x + 21x^2 + 50x^3 + 90x^4$ **(b)** $1 - 10x + 35x^2 - 40x^3 - 30x^4$

  (c) $81 - 216x + 324x^2 - 312x^3 + 214x^4$ **(d)** $128 + 448x + 672x^2 - 392x^3 - 1064x^4$

7 $a = -2, n = 5$ **8** $364x^2 : 135$ **9** $a = \frac{1}{9}, b = \frac{2}{9}$

10 $1 + 10x + 45x^2 + 120x^3 + 210x^4$; 1.105

11 $4096 - 24\,576x + 67\,584x^2 - 112\,640\,x^3$; 4022.88

12 $1 - 4x - 33x^2 + 118x^3$ **13** $96x^7$ **14** $\frac{115}{16}x^4$

Exercise 17.3

1 $1 - x + x^2 - x^3$; $-1 < x < 1$ **2** $1 - 4x + 12x^2 - 32x^3$; $-\frac{1}{2} < x < \frac{1}{2}$

3 $1 - \frac{1}{2}x - \frac{1}{8}x^2 - \frac{1}{16}x^3$; $-1 < x < 1$ **4** $1 - 3x + \frac{3}{2}x^2 + \frac{1}{2}x^3$; $-\frac{1}{2} < x < \frac{1}{2}$

5 $\sqrt{2}\left[1 + \frac{x}{4} - \frac{1}{32}x^2 + \frac{1}{128}x^3\right]$; $-2 < x < 2$ **6** $1 - x - x^2 - \frac{5}{3}x^3$; $-\frac{1}{3} < x < \frac{1}{3}$

7 $1 + \frac{1}{2}x^2$; $-1 < x < 1$ **8** $\frac{1}{4}[1 - x + \frac{3}{4}x^2 - \frac{1}{2}x^3]$; $-2 < x < 2$

9 $\frac{1}{4}\left[1 - \frac{x}{4} + \frac{x^2}{16} - \frac{x^3}{64}\right]$; $-4 < x < 4$ **10** $2 - \frac{1}{4}x^2 + \frac{5}{64}x^4$; $-1 < x < 1$

11 $1 + \frac{3}{2}x + \frac{7}{8}x^2 + \frac{11}{16}x^3 + \frac{75}{128}x^4$

12 **(a)** $1 + 2x + 2x^2 + 2x^3$ **(b)** $2 - 3x + 4x^2 - \frac{13}{2}x^3$ **(c)** $-\frac{1}{4} + \frac{3}{4}x - \frac{11}{16}x^2 + \frac{1}{2}x^3$

13 $8 - 3x + \frac{3}{16}x^2 + \frac{3}{128}x^3$;   7.97   **14** **(a)** $1 + x - x^3 - x^4$ **(b)** $1 - x - x^2 - x^3$

15 $1 - 8x + 24x^2 - 72x^3$; $|x| < \frac{1}{3}$ **16** $1 + x + \frac{1}{2}x^2 + \frac{1}{2}x^3 + \frac{3}{8}x^4 + \frac{3}{8}x^5 + \frac{5}{16}x^6 + \frac{5}{16}x^7$;   1.732

17 **(a)** $\dfrac{1}{2} + \dfrac{x}{4} - \dfrac{x^2}{8} + \dfrac{x^3}{16}$ **(b)** $\dfrac{3}{8}\left[1 - \dfrac{x}{4} - \dfrac{3x^2}{16} - \dfrac{7x^3}{32}\right]$ **(c)** $1 + 2x^3 - 3x^4 - 6x^5$

18 **(a)** $\dfrac{1}{x} - \dfrac{2}{x^2} + \dfrac{4}{x^3}$ **(b)** $1 - \dfrac{1}{x} + \dfrac{2}{x^2}$ **(c)** $-\dfrac{1}{x} - \dfrac{1}{x^3} - \dfrac{1}{x^5}$ **(d)** $\dfrac{1}{x^2} + \dfrac{5}{x^3} + \dfrac{31}{x^4}$

Exercise 17.4

1 **(a)** $\dfrac{1 - (n+1)x^n + nx^{n+1}}{(1-x)^2}$ **(b)** $\dfrac{1 - (4n-3)x^n}{(1-x)} + \dfrac{4x(1 - x^{n-1})}{(1-x)^2}$

2 $\dfrac{3 - (2n+1)x^n}{(1-x)} + \dfrac{2x(1 - x^{n-1})}{(1-x)^2}$;   **(a)** 3841   **(b)** $\dfrac{789}{512}$

3 **(a)** 765   **(b)** 472   **(c)** $n^2 - n + 1$

4 **(a)** $\dfrac{n}{3n+1}; \dfrac{1}{3}$ **(b)** $\dfrac{n}{2(n+2)}; \dfrac{1}{2}$ **(c)** $\dfrac{1}{24} - \dfrac{1}{6(3n+1)(3n+4)}; \dfrac{1}{24}$ **(d)** $\dfrac{1}{12} - \dfrac{1}{4(2n+1)(2n+3)}; \dfrac{1}{12}$

   (e) $\dfrac{1}{8} - \dfrac{(4n+3)}{8(2n+1)(2n+3)}; \dfrac{1}{8}$ **(f)** $\dfrac{17}{96} + \dfrac{1}{8}\left(\dfrac{1}{n+2} + \dfrac{1}{n+3} - \dfrac{5}{n+4} - \dfrac{5}{n+5}\right); \dfrac{17}{96}$

5 $\dfrac{3}{2} - \dfrac{n+2}{n(n+1)}$

Exercise 17.5

1 **(a)** 7381   **(b)** 1015   **(c)** 2025   **(d)** $n(2n-1)$ **(e)** $\frac{1}{6}n(n-2)(n-1)(2n-3)$

   (f) $(n+2)^2(2n+3)^2$ **(g)** 522   **(h)** 1235   **(i)** $\frac{3}{4}n^2(n+1)(5n+1)$ **(j)** n^2

   (k) $\frac{2}{3}n(n+1)(2n+1) - 4$ **(l)** $(n+1)^2[2n^2 + 4n + 1] - 1$

2 **(a)** $\frac{1}{3}n(n+1)(n+2)$ **(b)** $\frac{16}{3}n(n+1)(2n+1) - 8$ **(c)** $\frac{1}{3}n(2n+5)(2n+7)$ **(d)** $n(n+1)^2$

3 $\frac{1}{12}n(n+1)(3n^2 + 19n + 8)$ **4** **(a)** $\frac{1}{6}n(n+1)(2n+7)$ **(b)** $\frac{1}{6}n(3n^3 + 10n^2 + 6n - 7)$

5 $\frac{1}{3}n(2n+1)(12n^2 + 2n - 1)$ **6** **(a)** $\frac{1}{6}n(3n^3 + 28n^2 + 87n + 98)$ **(b)** $\frac{1}{12}n(n+1)(3n^2 + 7n + 8)$

7 $\frac{1}{12}n(3n^3 + 38n^2 + 153n + 238)$ **8** $\frac{1}{12}n(n+1)(3n^2 + 19n + 26)$

Chapter 18

Exercise 18.1

1 **(a)** Line $x = 2$ **(b)** Line $x = 2$ **(c)** Plane $x = 2$

2 **(a)**   **(b)** Line $y = 2x - 3$ **(c)** Plane $y = 2x - 3$

3 (a)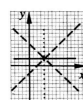

(b) Lines $y = x - 1$
$x + y = 3$

(c) Planes $y = x - 1$
and $x + y = 3$

4 (a)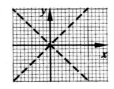

(b) Lines $y = x$
and $y = -x$

(c) Planes $y = x$
and $y = -x$

5 (a)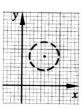

(b) Circle
$(x - 3)^2 + (y - 4)^2 = 4$

(c) Sphere
$(x - 3)^2 + (y - 4)^2 + z^2 = 4$

6 (a)

(b) Line $y = \frac{1}{2}x$

(c) Plane $y = \frac{1}{2}x$

7 (a)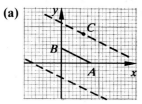

(b) Circle on AB as
diameter
$(x - 2)^2 + (y - 1)^2 = 5$

(c) Sphere

8 (a)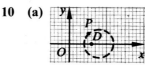

(b) Two major arcs
of circles, AB as
common chord.

(c) Apple-shaped

9 (a)

(b) Line through $C \parallel AB$
$x + 2y = 11$ and its
reflection in AB,
i.e. $x + 2y = -3$

(c) Cylinder

10 (a)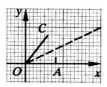

(b) Circle centre $(4, 0)$,
$r = 2$
$x^2 + y^2 - 8x + 12 = 0$

(c) Cylinder

Exercise 18.2

1 (a) $(x - 3)^2 + (y - 4)^2 = 4$; $x^2 + y^2 - 6x - 8y + 21 = 0$ **(b)** $(x - 3)^2 + (y - 4)^2 = 25$; $x^2 + y^2 - 6x - 8y = 0$

(c) $(x - 3)^2 + y^2 = 16$; $x^2 + y^2 - 6x - 7 = 0$ **(d)** $x^2 + (y - 2)^2 = 9$; $x^2 + y^2 - 4y - 5 = 0$

(e) $(x - 3)^2 + (y + 4)^2 = 4$; $x^2 + y^2 - 6x + 8y + 21 = 0$ **(f)** $(x + 3)^2 + (y - 2)^2 = 16$; $x^2 + y^2 + 6x - 4y - 3 = 0$

(g) $(x + 1)^2 + (y + 2)^2 = 9$; $x^2 + y^2 + 2x + 4y - 4 = 0$

(h) $(x - 5)^2 + (y - 12)^2 = 169$; $x^2 + y^2 - 10x - 24y = 0$

2 (a) $(-3, 2)$; $r = 5$ **(b)** $(3, 4)$; $r = 4$ **(c)** $(4, -5)$; $r = 13$ **(d)** $(0, -3)$; $r = 14$

3 (a) $(1, 2)$; $r = 1$ **(b)** $(3, 4)$; $r = 2$ **(c)** $(-2, -1)$; $r = 4$ **(d)** $(-\frac{1}{2}, -1\frac{1}{2})$; $r = 1\frac{1}{4}$

(e) $(2, -6)$; $r = 8$ **(f)** $(5, 6)$; $r = \sqrt{61}$ **(g)** $(-3, 0)$; $r = 4$ **(h)** $(0, 0)$; $r = 3$

Exercise 18.3

1 $(1, 1), (4, 4)$ **2** $(1, 1), (5, 1)$ **3** $(1, 3), (4, 0)$ **4** $(2, 4)$ **5** $(4, 1), (6, 3)$ **6** $(5, 10), (1, 2)$; $(2, 1), (10, 5)$

7 $T(3, 6), S(6, 3)$; $OC = 5\sqrt{2}$; $OT = OS = 3\sqrt{5}$ **8** SC is $y + 2x = 15$; Area $= 10$ **9** $(6, 3), (3, 4)$

10 $P(1, 1), Q(4, 4), R(2, 0), S(5, 1)$; $\dfrac{XR}{XQ} = \dfrac{\sqrt{10}}{5\sqrt{2}} = \dfrac{1}{\sqrt{5}}$; $\dfrac{RP}{QS} = \dfrac{\sqrt{2}}{\sqrt{10}} = \dfrac{1}{\sqrt{5}}$; $\dfrac{XP}{XS} = \dfrac{2\sqrt{2}}{2\sqrt{10}}$

Exercise 18.4

1 (a) $y = 3x - 7$; $3y + x = 11$ **(b)** $y = 2x - 2$; $2y + x = 1$ **(c)** $4y + x = 8$; $y = 4x - 15$

(d) $y = x + 1$; $x + y = 3$ **(e)** $y = 4$; $x = 2$ **(f)** $y + 3x = 2$; $3y = x - 4$

Answers

2 $m = 3.081$ or -0.081 **3** $m = -0.172$ or -5.828

4 **(a)** $y = x - 1$; $x + y = 3$ **(b)** $y = 2x - 2$; $2y + x = 6$ **(c)** $x + y = 4$; $y = x$

   (d) $2y = 3x - 1$; $3y + 2x = 5$ **(e)** $3y + 4x = 25$; $4y = 3x$

5 $y = 2x - 4\frac{1}{4}$; $16y + 8x = 7$ **(6)** 3.245 or -1.245

Exercise 18.5

1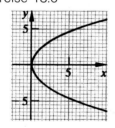

Parallel lines
reflects through
focus F.

2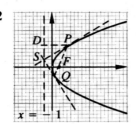

Directrix

Tangents at P and Q
(focal chord) meet
and $x = -1$ at S.
$D\hat{P}S = S\hat{P}F$

3

Focus $(\frac{1}{4}, 0)$

Directrix $y = -\frac{1}{4}$

4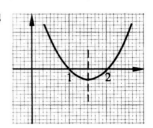

Axis $y = 1\frac{1}{2}$

Focus $(1\frac{1}{2}, 0)$

Directrix $y = -\frac{1}{2}$

5 $y = x + 1$; $x + y = 3$

6 **(a)** $y^2 = 8x$ **(b)** $y = \frac{x}{4}(x - 4) = \frac{1}{4}x^2 - x$ **(c)** $(y - 1)^2 = 4(x - 1)$ **(d)** $y = 2 - \frac{1}{4}(x - 2)^2$

7 $(y - 2)^2 = 2(x - 1)$

Exercise 18.6

1

$y^2 = 4x$

$ty = x + t^2$

$y + tx = 2t + t^3$

$y = x + 1$

$y + x = 3$

2 Grad $PQ = \dfrac{2}{p + q}$

Equation of PQ is $(p + q)y = 2x - 2apq$

Tangent at P is $py = x + ap^2$

Tangent at Q is $qy = x + aq^2$

F on PQ $\Rightarrow$ $pq = -1$ $\Rightarrow$ tangents at right angles.

Tangents intersect on $x = apq = -a$.

3 Tangent at $(1, 2)$ **4** $(1, 2), (1, -2)$ **5** $c = 2$; $(4, 4)$

6 $(1, 2), (1, -2)$; Curves touch (tangents to each other).   **7** $(0, 0), (2, 2\sqrt{2}), (2, -2\sqrt{2})$

8 **(a)** Circle centre $(0, 0)$ radius 4; $x^2 + y^2 = 16$ **(b)** Ellipse centre $(0, 0)$; $\dfrac{x^2}{36} + \dfrac{y^2}{4} = 1$

   (c) Ellipse centre $(0, 0)$; $x^2 + \dfrac{y^2}{4} = \dfrac{64}{9}$

9 $3y = x + 5$ **10** Circle $(4\frac{1}{2}, 0)$ radius $1\frac{1}{2}$.   **11** Concurrent at $(4, 2)$ almost; $(4\frac{1}{6}, 1\frac{5}{6})$ exactly.

12 ABC is isosceles; $x + y = 6$ is line of symmetry. Hence proof. Orthocentre $(3\frac{2}{3}, 2\frac{1}{3})$.

13 Ellipse centre $(0, 0)$; $\dfrac{x^2}{25} + \dfrac{y^2}{7} = 1$

14
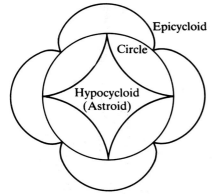

15
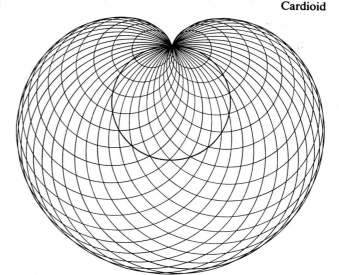

Chapter 19

Exercise 19.1

1 $13.5\,\text{rad}\,\text{s}^{-1}$ **(2)** $8.17\,\text{rad}\,\text{s}^{-1}$; $2.07\,\text{m}\,\text{s}^{-1}$ **3** $0.47\,\text{m}\,\text{s}^{-1}$ **4** 5.45 mins after 1 pm

5 Yes, 1 mile in $3\frac{1}{3}$ minutes **6** $1.51\,\text{m}\,\text{s}^{-1}$; $15.16\,\text{m}\,\text{s}^{-2}$

7 $0.314\,\text{m}\,\text{s}^{-1}$; $0.987\,\text{m}\,\text{s}^{-2}$. If anything, the pea should move away from the centre, but is prevented by friction and/or plate rim. Microwave cooks from outside, inwards.

8 $\frac{\pi}{3}\,\text{m}\,\text{s}^{-1}$; $\frac{\pi^2}{18}\,\text{m}\,\text{s}^{-2}$; $\frac{\pi}{6}\,\text{rad}\,\text{s}^{-1}$ **9 (a)** $3\,\text{rad}\,\text{s}^{-1}$ **(b)** 0 **(c)** $\frac{2}{4+t^2}$ **(d)** $\frac{1}{1+t^2}$

10 $465\,\text{m}\,\text{s}^{-1}$ ($=207\,\text{m.p.h}$); London: $v=290\,\text{m}\,\text{s}^{-1}$; N. Pole: standing still, but rotating.

Exercise 19.2

1 (a) $a=2\pi\simeq6.28\,\text{rad}\,\text{s}^{-2}$ **(b)** $\frac{\pi}{15}\,\text{rad}\equiv12°$ **(c)** $a=\pi\simeq3.14\,\text{rad}\,\text{s}^{-2}$ **(d)** $\frac{2\pi}{15}\,\text{rad}\equiv24°$

2 $2250\,\text{rad}=358\,\text{revs}$ **3** $360\,\text{revs}\,\text{min}^{-2}$; $\frac{\pi}{5}\,\text{rad}\,\text{sec}^{-2}$ **4** $\frac{8\pi}{3}\,\text{rad}\,\text{s}^{-2}$; $\omega=8\pi\,\text{rad}\,\text{s}^{-1}$

5 (a) $-0.5\,\text{rad}\,\text{s}^{-2}$ **(b)** $10\,\text{s}$ **(c)** $25\,\text{rad}=\frac{25}{2\pi}\,\text{revs}\simeq3.98\,\text{revs}$ **6** $\frac{3\pi}{2}\,\text{rad}\equiv270°$; $\frac{5\pi}{2}\,\text{rad}\equiv450°$

Exercise 19.3

1 (a) $\omega=1.5\,\text{rad}\,\text{s}^{-1}$; $45\,\text{N}$ **(b)** $12\,\text{m}\,\text{s}^{-1}$; $360\,\text{N}$ **(c)** $v=2\,\text{m}\,\text{s}^{-1}$; $\omega=0.4\,\text{rad}\,\text{s}^{-1}$

2 $36\,000\,\text{N}$; $10\,\text{m}\,\text{s}^{-1}$ **3** $\mu=0.002\,19$ **4** $\mu=\frac{\pi}{6}=0.52$

5 $v=26.5\,\text{m}\,\text{s}^{-1}$; $F=5091\,\text{N}$ ($2546\,\text{N}$, assuming $r=2\,\text{m}$) **6** $v=24.5\,\text{m}\,\text{s}^{-1}$; $F=1200\,\text{N}$ (radius 1 m)

Exercise 19.4

1 (a) $v=2.92\,\text{m}\,\text{s}^{-1}$; $T=66\frac{2}{3}\,\text{N}$ **(b)** $\theta=68.2°$; $T=107.7\,\text{N}$ **2** $2.74\,\text{m}\,\text{s}^{-1}$

3 $v=22.4\,\text{m}\,\text{s}^{-1}$; $T=50\,\text{N}$ **4** $v=\sqrt{3.6}=1.9\,\text{m}\,\text{s}^{-1}$; $48.5\,\text{N}$; $48\,\text{N}$ **5** $\frac{10}{9}=1\frac{1}{9}\,\text{m}$

6 Circle $x^2+y^2=9$ in plane $z=4$; angular velocity $2\,\text{rad}\,\text{s}^{-1}$

7 Helix whose section is a circle radius 2, centre $(0,0,0)$, angular velocity $1\,\text{rad}\,\text{s}^{-1}$, of depth (1 revolution) 4 m.

Exercise 19.5

1 $14.64\,\mathrm{m\,s^{-1}} \simeq 33\,\mathrm{m.p.h.}$ **2** $25.6°$ **3** $15.8\,\mathrm{m\,s^{-1}} \simeq 35.6\,\mathrm{m.p.h.}$ **4** $\mu = 0.5$; $32.5\,\mathrm{m\,s^{-1}} \simeq 73\,\mathrm{m.p.h.}$

5 $13\,500\,\mathrm{N}$; $2.7\,\mathrm{cm}$ **6** $40.97\,\mathrm{m\,s^{-1}}$; $35.05\,\mathrm{m\,s^{-1}}$

7 **(a)** $8990\,\mathrm{N}$ inwards   **(b)** $11\,590\,\mathrm{N}$ outwards   **(c)** $20\,560\,\mathrm{N}$ inwards

Index